Processes for Major
Addition-Type Plastics
and Their Monomers

Complex of Chemplex Co. for production of ethylene and polyethylenes. The ethylene plant is on the left side. High-density polyethylene units are behind and slightly to the right. Low-density polyethylene units are on the right. (Chemplex Co.)

Processes for Major Addition-Type Plastics and Their Monomers

Lyle F. Albright

Professor of Chemical Engineering
Purdue University, Lafayette, Indiana

McGRAW-HILL BOOK COMPANY

New York   St. Louis   San Francisco   Düsseldorf   Johannesburg
Kuala Lumpur   London   Mexico   Montreal   New Delhi
Panama   Rio de Janeiro   Singapore   Sydney   Toronto

Library of Congress Cataloging in Publication Data

Albright, Lyle Frederick, date.
  Processes for major addition-type plastics and their monomers.

  Includes bibliographical references.
  1. Polyolefins. I. Title.
TP1180.P67A42   668.4′23   73-12991
ISBN 0-07-000965-1

1 2 3 4 5 6 7 8 9 0 KPKP 7 6 5 4 3 2 1

*The editors for this book were Jeremy Robinson and Stanley
E. Redka, the designer was Naomi Auerbach, and its production
was supervised by George E. Oechsner.   It was set in Caledonia
by The Maple Press Company.*

It was printed and bound by The Kingsport Press.

Contents

Preface

Processes for major plastics produced by addition-polymerization techniques and for the monomers of these polymers are described and discussed in this book. These plastics are defined here as polyethylene, polypropylene, polyvinyl chloride, polystyrene, and related copolymers. In planning and organizing the present book, the response to a series of process-oriented papers by the author published in 1966 and 1967 in *Chemical Engineering* was of considerable help. Calvin Cronan and Steven Danatos of that magazine provided valuable assistance and advice in preparing that series. The polyolefins chosen for inclusion in this book account for about 70 percent by weight of the total plastic production in this country. Processes described here are often similar to those used for other plastics.

In the author's opinion, a process-oriented book should contain, besides process descriptions, background and economic information; a critical analysis and review, indicating key features and bottlenecks; a discussion of the chemistry and fundamental information required for an understanding of the key steps, including polymerization; comparison of alternate processes; and expected process and sales trends.

It is hoped that this book will be even more useful than the *Chemical Engineering* papers. They were used as reference material in undergraduate and graduate design courses for chemical engineers; as training material for new employees in several industrial plants; descriptive material employed by salesmen of plastics and equipment; and as general reading for industrial executives, plant and research personnel, and government planners. Representatives of at least two foreign countries have used the series for planning industrial units in their country. The author has found that with the aid of the material in this book senior chemical engineering students can do a good job of designing and costing specific processes described here.

Process improvements of the last 5 years have been significant for all the polyolefins and monomers discussed in this book. Process descriptions included in the previous series of papers have all been drastically revised and updated.

The author's objective has been to include all available nonproprietary information of importance but to exclude information that is definitely proprietary. In a few cases, this has necessitated judgments on the part of the author since certain information is considered proprietary by one company but not by other companies that discussed it openly.

How the information for this book was obtained may be of interest.

1. A thorough and *critical* review was made of the technical papers, patents, and other literature. Most information was obtained in this manner.

2. Several companies invited the author to visit their facilities and to meet their personnel. Considerable information and perspective were gained in each case. The author sincerely appreciates this cooperation.

3. First and some second drafts of chapters were sent to key authorities for review and suggestions.

The author is most grateful to the many individuals who have provided information, suggested literature to read, reviewed drafts, and/or offered encouragement. Individuals to whom special thanks is given include W. W. Albright, Jr., E. T. Black, L. G. Brackeen, K. E. Coulter, R. A. Darling, H. G. Davis, K. D. Demarest, K. W. Doak, J. R. Fair, E. Farber, M. R. Frederick, S. A. Gardner, C. A. Heiberger, C. W. Johnston, R. G. Keister, I. Leibson, D. W. McDonald, J. A. Moore, E. A. Rose, H. E. Ross, K. Shoenemann, R. N. Shreve, E. S. Smith, R. E. Stack, J. R. Underhill, and H. B. Zasloff.

Several other people asked that their names not be publicized because of company policy.

If the author has erred in any way, he will be most grateful to any reader who will let him know. He hopes that this book will help develop an even better plastics industry.

Lyle F. Albright

Processes for Major Addition-Type Plastics and Their Monomers

CHAPTER ONE

Introduction

Many process improvements have been realized in the last 10 years for the commercial production of major plastics produced by addition polymerization of olefinic, or unsaturated, hydrocarbons. These major plastics are defined here as polyethylene, polypropylene, polyvinyl chloride, polystyrene, and related copolymers (including ABS plastics). As shown in Table 1-1, these plastics account for about 70 percent by weight of the production of plastics in this country. Recent process improvements have resulted in lower polymerization costs and/or plastics with improved properties. The improved technology has already been adopted to some extent in the production of other plastics.

The most important family or group of synthetic high polymers is the plastics that are often produced from unsaturated hydrocarbons. Other groups are synthetic rubbers, or elastomers, and synthetic fibers, adhesives, and surface coatings (paints and varnishes). Natural high polymers of importance include natural rubber, natural fibers (cotton, wool, silk, linen, etc.), and wood. Some natural high polymers that are chemically modified include paper, rayon, and various cellulosic plastics. Synthetic high polymers are generally produced from petrochemicals.

Over 14 billion lb of polyethylene, polypropylene, polyvinyl chloride,

TABLE 1-1 U.S. Plastic Consumption, 1960–1980 *

Notes

1. These estimates are based on trends and industry appraisals with emphasis on new as well as expanding major market penetrations of plastics.

2. Totals for all Materials, Markets and Processes include those categories listed as well as other categories not listed.

3. All materials listed include their copolymers except that ABS is separated from polystyrene.

4. Powder process includes electrostatic, fluidized, rotational, slush and static.

5. Blank spots, where no numbers are listed on the chart, indicate no significant consumption.

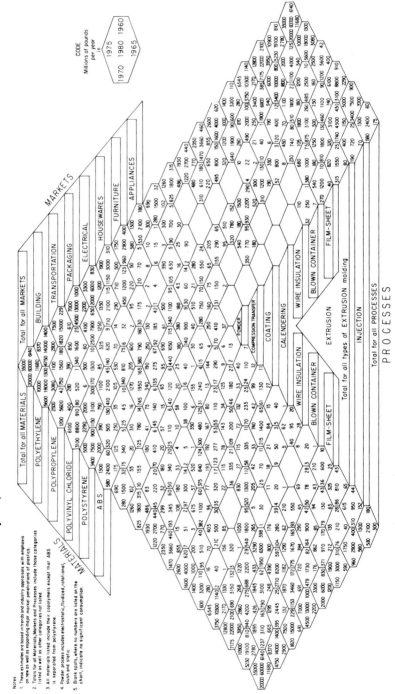

Percentage of plastics consumed by market, process, and material

By market

	1960	1965	1970	1975	1980
Building	22.8	24.8	23.8	23.9	23.3
Transportation	3.7	5.2	9.1	21.4	25.0
Packaging	16.3	18.8	22.5	20.0	21.7
Electrical	13.5	10.3	10.0	9.4	10.0
Housewares	5.0	5.1	5.8	5.4	5.0
Furniture	6.5	5.0	4.8	5.0	4.8
Appliances	2.9	2.6	2.9	3.7	3.5
Others	29.3	28.2	21.1	11.2	6.7
	100.0	100.0	100.0	100.0	100.0

By process

	1960	1965	1970	1975	1980
Injection	13.2	23.8	26.2	31.1	31.8
Extrusion:					
Film and Sheet	18.5	18.6	18.8	18.7	17.2
Blown	1.8	2.4	3.0	4.0	5.3
Wire Insulation	10.1	8.6	8.0	7.4	6.7
Others	3.8	4.2	6.1	6.5	7.5
Calendering	7.5	7.3	7.4	6.7	6.1
Coating	7.2	6.6	6.1	5.5	4.5
Compression-transfer	5.1	4.2	4.1	3.6	3.0
Powder	0.5	0.9	1.4	2.0	2.5
Others	32.3	23.4	18.9	14.5	15.4
	100.0	100.0	100.0	100.0	100.0

By material

	1960	1965	1970	1975	1980
Polyethylene	21.2	26.4	28.0	28.6	30.0
Polypropylene	0.7	3.4	6.0	7.1	9.3
Polyvinyl chloride	14.8	19.6	15.5	17.4	14.7
Polystyrene	14.7	17.1	17.0	14.3	12.5
ABS	1.0	1.5	2.9	3.7	4.0
Phenolic	8.8	7.8	6.5	5.7	5.0
Polyester	2.5	3.4	3.5	4.0	3.5
Others	36.3	20.8	20.6	19.2	21.0
	100.0	100.0	100.0	100.0	100.0

* Abridged and published with permission of *Plastics World* (2).

polystyrene, and related copolymers were produced in the United States in 1971 (1).* Each family accounted for over 1 billion lb by 1970, and each finds multiple uses and markets, as shown in Table 1-1. Although these polyolefins account for about 70 percent of the American plastics market, their share of the market has been steadily growing. Production of these polyolefins throughout the remainder of the world is several times greater than in the United States. Based on American sales, they all rate in the group of most important chemicals sold. The sales value of all polyethylenes is now higher than that of any other chemical. The polyolefins discussed in this book are obviously no longer speciality chemicals but now must be considered as "heavy chemicals" of the petrochemical industry.

Although various polymerization processes, including techniques of bulk, suspension, emulsion, and solvent polymerization, have been widely used for many years to produce major polyolefins, significant process improvements have recently resulted because of a better understanding and application of polymerization fundamentals. Furthermore, some new polymerization techniques have been developed, e.g., processes using new reactor concepts, such as a gas-solid polymerization unit, and improved catalyst systems. Polyolefins can now be better "tailored" with modern units for specific applications and improvement of selected properties.

Since the major cost for production of all major polyolefins is the cost of the monomer (ethylene, propylene, vinyl chloride, or styrene), and since these monomers are used in large amounts, considerable effort has also been devoted in the last several years to finding improved manufacturing techniques for these monomers. It should be recognized that both ethylene and propylene are used for the production of numerous petrochemicals in addition to polyethylene and polypropylene. Significant progress has been made in developing improved processes for the monomers, as later chapters will show. In general, new monomer plants have much larger capacities than older plants, and reduced operating expenses have resulted. Scale-up of the plants has sometimes necessitated new design and operating techniques.

There is increased interest and awareness on the part of the general public of the need for planning and controlling the utilization of our natural resources. In regard to petroleum and natural-gas supplies, competition for hydrocarbons to be used for petrochemicals or fuels is already severe and is likely to intensify. Since plastics (and other synthetic high polymers) are in competition with other materials of

---

* References, indicated by numbers in parentheses, are listed in Literature Cited at the end of the chapter.

construction (metals, wood, glass, cement, etc.), the amount of plastics used affects the rate of utilization of ore deposits, forest reserves, etc. The level of plastics (and other high-polymer) production is now sufficiently large to be an important factor in American economic and resource planning.

In the discussion of various polyolefins and monomers in the subsequent chapters of this book, emphasis will be placed on the commercial processes. In addition, attention will be given to background and economic information and to chemical and fundamental polymerization considerations.

Literature Cited

1. *Chem. Eng. News*, June 5, 1972, pp. 12–14.
2. U.S. Plastic Consumption: 1960 to 1980, *Plastics World*, 1971.

CHAPTER TWO

Pyrolysis Processes for Production of Ethylene and Propylene

I. INTRODUCTION AND GENERAL TRENDS

Ethylene and propylene are the first and second most important petro-chemicals, respectively, both in the United States and the rest of the world. They are produced mainly by the high-temperature pyrolysis (or decomposition) of various hydrocarbon feedstocks that are often paraffinic in nature. Important by-products of pyrolysis processes frequently include butadiene, aromatics, blending stocks for gasoline, and other olefins and diolefins. Tubular reactors operated at relatively steady-state conditions are employed almost exclusively for processes involving only thermal decomposition (or cracking) reactions. Some processes that include a combination of decomposition and oxidation reactions, however, employ other types of reactor (19, 22).

Production Trends

Table 2-1 indicates past, present, and predicted future production information for ethylene and propylene. Although the growth rates for both olefins have been healthy in the United States (6, 7, 15, 18, 26, 27, 29), even larger rates have occurred and are expected to occur in the next several years in western Europe and Japan (3, 8, 9, 12, 24). Por-

TABLE 2-1   Free World's Ethylene and Propylene Production (Million Metric Tons per Year) and Each Area's Percentage of Total (4)

	1960		1965		1970		1975		1980	
	Amount	%	Amount	%	Amount	%	Amount	%	Amount	%
Ethylene										
U.S.	2.6	75	4.6	61	7.5	45	12.0	40	18.0	37
Western Europe.....	0.7	21	2.0	26	6.0	36	12.0	40	19.0	38
Japan...............	0.1	3	0.8	11	2.4	15	4.4	14	7.0	14
Rest of free world...	Neg.	1	0.1	2	0.7	4	2.1	6	5.0–6.0	11
Total...........	3.4		7.5		16.6		30.5		49.0–50	
Propylene										
U.S. and Canada....	1.2	67	2.3	47	3.6	39	5.6	36	8.2	35
Western Europe.....	0.5	28	1.8	37	3.6	39	5.6	36	8.7	37
Japan...............	0.1	5	0.6	12	1.6	17	3.0	20	4.5	18
Rest of free world...	Neg.	...	0.2	4	0.5	5	1.2	8	2.4	10
Total...........	1.8		4.9		9.3		15.4		23.8	

tions of the world not included in Table 2-1 have also experienced high growth rates, but detailed information is unavailable for them.

Ethylene and propylene are the starting materials for numerous major petrochemicals, including polyolefinic-type high polymers employed as plastics, synthetic fibers, and synthetic elastomers. Table 2-2 indicates the major uses or outlets for ethylene in the United States. About 50 percent of the ethylene is used directly or indirectly for the production of plastics, since ethylene is employed for the production of polyethylene, vinyl chloride (used to produce polyvinyl chloride polymers), and styrene (used to produce polystyrene). Production processes for each are discussed in later chapters of this book.

TABLE 2-2   Uses for Ethylene (21)

Direct use	Ethylene use, %	Use in plastics, %
Ethanol................	12	
Ethylene oxide..........	20	Small, but growing
Polyethylene............	35	100
Styrene................	9	60
Vinyl chloride..........	10	100
Chlorinated solvents and other compounds.......	14	Small
Total................	100	

TABLE 2-3   Propylene Demand in 1968 (6)

Product	Propylene demand, millions of pounds
Isopropanol	1,550
Acrylonitrile	1,300
Polypropylene	900
Propylene oxide	800
Heptene	750
Cumene	500
Butyraldehyde	450
Dodecene	350
Nonene	300
Glycerin	130
Epichlorohydrin	100
Isoprene	100
EPDM rubber	30
Acrylic acids	10
Miscellaneous	50
Total	7,320

Propylene ranging in purity from about 91 to 99.5 percent is used in large quantities for production of numerous petrochemicals, as shown in Table 2-3. Polymer-grade propylene with purity often as high as 99 to 99.5 percent is employed for the production of polypropylene plastics and of ethylene-propylene (EP) elastomeric copolymers. Propylene is a feedstock for production of acrylonitrile and isoprene, both of which are polymerized in large amounts. The growth rate for polypropylene is high, and the propylene demand for this product will probably be the largest one before 1980.

Significant amounts of propylene are formed in catalytic crackers of a refinery. The resulting hydrocarbon streams are sometimes fractionated to obtain high-purity propylene, but in many cases only partial separations are made. Refinery-grade (rather low purity) propylene is used in large quantities for the manufacture of rather high-quality gasolines. Alkylation of isobutane with olefinic mixtures containing significant amounts of propylene and polymerization (to produce primarily dimers and trimers) are important refinery operations. Over 20 billion lb/year of this refinery-grade propylene is used in such refinery processes, and a relatively small increase is expected in the future (7). The growth rates for such uses are not expected to be large since higher-quality gasolines can be produced by other methods.

As the production rates of both ethylene and propylene have increased rapidly within the last several years, several trends in the manufacture have allowed decreased operating expenses and/or improved product quality.

1. Plant sizes have increased significantly (28). In 1956, the average plant capacity was less than 150 million lb/year of ethylene. By 1968, it had doubled, and several newer plants now have capacities up to 1.25 billion lb/year.

2. Energy and utility requirements are less, based on a given amount of product. Considerable energy is required to heat the feedstock to reaction conditions, which are generally in the 800 to 900°C (1472 to 1652°F) range. Separation of the products includes liquefaction of the methane and requires temperatures which may be as low as −130°C (−202°F). Numerous heat exchangers are provided to minimize energy requirements for the plant. Process modifications have resulted in better utilization of the energy supplied to the unit.

3. The higher-purity ethylene and propylene currently needed, especially for polymerization purposes, have necessitated more fractionation. In addition, more complete recovery and separation of products from the pyrolysis furnace salvage valuable by-products.

4. Higher-severity furnaces are now employed. These furnaces operate at higher temperatures, and greater conversions of feedstocks are then realized. This trend has increased the amount of ethylene produced and reduced the amount of feedstock required.

5. Improvements have been made in all or almost all pieces of equipment in the complicated plant, in particular in the pyrolysis furnaces, compressors, and distillation sections.

The manufacturing costs for ethylene and propylene depend on many factors, including the cost of the feedstock, specific process being used, size of plant, purity of product olefins, level of conversion employed, and desired flexibility of operations. Choosing the "best" plant and feedstock for a specific plant is not easy and requires careful analysis of many factors.

Feedstocks for Pyrolysis Processes

Many feedstocks are currently used throughout the world for production of ethylene and propylene. Important ones arranged in order of increasing molecular weight (or decreasing ratios of hydrogen to carbon atoms in the feedstock) include ethane, propane, n-butane, naphthas, gas oils, and even crude oils (1, 2, 5, 10, 11, 13, 14, 16, 17, 23).

Higher yields of ethylene (expressed as weight of ethylene produced per weight of feedstock used) can generally be obtained by using lower-molecular-weight feedstocks, but higher-molecular-weight feedstocks commonly result in higher propylene-to-ethylene ratios. Normal paraffins with even-number carbon atoms often produce more ethylene than those with odd-number carbon atoms. Higher-molecular-weight feeds generally produce more by-products, including C_4 and C_5 olefins and diolefins (including butadiene and isoprene), high-quality gasoline cuts, and aromatics. The off-gas of a pyrolysis unit is primarily hydrogen and methane.

Normal paraffins in general produce more ethylene than isoparaffins of similar molecular weight (16, 29). Isobutane, which is of commercial interest as a feedstock, produces little ethylene but does produce considerable amounts of propylene and isobutylene.

To maximize ethylene production, the heavier olefins initially produced are sometimes recovered and recycled to the pyrolysis furnaces. Propylene and heavier olefins produce appreciable amounts of aromatics. Naphthenic hydrocarbons can be used as feedstocks, but paraffins are preferred. Aromatic hydrocarbons are generally of little value for the production of olefins. Hence naphthas and gas oils containing mainly paraffinic hydrocarbons are preferred high-molecular feedstocks.

Table 2-4 indicates the approximate amount of several feedstocks required and the amount of by-products obtained in producing 1.0 billion lb/year (450,000 metric tons) of ethylene. The amounts of feedstocks used are in all cases large, ranging from about 28,000 to 45,000 bbl/day, depending on the feedstock. Such a quantity is greater than that

TABLE 2-4   Feed and Product Summary for Various Feedstocks (16)

1 billion lb/yr ethylene production

| | | | | Full-range naphtha | Gas oil | |
Feedstock	Ethane	Propane	n-Butane		Light	Heavy
Feed rate, million lb/yr.	1,311.8	2,379.6	2,497.4	3,204.7	3,801.4	4,286.5
Products, million lb/yr:						
Off-gas.	211.7	713.7	626.4	551.2	431.5	428.6
Ethylene (polymer grade).	1,000.0	1,000.0	1,000.0	1,000.0	1,000.0	1,000.0
Propylene (chemical grade).	38.4	385.0	521.5	517.3	573.3	613.9
C_4 fraction:						
Butadiene.	17.6	76.1	73.4	144.2	151.3	173.1
Butylenes.	7.5	32.6	171.6	144.2	186.1	187.9
C_5/400°F naphtha.	36.6	142.9	92.2	703.4	575.9	591.1
400°F + fuel.	. . . .	29.3	12.3	144.4	883.3	1,291.9
Total.	1,311.8	2,379.6	2,497.4	3,204.7	3,801.4	4,286.5
Aromatics in C_5/400°F naphtha, million lb/yr.	. . . .	80.9	58.2	387.0	288.0	295.7
Hydrogen in feed, wt. %.	20.0	18.2	17.3	15.0	13.3	13.0

needed by some small refineries. The feedstocks required for ethylene-propylene plants in the United States represent a significant and growing fraction of the total amount of hydrocarbons used.

When higher-molecular-weight feedstocks are used, a much more complex product mixture is obtained, as indicated by Table 2-4. In the past, it was not always economically feasible to recover and purify all by-products; instead they were sometimes used only for their fuel value or were recycled. Current economics generally dictates the recovery of most by-products, requiring a more complicated plant. Capital costs are thus greater when higher-molecular-weight feedstocks are used.

In the United States, ethane, propane, and n-butane are recovered in appreciable quantities from natural gas. The availability of each paraffin depends on several complicating factors (16). Some ethane is normally added to the natural gas or left in it to increase its heat content. If the ethane were removed, it could be made available for ethylene production. Ethane prices are currently about 2.5 to 3.25 cents/gal on the Gulf Coast of the United States (29).

Propane and n-butane are used in large amounts for home and industrial heating. Propane makes a desirable antipollution fuel for automotive engines, but its major fault is the high pressures required for storage as a liquid in the fuel tank. Pyrolysis units then must compete for propane with these other uses. Propane was available several years ago for about 3.25 to 3.75 cents/gal on the Gulf Coast (29), but currently it is in the range of 5 to 6 cents/gal.

n-Butane can possibly be recovered in larger amounts in the future from several sources.

1. As more natural gas is processed, more n-butane can be recovered.
2. More n-butane can be recovered from various streams in some refineries.
3. n-Butane is blended in large amounts in gasolines and various liquefied hydrocarbons. Stricter volatility requirements to reduce air pollution may restrict such uses in future.

Isobutane may be of considerable interest as a feedstock in the future especially if the demand for propylene and isobutylene increases. Hydrocracking results in higher amounts of isobutane than catalytic cracking. Isobutane is relatively cheap, being about 5 to 7 cents/gal or even less.

In western Europe and Japan, naphthas were delivered in 1970 for about 5 to 6 cents/gal (24, 29). Even lower prices may occur in the Caribbean area. Because of government restrictions, relatively little foreign naphtha can enter the United States on a tariff-free basis.

American naphthas and gas oils are considerably more expensive, and they are normally utilized almost exclusively for gasoline manufacture. The establishment of some free-trade zones allows some cheap naphthas and gas oils to be imported for petrochemical purposes.

Refinery gases produced by the cracking of heavier hydrocarbons contain considerable amounts of ethylene and propylene. When such a gas can be obtained at essentially fuel-gas prices, the olefins, ethane, and propane can be recovered and used to produce high-purity ethylene and propylene. A major disadvantage of this method is that modifications in the operation and feedstocks to the refinery change the refinery gas. Hence the recoverable ethylene and propylene may vary significantly with time.

Freiling, King, and Newman (17) have shown that the production costs for ethylene vary significantly with the type and cost of feedstock used (Fig. 2-1). Major by-products are recovered for sale to maximize profits. The costs shown in Fig. 2-1 should be considered only approximate since the assumptions may not be applicable in all plants.

At present, about 80 percent of the feedstocks used in the United States for production of ethylene are light hydrocarbons, mainly ethane, propane, and n-butane. It is predicted, however, that 43 percent of

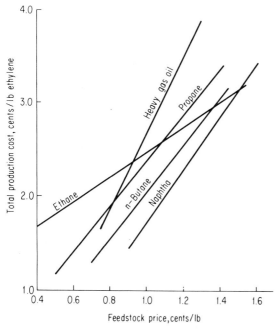

Fig. 2-1 Effect of feedstock prices on United States ethylene-production costs (by-products are recovered and sold) (17).

the ethylene will be produced from heavier feedstocks by 1980 (10). Obtaining adequate supplies of feedstocks for the expected increased demand of ethylene, propylene, butadiene, aromatics, and other valuable hydrocarbons will require careful planning.

Plants using particularly heavier feedstocks should not as a rule be called ethylene plants. In a broad sense, they are olefin or basic petrochemical plants since they yield such a wide variety of products. To an increased extent, these plants are being designed (at least in part) to obtain many products that were previously considered rather undesirable. Butadiene, for example, will be produced in the future more frequently in pyrolysis plants than by the dehydrogenation of n-butane and n-butenes (20).

Computer programs are widely used to evaluate major changes in new or existing plants, but relatively little has been published on these programs. Chem Systems, Inc. has publicized (25) a linear program that evaluates changes of the feedstock, conversion levels, products recovered, and investment. Considerable effort can be expected in the future in the development and use of improved programs.

II. FUNDAMENTALS OF PYROLYSES

Relatively limited fundamental information is currently available for the pyrolyses of several hydrocarbons of commercial importance. The need for additional information and a review of how such information could be used in designing commercial pyrolysis units will now be discussed.

Mechanism of Pyrolysis

Commercial pyrolyses of hydrocarbons to produce ethylene and propylene occur primarily by a highly complicated sequence of consecutive and simultaneous free-radical steps that occur in the gas phase (35, 37, 39–43, 46, 47, 49, 50, 56). In addition, *wall effects* are often quite important to the composition of the product stream and the kinetics of the overall reaction sequence (34, 47, 48, 53, 54). A significant number of different reactions and phenomena are occurring at the walls of the reactor and will be discussed in more detail later.

In spite of the commercial importance of the various pyrolysis processes, relatively little information has been reported in the literature for the reaction sequence and the kinetics of these reactions. Several companies are known, however, to have unpublished information in their files. The most complete information currently available is for the pyrolysis of ethane, propane, ethylene, and propylene.

The following reaction steps have been suggested by Snow et al. (51, 52) for the pyrolysis of ethane:

(1) $CH_3CH_3 \rightleftharpoons 2CH_3\cdot$
(2) $CH_3\cdot + C_2H_6 \rightleftharpoons CH_4 + CH_3CH_2\cdot$
(3) $CH_3CH_2\cdot \rightleftharpoons CH_2{=}CH_2 + H\cdot$
(4) $H\cdot + CH_3CH_3 \rightleftharpoons CH_3CH_2\cdot + H_2$
(5) $H_2 + CH_3\cdot \rightleftharpoons H\cdot + CH_4$
(6) $H\cdot + CH_3CH_2\cdot \rightarrow CH_3CH_3$
(7) $CH_3\cdot + CH_3CH_2\cdot \rightarrow CH_3CH_2CH_3$
(8) $2CH_3CH_2\cdot \rightarrow CH_3CH_2CH_2CH_3$
(9) $2CH_3CH_2\cdot \rightarrow CH_3CH_3 + CH_2{=}CH_2$

Steps 1 to 5 are reversible, and 14 chemical steps are listed. The production of ethyl radicals by steps 1 and 2 initiates the reaction. Ethylene and hydrogen are the major products, and steps 3 and 4 involve a chain for production of these two components. The model is certainly an oversimplification of the actual sequence, and other reaction steps are occurring to at least some extent, including pyrolysis of ethylene (39). Both propane and n-butane react thermally more readily than ethane, yet no provisions are made in the model for the reaction of these two paraffins or for production of heavier hydrocarbons. The amounts of propane and n-butane present, however, are relatively low, especially at lower conversion levels of ethane.

Several investigators (32, 34, 37, 38, 40, 41) have obtained extensive data for the pyrolysis of propane and propylene. Conditions investigated (37) for propane include temperatures from 650 to 850°C; steam-to-propane feed ratios of 0:1, 1:3, and 1:1; propane conversions to about 95 percent; nitrogen and helium as diluents instead of steam; different flow velocities; and pretreatment of the inner walls of the reactor.

Table 2-5 indicates a simplified chemical model for propane (37) that predicts the product and kinetics of the reaction reasonably well at propane conversions up to about 70 percent. The main initiation step for the production of radicals is the fragmenting of propane (step 1) to produce methyl and ethyl radicals. Ethane is harder to fragment, and radical formation from ethane, which is undoubtedly quite unimportant during propane pyrolysis, is ignored in Table 2-5.

Propyl radicals are produced primarily by steps 2, 3, and 6 of Table 2-5. Isopropyl and n-propyl radicals probably isomerize rapidly to form essentially an equilibrium mixture. This isomerization probably is not a simple first-order reaction but possibly occurs by forward and reverse reactions, as shown in step 5 of Table 2-5. Ethylene and methane are produced mainly by a chain sequence involving n-propyl radicals (steps

TABLE 2-5   Main Reaction Steps for Propane Pyrolysis (37)

Reaction	A s^{-1} or $cm^3/(g\ mol)(s)$		E, kcal/ g mol
(1) $CH_3CH_2CH_3 \rightleftharpoons CH_3CH_2\cdot + CH_3\cdot$	6×10^{14}	4.2×10^{13}	75 0
(2) $CH_3CH_2\cdot + CH_3CH_2CH_3 \rightleftharpoons CH_3CH_3 +$ $CH_3\dot{C}HCH_3$ (or $CH_3CH_2CH_2\cdot$)	4×10^{10}	1×10^{11}	5 4
(3) $CH_3\cdot + CH_3CH_2CH_3 \rightleftharpoons CH_4 +$ $CH_3\dot{C}HCH_3$ (or $CH_3CH_2CH_2\cdot$)	1.5×10^{12}	1.1×10^{10}	4.9 5
(4) $CH_3CH_2CH_2\cdot \rightarrow CH_2{=}CH_2 + CH_3\cdot$	4×10^{10}	3.6×10^{11}	32 2
(5) $CH_3\dot{C}HCH_3$ (or $CH_3CH_2CH_2\cdot$) $\rightleftharpoons$ $CH_3CH{=}CH_2 + H\cdot$	3.6×10^{10}	1.8×10^{12}	30 1
(6) $H\cdot + CH_3CH_2CH_3 \rightleftharpoons H_2 +$ $CH_3\dot{C}HCH_3$ (or $CH_3CH_2CH_2\cdot$)	1.8×10^{11}	5.6×10^{9}	4.6 8
(7) $C_2H_5\cdot \rightleftharpoons CH_2{=}CH_2 + H\cdot$	3×10^{11}	7.5×10^{11}	35 5.4
(8) $CH_3\dot{C}HCH_3$ (or $CH_3CH_2CH_2\cdot$) $+ CH_3\cdot \rightarrow$ $CH_3CH{=}CH_2 + CH_4$	1×10^{14}		0

3 and 4), and propylene and hydrogen are produced by a sequence involving either n-propyl or isopropyl radicals (steps 5 and **6**).

At low conversions, essentially equal numbers of moles of ethylene, methane, propylene, and hydrogen are produced, as shown in **Fig. 2-2**, but the ratio of ethylene to propylene increases significantly as the propane conversion increases. Increased temperatures promote ethylene production but lower it for propylene. All olefins (but especially propylene and heavier olefins) tend to react. Although not shown in Fig. 2-2, yields up to 1 or 2 percent occur for butanes at low conversions, for butenes at 30 to 70 percent conversions, and for butadiene at 70 percent conversions and higher. Significant amounts of aromatics are produced at 90 percent conversions and higher. The reaction scheme for propane pyrolysis is obviously more complicated than the one shown in Table 2-5, especially at higher propane conversions. Additional reaction steps that may be of some importance are

$$(10)\quad 2CH_3\cdot \rightleftharpoons 2C_2H_6$$
$$(11)\quad H_2 + CH_3\cdot \rightarrow CH_4 + H\cdot$$
$$(12)\quad H_2 + C_2H_5\cdot \rightarrow C_2H_6 + H\cdot$$

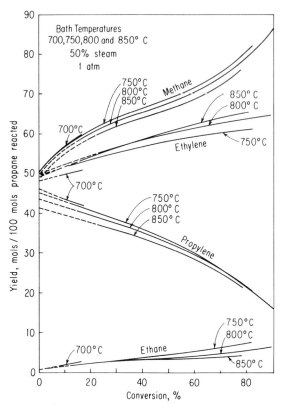

Fig. 2-2   Product yields vs. conversions of propane for various temperatures using 50 percent steam. [*From (37) with permission of AIChE Journal.*]

The forward reaction of Eq. (10) is of importance only when a fairly high concentration of methyl radicals is present. Hydrogen is known to be a reactive intermediate in pyrolysis reactions. When hydrogen is added as a diluent to ethane, there is relatively little net formation of hydrogen during the subsequent pyrolysis. The hydrogen probably reacts with methyl and ethyl radicals, as shown above.

Ethane and (especially) methane are relatively unreactive molecules in propane pyrolysis since the C—H bonds are stronger in these compounds than the C—H bonds in propane or propylene. Reactions of either ethane or methane with various radicals, such as methyl, propyl, or hydrogen radicals, may then be of lesser importance. It is quite probable that some allyl radicals are formed from propylene since the allylic hydrogens of propylene are rather weakly bound. These allylic radicals may be precursors to aromatics. When propylene is added to propane, the pyrolysis reaction is inhibited to some extent, apparently because the allylic radical does not propagate the chain mechanism.

The overall pyrolysis-reaction sequence is endothermic, but some of the free-radical reactions occurring during pyrolysis are exothermic. Combination of free radicals, e.g., those which reform ethane or propane or those which produce higher-molecular-weight products, are examples of such reactions. The question has been raised how the exothermic heat can be dissipated to allow the formation of relatively stable molecules. One suggestion has been that the energy is transferred to a *third body* (46, 50), such as steam or a relatively heavy hydrocarbon, but as will be discussed later, steam does not appear to act as a third body.

One of the main purposes of using steam as a diluent for the hydrocarbon feedstock is to lower the partial pressure of the hydrocarbons, and other "inert" gases might be used. Methane, being essentially inert at reaction conditions, has been considered. Less heat is required to heat methane to reaction temperatures than is needed for steam. As a result, the energy requirements for a pyrolysis furnace would be less using methane than steam. The main disadvantage of methane would be the cost of separating the additional methane from the product mixture. Demethanization accomplished by low-temperature distillation is always relatively difficult and expensive.

Another possible diluent is ethane. Some ethane of course reacts, but the ethane conversion could be kept sufficiently low to ensure a partial pressure of products no higher than currently practiced using a hydrocarbon-steam mixture. Additional costs for the large deethanizing columns probably are greater in all cases than the savings in the energy requirements for the pyrolysis furnaces.

Some limited pyrolysis information is available in the literature for butane (30, 31), isobutane (33), and heavier hydrocarbons (36, 45). More is needed for effective design of pyrolysis units using these heavier hydrocarbons. Recently Kunugi and coworkers have reported extensive data for the pyrolyses of ethylene (39) and propylene (40, 41). They have proposed a total of 60 and 48 reaction steps, respectively, for these two olefins. It is obvious that many undesired reactions can occur to lower the yields of ethylene and propylene.

Kinetics of Pyrolysis Reactions

At the conditions employed commercially, the kinetics of pyrolysis reactions is very fast. In modern high-severity furnaces operated at high temperatures, 80 to 95 percent conversions are sometimes obtained in 0.5 s or less. In older furnaces operated at lower temperatures, relatively much longer times were required to obtain the desired conversion. Figure 2-3 shows the important effect of temperature on the kinetics

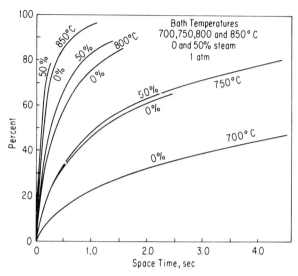

Fig. 2-3 Propane conversion vs. space-time for various temperatures and steam concentrations. [*From* (37) *with permission of AIChE Journal.*]

of propane pyrolysis.   An equimolar mixture of propane and steam has a slightly higher rate of propane conversion than pure propane.

Figure 2-3 indicates that the kinetics of pyrolysis is initially rapid but decreases significantly as the conversion progresses.   Attempts have been made to represent the overall rate of disappearance of the primary reactant by first-order, ⅔, and second-order kinetic equations.   Kunugi et al. (40), for example, reported a ⅔ order as an approximate fit of the kinetic data for propylene.   Such equations can be considered at best as only an approximation which should be employed over only a limited range of operating conditions (32, 34).   Such a conclusion is obvious when it is realized that the overall reaction sequence involves numerous specific reaction steps most of which are reversible.   Furthermore these specific reactions follow either first-order kinetics, e.g., the forward reaction of step 1 of Table 2-5, second-order kinetics, e.g., the reverse reaction of step 1, or possibly even third-order kinetics.   Reactions that follow third-order kinetics might be those that involve a third body.

Snow et al. (51, 52) and Herriott, Eckert, and Albright (37) have had considerable success in modeling both the product distribution and the kinetics for ethane and propane pyrolysis, respectively.   In both cases, the most probable free-radical reaction steps were used.   Reaction rate constants for each of the reaction steps were correlated using the

Arrhenius equation ($k = Ae^{E/RT}$). The energies of activation E as used agreed reasonably well with those rather widely accepted in the literature (50).

Kinetics equations modeled using the most important reaction steps (such as those shown in Table 2-5) require a large computer to solve. The models used by Snow et al. and by Herriott et al. are considered to be simplified based on the actual sequence of the numerous re-action steps. Although their approach has been checked to only a limited extent, it seems to be basically a much sounder and more theo-retical approach for the development of reliable kinetic equations than any other reported to date.

All reaction steps for ethane pyrolysis are also occurring during pro-pane pyrolysis to at least some extent, and vice versa. A similar observa-tion is true for the pyrolysis of heavier hydrocarbons starting with n-butane, isobutane, pentanes, etc. Probably the E and A terms would be relatively constant regardless of the feedstock and temperature, and a generalized kinetic equation could be developed for any combination of hydrocarbons in the feed. If third-body effects are important, the A and E values might change somewhat, however. Possibly additional reactions should be added to the generalized model for third-body effects.

As a first step toward the development of such a generalized kinetic equation, additional and more reliable pyrolysis data for ethane are apparently still needed. Propane-pyrolysis data have been reported over a wide range of conditions except for pressures. A combination of ethane, propane, propylene, and ethylene data should allow development of a rather complete kinetic equation for all feed mixtures of ethane and propane. Later this model could be expanded as data for butanes and high hydrocarbons are obtained.

The rate-controlling step is often the initiation step to form free radi-cals to start the chain sequence. Various initiators such as oxygen or halogens have been tested for obtaining free radicals at lower reaction temperatures. Initiators that are somewhat effective in lowering the reaction temperature have been found, but apparently they are not used commercially. The cost of the initiator or catalyst, increased cost of product separation, and other process complications have prevented their industrial use.

Important Operating Variables

Temperatures and steam-to-hydrocarbon ratios in the feed stream are particularly important variables. Both variables have significant effects on the product composition and the kinetics of pyrolysis, as shown in Figs. 2-2 to 2-4. Increased amounts of steam dilution (or higher

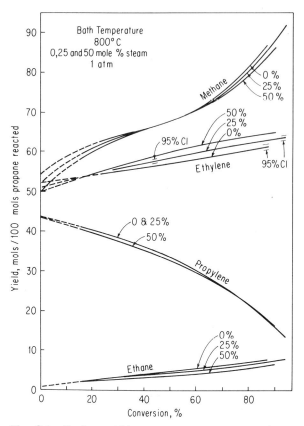

Fig. 2-4 Product yields vs. propane conversions for various steam concentrations at 800°C. [*From* (37) *with permission of AIChE Journal.*]

steam-to-hydrocarbon ratios) reduce the partial pressures of the hydrocarbons in the reaction mixture. Decreased partial pressures decrease the rates of the second-order reaction steps to a greater degree than those of the first-order steps. Higher reaction temperatures and lower partial pressures promote ethylene production. Propylene formation is favored by lower temperatures and is affected to only a small extent by pressure.

Table 2-6 indicates common levels of steam used for various feedstocks. In addition to lowering the partial pressure of the hydrocarbons, steam acts as an energy source to promote more isothermal conditions in the highly endothermic reaction mixture, but it also reacts with the coke deposits on the reactor walls and with the metal walls themselves to form metal oxides, as will be discussed in more detail later.

TABLE 2-6   Steam Used Commercially for Various Feedstocks (29, 44)

Feed	Weight steam / Weight feed
Ethane	0.25–0.4
Propane	0.25–0.5
Naphthas	0.5 –0.7
Gas oils	0.7 –1.0

Less coking occurs at lower partial pressures. Presumably at least part of the coking sequence involves the initial adsorption of a hydrocarbon on the wall.   Less volatile hydrocarbons, such as aromatics, are likely candidates for adsorption, and adsorption is more probable at higher partial pressures and lower temperatures.   Adsorbed hydrocarbons probably react readily since the reactor walls are hotter than the main body of the gas.

The partial pressures of hydrocarbons vary in a tubular reactor as a function of tube length, for at least two reasons.   The pressure drop of the gases as they pass through the tube tends to reduce the partial pressures, but pyrolysis increases the number of moles of hydrocarbons.   The partial pressures of the least volatile hydrocarbons are generally highest at the reactor outlet, and these pressures may be critical design considerations.   More steam is needed as a feed diluent for higher-molecular-weight feedstocks (see Table 2-6) to prevent the adsorption on the wall of the heavier hydrocarbons.

Steam potentially could act as a third body in the free-radical reactions that occur during pyrolysis (46, 50).   Propane runs made using helium as diluent, however, resulted in production distribution and kinetics similar to those of runs made with steam (37).   It would appear that third-body effects involving steam are of minor importance during the pyrolysis of propane, at least.

The mass velocity of reactants in a tubular reactor affects the heat transfer through the boundary layer at the wall.   Linde (44) specifies that the mass velocities should be in the range from 16 to 29 lb/(ft^2)(s) [80 to 145 kg/(m^2)(s)].   Higher velocities promote higher heat-transfer coefficients (involving mainly conduction and convection) in the boundary layer.   Higher velocities, however, cause increased pressure drops in the gas stream as it flows through the tube and more erosion of the hot tubes, especially at bends.   Higher inlet pressures for the feedstock are required when higher mass velocities are used.

Plug flow of the gas mixture was apparently approximated in laboratory reactors even though they were operated at reasonably low velocities and hence nominally low Reynolds number (34, 37). Conversion and product distribution were found to depend only on the residence time of the reactants at reaction temperatures and to be independent of the flow rates. High diffusivity coefficients for the hydrocarbons at the reaction temperatures apparently promote rapid mixing. Plug flow probably is also a reasonably good approximation for commercial tubular reactors.

Factors Affecting Reactor Walls

The condition of the reactor walls and methods of treating them have long been known to be of importance for pyrolysis reactions. Surface reactions occurring during pyrolysis have been investigated by Albright and associates, who have published part of their findings (34, 37). Table 2-7 summarizes the reactions for producing and destroying surface coke and surface oxides. At least part of the coke (or carbon) deposition during pyrolysis reactions occurs because of surface reactions. Propylene, dienes, and aromatics frequently are significant coke producers.

TABLE 2-7   Surface Reactions Occurring during Pyrolysis

(A) *Formation of Coke:*
    (1) Hydrocarbons → coke + H_2
        Hydrocarbons that are high coke producers include aromatics, diolefins, and propylene
(B) *Destruction of Coke:*
    (1) Coke + H_2O → CO + CO_2 + H_2
    (2) Coke + O_2 → CO + CO_2
        Little water is produced, indicating that coke contains few or no hydrogen atoms
    (3) Coke + H_2 → CH_4
        This reaction is probably of minor importance
(C) *Production of Metal Oxides on surface:*
    (1) Metal + H_2O → metal oxides + H_2
    (2) Metal + O_2 → metal oxides
        The metal oxides are probably higher oxides, whereas the "metal" is most probably lower oxides of the metal
(D) *Destruction (or Reduction) of Metal Oxides:*
    (1) Metal oxides + CO → metal + CO_2
    (2) Metal oxides + H_2 → metal + H_2O
    (3) Metal oxides + hydrocarbons → metal H_2O + CO + CO_2
    (4) Metal oxides + H_2S → metal sulfides + H_2O
        Sulfides, mercaptans, and sulfur probably also produce metal sulfides
(E) *Shift Reactions with Steam:*
    (1) CO + H_2O → CO_2 + H_2
    (2) Hydrocarbons + H_2O → CO or CO_2 + H_2
        These shift reactions may actually involve a combination of reactions C(1), D(1), and D(3) above

Surface oxides, which are always present on the surface of commercial reactors, are effective for oxidizing carbon monoxide to carbon dioxide, hydrogen to water, and hydrocarbons to carbon oxides and water. Propylene and possibly ethylene are sometimes oxidized in significant amounts. Such oxidation steps reduce the metal oxides, in general, to lower oxides. The metal oxides are produced in commercial reactors by at least two methods. Steam reacts at the high temperatures employed with the metal walls to form metal oxides and hydrogen. Metal oxides are also produced when either oxygen or air is employed to burn out the coke in the so-called decoking operation.

Coke is also destroyed to some extent while pyrolysis is occurring by reactions with steam. Shift reactions with carbon monoxide and adsorbed hydrocarbons may also occur. Most of the carbon oxides (if not all) and part of the hydrogen produced result because of these steam reactions. A small amount of methane is also produced by reactions between hydrogen and surface coke.

Albright and associates (34, 37) have investigated several methods of treating the inside walls of a tubular reactor. Their conclusions, based on ethane and propane pyrolyses, are at least qualitatively applicable to other feedstocks. Hot reactor tubes contacted with oxygen for several minutes or with steam for several hours become activated because of the metal oxides formed. These activated reactors cause a higher rate of pyrolysis, lower yields of ethylene and propylene at a given hydrocarbon conversion, more hydrogen formation, and increased coking.

Surface oxides can be reduced by several techniques:

1. Operation of reactor for perhaps ½ h. During this period, the metal oxides react with the hydrocarbons or the hydrogen formed, and relatively large amounts of carbon oxides are produced. This operation may not always be successful; small laboratory reactors sometimes plug before the surface can be sufficiently reduced.

2. Treatment of the reactor with hydrogen or carbon monoxide.

3. Treatment of the reactor with sulfur-containing compounds such as H_2S, mercaptans, sulfides, or elemental sulfur, to form relatively stable metal sulfides.

Some hydrocarbon feedstocks, particularly naphthas and gas oils, contain sulfur compounds. When ethane, propane, or butanes are used as feedstocks, hydrogen sulfide or mercaptans are deliberately added, however, at least in some plants. Although more information is needed on the role of these compounds, a sufficient amount should be added to keep the level of metal oxides on the tube wall low. Too much, however, may lead to coke production. One postulate is that thiophene may be a by-product and that this compound adsorbs on the tube walls, contributing to coking. A surface that has been passivated with hydrogen sulfide is activated only when treated with oxygen for an extended time (34, 37).

During a pyrolysis run, an essentially steady-state condition is generally reached on the walls of a reactor within several hours. At any position in the tube reactor, the rate of coke deposition is just slightly greater (at most) than the rate of coke removal. In addition, the rate of formation of surface oxides approximates that for their reduction. It should be recognized that the rates of these surface phenomena vary as a function of tube length and of time of cycle (after a coking operation).

Two techniques are used commercially to remove the coke deposits that eventually form on the reactor walls. In the first method, air is added in small quantities to steam, which is fed to the tube at reaction temperatures. The oxygen "burns" out the carbon relatively quickly, depending in part on the amount of oxygen present in the mixture. The second method consists of treating the reactor for perhaps 2 or 3 days with steam. Since both methods tend to activate the surface, a passivating technique such as hydrogen sulfide treatment is often needed.

The types of metal oxides formed on the surface of a reactor obviously depend to some extent on the material of construction. The nickel content of the metal, for example, has some effect on product yield.

Termination steps for the free-radical chain reactions probably occur to a significant extent at the walls of the reactor. Some initiation steps have also been postulated as occurring there (48, 54).

Methods of scaling-up a reactor have improved significantly in the last few years as more fundamentals of the pyrolysis reaction have become known, but scale-up is still partly an art. The temperature of the wall of the reactor (which may be appreciably hotter than the reacting gases), composition of the wall, and roughness are all important design factors. Wall temperatures tend to be higher as the diameter of the tube reactor increases since a proportionally greater amount of heat transfer per given surface area is required then. Temperature and flow characteristics of the reacting gases may vary, depending on both axial and radial positions in the tube reactor. When larger reactors are built, these factors may change significantly. Mathematical modeling has been of considerable help in designing new reactors.

Conclusions

Although the mechanisms of the pyrolyses of light paraffins have been clarified in the last few years, more information is still needed. Considerably less information is available for the reactions and products of heavier feedstocks. Design of a commercial pyrolysis furnace is now considerably more scientific than it was several years ago, and an even more scientific approach seems possible when more mechanistic information has been obtained.

III. PROCESS DETAILS

A modern pyrolysis plant for the production of ethylene, propylene, and other hydrocarbons is complex and huge (29, 44, 102, 115); Fig. 2-5 shows the 0.5 billion lb/year ethylene plant of Continental Oil Company, Lake Charles, Louisiana. A pyrolysis plant always contains pyrolysis furnaces, where the actual pyrolysis operation occurs; compressors, both for the product stream and for refrigeration streams; distillation columns, for the separation of the various products and by-products; numerous heat exchangers; and units for separating oil and water, drying

Fig. 2-5 Ethylene plant of Continental Oil Company. Distillation columns are in the right foreground, and cracking furnaces are in the left background. (*Continental Oil Company and The Lummus Company.*)

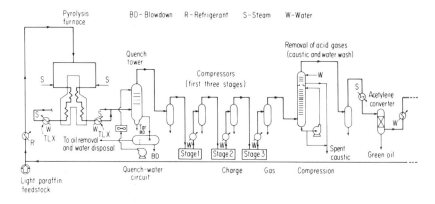

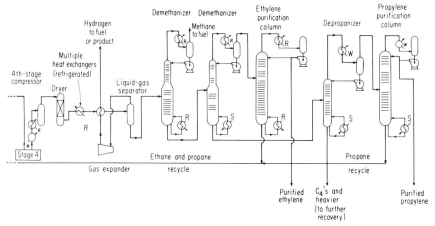

Fig. 2-6   Simplified flowsheet for pyrolysis plant for light paraffins.

gases, removing acid gases, and removing or reducing acetylenic compounds. Figure 2-6 is a basic flowsheet for an ethylene plant that uses light paraffins, including at least some propane or *n*-butane as well as ethane, for a feedstock. The flowsheet for other pyrolysis processes (44, 69, 88, 90, 115) sometimes differs from that shown here, and such variations will be discussed later.

Pyrolysis Systems and Auxiliary Equipment

The sequence of steps in and around the pyrolysis furnace includes vaporizing feedstock, mixing gaseous feedstock and steam, preheating gas mixture to reaction temperature, and providing enough energy to obtain the desired reactions.

The method of vaporization used for the feedstock depends on the type of feedstock used. When a light hydrocarbon such as ethane and/or propane is used, it may be vaporized by heat exchange with condensing propylene, which is often used as a refrigerant in the recovery section of the process (57). Vaporization of naphthas and gas oils, however, requires higher temperatures. Such hydrocarbons are generally partially preheated outside the furnace and then vaporized as they flow through tubing positioned in the convection zone of the furnace. The tubing is constructed of several straight sections joined by U bends, and it is normally positioned horizontally in the convection zone, as in the furnace shown in Fig. 2-7. Steam is also superheated in other tubing positioned in this zone, and the steam and gaseous hydrocarbons are then mixed and heated to about 650 to 700°C in the convection section. The mixture then enters the tube reactors in the radiant zone of the furnace, where the mixture is heated to reaction temperatures and the main pyrolysis reactions occur.

Design of the pyrolysis furnaces, including the tube reactors, is a combination of science and art. Not all design considerations are well understood. The temperature, total pressure, partial pressure of hydrocarbons (or steam-to-hydrocarbon ratio), flow rate of reaction mixture, and residence time of the reactants at reaction conditions must all be controlled to obtain optimum results.

Parameters of the tube reactor include the length, diameter, wall thickness, and configuration in the furnace. Designing the furnace to heat

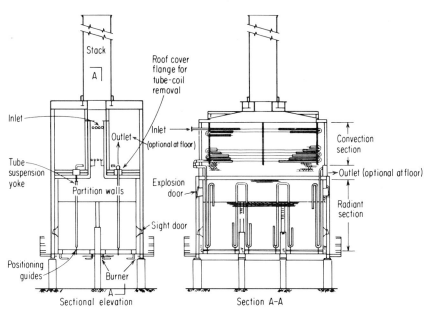

Fig. 2-7 Multizone pyrolysis furnace. (*Foster-Wheeler Corp.*)

the tube reactor involves consideration of the geometry of the basic shell, partitions, materials of construction, location and design of the burners, provisions for fuel and air to the burners, and the methods of supporting the reactors in the furnace. Some literature (65) has been published concerning the rather different designs of The Lummus Company (106, 107, 113, 114), M. W. Kellogg Company (57, 97), Foster-Wheeler Corporation (76, 101), Selas Corporation of America (5), Stone and Webster Engineering Corporation (29, 90), and Linde of Germany (44).

Combustion Zone of Furnace All furnaces are divided into radiation and convection sections, and Fig. 2-7 indicates the major features of a furnace designed by Foster-Wheeler Corp. (76). Actual combustion of the fuel is in the radiation section, which is the hotter of the two sections. Temperatures of the combustion gases in this section may range from about 1150°C (2100°F) to at least as high as 1260°C (2300°F) in some furnaces. Heat transfer from the gases and the furnace walls to the several reactors positioned in the furnace is largely by radiation. Gases that radiate are mainly carbon dioxide and steam. Because of the high rates of heat transfer through the tube walls, the temperatures of tube walls are significantly lower than those of the combustion gases or even the furnace walls. Refractory brick and any other construction materials for this zone of the furnace must be carefully chosen in order to realize long and dependable furnace operation.

The combustion gases from the radiant zone pass into the convection zone, and sensible heat from the combustion gases is used here to preheat both the hydrocarbon feedstock and the steam and possibly to produce high-pressure steam needed for operation of the overall unit. Heat transfer in this section is mainly by convection and by conduction through the boundary layers surrounding the tubes. The relatively cool combustion gases from the convection zone then pass to the stack.

The radiant and the convection sections are generally separated by a baffle or a semiwall. The radiant zone is sometimes compartmentalized by walls or partitions, as shown in Fig. 2-7. A tube reactor passes successively in this furnace through separate compartments, or zones, in each of which a different firing rate, and hence different temperatures, can be employed (76). Such an arrangement is used to give better control of the temperatures for each reactor tube.

Fuels for Pyrolysis Furnace and Burners Choice of the fuel for a pyrolysis furnace is an important consideration because energy requirements are high. In modern large furnaces, the requirements are often 50 to 140 million Btu/h (12.5 to 35 million kcal/h), and fuel costs are obviously large. Burner design and the furnace details depend on the fuel used. Clean gaseous fuels of constant composition are preferred to liquid fuels, for several reasons.

1. The flow rate of a gas to a burner can be adjusted more easily than that of a liquid with consequently more uniform heat release rates for burner.

2. The gaseous fuels are premixed with air before combustion, whereas atomizing steam is required in many cases with a liquid fuel.

3. Liquid fuels which contain aromatic or unsaturated hydrocarbons often produce luminous flames and are particularly troublesome in relation to burner fouling and soot or smoke formation.

Gaseous fuels frequently are required for high-severity furnaces in order to obtain the high flame temperatures needed.

Pyrolysis furnaces, like the other types, contribute to air pollution. Nitric oxide is formed, especially at high temperatures, from oxygen and nitrogen. Incomplete combustion of the fuel is more common with liquid fuels. Methods to minimize or essentially eliminate the pollution problems can probably be developed in the near future.

The desired combustion of the fuel results in a short nonluminous flame, and a large number of radiant wall burners are used for providing the heat to a pyrolysis furnace. Better heat distribution in the furnace can be obtained by increasing the number of burners and in locating the smaller burners in more strategic positions. Both capital and operating costs, however, increase as the number of burners and auxiliary control devices are increased. The optimum size of burner for pyrolysis furnaces is often 0.5 to 1.0 million Btu/h. Variations occur in the methods of positioning the burners in the furnace. Some manufacturers position the burners mainly in the walls of the furnaces, but others place a significant number of burners in the floor, as shown in Fig. 2-7.

Excess air, perhaps 5 to 10 percent, is generally provided in order to ensure complete combustion of the fuel. All furnaces have several openings, peepholes, etc. A high wind can sometimes cool the windward side of the furnace, causing temperature changes in it.

Tubular Reactors   Each tube reactor generally is manufactured by welding alternately straight sections of tubing and tubular U bends of the same internal diameter. In some cases, the U bends have slightly thicker walls. The final reactor consists of several hairpin-type coils joined in series so that the straight sections of tubing are parallel to each other. Portions of a tube reactor, as it is suspended in the radiation zone, are shown in section A-A of Fig. 2-7. Such a reactor, providing a long flow path for the reactants, can be positioned inside a furnace of reasonable size.

About 4 to 24 tube reactors are provided in a furnace (29). One or more common headers are provided to give parallel feed to the various tubes in the furnace. Product headers are used for the product gases from the tube reactors. The header and its inlet and outlet lines should be designed to minimize pressure drops and the residence time of the product in them. Coking is sometimes a problem in the headers.

Equal flow rates and identical temperature and pressure conditions are desired in each tubular reactor. In modern furnaces, the flow rate to a reactor is measured and controlled. The temperature pattern in each tube can be controlled within limits by adjusting the various burners (often located in both the wall and floor of the furnace). Measuring the exact temperature of the tube is not easy and requires consideration of several variables (102). Often optical pyrometers are used. Sampling and analyzing the exit gas from a reactor involve careful attention to details if accurate analyses are to be obtained (95).

Significant temperature and pressure variations occur as the gases flow through a tube reactor. The highest rates of reaction, and hence the highest energy demand, occur relatively near the inlet, where the feedstock concentrations are high. In general, the temperature increases steadily but not necessarily uniformly in the reactor, so that the highest temperature is at the exit end of the reactor. A rising temperature profile in a tube is often thought to minimize coking, but whether it is the ideal one relative to yields of ethylene and propylene has not been completely determined.

The mass velocity of the reaction mixture (hydrocarbons and steam) in a tube reactor varies from about 16 to 29 lb/(ft^2)(s) [80 to 145 kg/(m^2)(s)] (44). This velocity is sufficiently high to provide high heat-transfer coefficients between the tube wall and the gas mixture. Too high a velocity, however, may cause serious erosion of the hot tubes and large pressure drops.

The linear velocities of the reaction mixture as the gas mixture flows through the reactor sometimes increase to as high as 1,000 ft/s because of the decrease in the gas densities resulting from pressure drops in the reactor, the higher temperatures at the reactor exit, and the increased number of moles in the product stream. An increase in the number of moles occurs with higher-molecular-weight feedstocks especially. The pressure drop in a tube is quite nonlinear relative to tube length. In general, smaller pressure drops occur in reactors with fewer U bends and longer lengths of straight tubing.

The tube reactors have diameters varying from 2.5 to 5.5 in and tube lengths in the radiation zone of 100 to 500 ft. The tube wall thickness is not critical from an operational standpoint, and it is set primarily by manufacturing considerations. Smaller-diameter tubes provide a larger surface-to-volume ratio and more surface for heat transfer. Hence, the differences in temperature between the wall (or skin) of the tube and the gases inside tend to be smaller. Higher reactant temperatures are permissible with smaller-diameter tubes, and such tubes have become more common with the increased interest in high-intensity furnaces. Zdonik et al. (29) have reported that the length-to-diameter ratio of a tube reactor is an important parameter for correlating tube-metal (or skin) temperature, firebox temperature, and pressure drop.

Mitsubishi Petrochemical Co. indicated in 1972 that tubes with oval cross sections provide better heat transfer, allowing up to 20 percent reduction in the capital cost of a pyrolysis unit compared with a conventional one having tubes of circular cross section. They also claim that the time between decoking is generally lengthened by as much as 10 percent. Constricted-waist or peanut cross sections are also being considered for the tubes.

Heat transfer from the wall of the tube reactor to the gas mixture inside involves both convection and radiation, but convection is usually considered to be more important. The relative importance of the two can only be approximated because of the lack of reliable physical-property data, including the emissivity values for various compounds at high temperatures.

Stainless steels containing considerable amounts of nickel and chromium are used in tubular reactors. Extruded Incoloy (32 to 35 chrome, 20 nickel) is used for gas temperatures up to at least 815°C, HK-40 (25 chrome, 20 nickel) for temperatures slightly beyond 875°C, and a 25 chrome–35 nickel steel for even higher temperatures (29, 44, 104). These steels frequently contain small amounts of manganese and molybdenum. HK-40 tubes are often centrifugally cast, bored out, and then polished on the inside wall to provide a smooth and uniform diameter. These tubes are more brittle than Incoloy tubes.

The reactor tubes should provide long life under the most severe conditions. The outside wall of the tube is exposed to the intense radiation of the radiant zone and simultaneously to an oxidizing atmosphere, since some excess oxygen is always present in the combustion mixture. The outside wall (or skin) temperatures of the tubes vary with the specific furnace, but the maximum skin temperature in a furnace frequently ranges from 980°C for a low-severity furnace up to at least 1050°C for high-severity furnaces. The inside wall of a tube has slightly lower temperatures, and it is exposed to a reducing atmosphere. In addition, small amounts of sulfur compounds are generally added in the reaction mixture to suppress carbon monoxide formation and coking and hence to promote improved pyrolysis reactions (37). A factor of concern is the buildup of appreciable amounts of coke, often at least as much as $\frac{1}{4}$ in thick. If a reactor tube which is highly coked is shut down quickly, differences in the thermal expansion of the coke and steel may lead to serious problems, including cracking of the metal.

Destructive carburization is a frequent cause of tube failure (104, 139). When a tube cracks, the issuing gas merely burns in the combustion zone. As a part of preventive maintenance, whenever a furnace is shut down, the tubes are inspected. Tubes that have become quite magnetic (caused by the partial loss of chromium) are replaced, since they are likely to fail shortly. Carburization, various other changes of the crystal structure of the metal, and problems associated with high-temperature operation of the tubes have been discussed by Zeis and Heinz (139).

At operating temperatures, all the stainless steels used creep. Normally a tube is designed to creep no more than 0.01 percent per 1,000 h, and its life expectancy would be about 10 to 12 years (104). In many plants, a tube is heated higher than design values for at least short periods of time. Tubes frequently last several years before they need to be replaced.

When a tube reactor fails, generally the faulty section of the tube is cut out and a new one electrically welded in place. The tube is then pressure-tested for leaks, and the weld is x-rayed to test for soundness. If the weld is too thin, too thick, or in any way faulty, the job must be repeated. Pinhole leaks are sometimes noted after a welding job, as small amounts of the reaction mixture leak into the combustion zone, where they burn. As the tube cokes up, such leaks are generally sealed.

As a rule, the tubes must be decoked every 1 to 3 months, preferably on a scheduled basis. Normally steam is passed through the hot tubes for 1 to 2 days with the production of carbon oxides and hydrogen. To obtain even faster decoking, a small amount of air is sometimes added to the steam. The oxygen removes the coke by burning. Care must be taken when air is used not to overheat the tubes or to oxidize the metal surfaces. Nitriding of the surfaces because of reactions with nitrogen can sometimes be a problem, especially if steam is not used.

The method of positioning the tubular reactors in the furnace differs significantly with various manufacturers. The reactor can be positioned so that the straight parallel passes of the reactor coils are either in a vertical or a horizontal position. When the reactor is positioned vertically, the supports for it are either in the roof of the radiation zone, as shown in Fig. 2-7, or in the floor of the furnace. Figure 2-7 indicates the locations of the tube suspension yokes and positioning guides. Reasonably common metals can be used as supports since they are not exposed to the high temperatures of the radiant zone.

Horizontally positioned tubes tend to buckle and/or sag at the high temperatures of the radiant zone unless they are supported at rather frequent intervals. These supports are subjected to the temperatures of the radiant zone. High-chrome supports tend to fail unless some method is made to cool them. Cradling the tubes in firebrick or other refractory material is sometimes used, but each support interferes with heat transfer and hence with uniform temperature control.

Furnaces using vertical coils can be adapted to conditions of higher severity because of higher heat fluxes and better control of the temperature profiles. Claims have been made that horizontal tubes result in smaller pressure drops for the reaction gases since longer passes of straight tubing are possible and fewer U bends are needed. Decoking is also claimed to be easier in horizontal tubes since loose coke can be blown out of the tubes relatively easily.

Both horizontal and vertical tubes are currently used by major ethylene and propylene manufacturers. A user of horizontal tube furnaces admits that their tubes sag but claims this does not seriously affect the reactions or the longevity of the tubes. Major furnace manufacturers, however, appear to be switching to vertical-tube furnaces whenever high-severity conditions are required.

Disagreement has been expressed among proponents of vertical tubes over the preferred location for the product-gas outlet from the furnace. In some cases, the exit tube from the product header leaves at the top of the radiation zone, as shown in Fig. 2-7, or at the bottom. The transfer-line exchanger (TLX or TLE) used to recover sensible heat must be located close to the product header (not shown in Fig. 2-7) since the connecting line must be as short as possible. It is desirable from a maintenance standpoint to have these exchangers located near ground level, but when the product line is through the floor of the furnace, the vertical coil must be supported by the floor. Thermal expansion of the vertical tubes is in an upward direction and is about 6 to 8 in. The tube support system is more complicated as a result. In addition, the lines to the TLXs tend to be longer when they are on the bottom of the furnace. Both top and bottom exit systems are currently used, but the top system is said to be more common (29).

Spacing of the vertical or horizontal rows of tubes often differs for commercial furnaces. Some furnaces use only a single row of tubes positioned at the centerline of the furnace (106). Others use two rows of tubes, one positioned on either side of the centerline (5). In each case burners are positioned on both sides of the tubes, so that radiant energy reaches the tube from several directions and (in theory) rather uniformly. Design features of the latest Lummus furnaces have been described (107, 113). More uniform reactant temperatures, shorter residence time for reactants, and less coking are claimed. A single furnace containing four tube reactors can be designed for production of 100 million lb of ethylene per year.

Several furnaces are needed in an ethylene-propylene plant in order to provide reasonably constant production levels at all times. Each furnace has to be decoked at intervals of 1 to 2 months for ethane or propane feeds and 2 to 3 months for naphtha feedstocks. The decoking sequence and any maintenance repairs on a furnace should be programmed if possible so that only one furnace is down at a time. The remainder of the plant, namely compression, separation of products, etc., is sufficiently flexible to be satisfactorily operated with the product streams of the remaining furnaces. The minimum number of furnaces for a plant has been suggested as four or five (44, 29) although several small plants have only three.

A tube-wall temperature of $1000°C$ ($1832°F$) is common in current high-severity furnaces with clean tubes. As coking occurs, the temperature increases to an upper temperature of about $1050°C$ ($1920°F$), as measured by an optical pyrometer. An experienced plant operator can also tell by visual observation of the white-hot tubes when one is heating up because of coking. Reaction-gas temperatures as high as 750 to $875°C$ are common in the newer furnaces. The higher temperatures are normally used for ethane or propane feedstock and the lower ones for higher-molecular-weight feedstocks.

The total pressure of the hydrocarbon-steam mixture varies from about 40 to 80 lb/in^2 gage at the reactor inlet to 10 to 20 lb/in^2 gage at the outlet. In one furnace, an inlet pressure of 65 lb/in^2 gage is used when the tubes are clean, and this pressure gradually has to be raised to 75 to 78 lb/in^2 gage as the tubes become coked. The exit pressure is maintained at 15 to 20 lb/in^2 gage regardless of the coke on the tubes. Steam, of course, reduces the partial pressure of the hydrocarbons, and the amount of steam normally used varies from less than 0.3 to 1.0 lb per pound of hydrocarbon feedstock, depending on the feedstock.

Several features common to modern commercial units may be outlined. Fluxes as high as 20,000 to 30,000 $Btu/(h)(ft^2)$ are common through the tube wall of high-severity furnaces. Appreciably lower fluxes are present in lower-severity units. Residence times of the reactants in the reaction zone range from about 0.3 to 0.5 s in higher-severity furnaces. With lower-severity furnaces, residence times as high as 2 or 3 s are sometimes employed. The conversion level varies significantly with the hydrocarbon feedstock, and common levels are as follows: ethane, 60 to 65 percent; propane, 85 to 95 percent; and n-butane, 85 to 96 percent. Measuring the conversion levels for naphthas, gas oils, and crude oils accurately is impossible since each of the above hydrocarbon cuts is a large mixture of hydrocarbons which are pyrolyzed at quite different rates.

Nontubular Reactors   Tubular reactors are not always practical, especially for high-molecular-weight feedstocks such as crude oils and some types of gas oils, since rapid coking of the tubes would occur. Hence different types of reactors must be used in such cases.

Fluidized-Bed Reactors: The Badische Anilin- und Soda-Fabrik (BASF) process (61, 125), now commercialized, and the K-K process (5), being developed at the University of Tokyo, employ a fluidized-bed reactor. Ceramic and coke particles, respectively, are used in these processes, which crack heavy-hydrocarbon feedstocks.

Figure 2-8 is a simplified diagram of the reactor system for these two processes. Steam and hydrocarbon feedstock are continuously introduced with hot, "clean," solid particles to the reactor, where the solids are fluidized. Heat is transferred from the solids to the hydrocarbons, which react to form ethylene, propylene, and numerous C_2 to C_6 byproducts. In addition, by-product coke is produced and deposits mainly on the fluidized solids.

Solid particles coated with coke are withdrawn continuously from the reactor and transferred to the regenerator. Fuel and hot air are

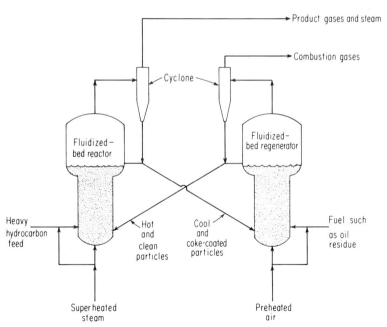

Fig. 2-8   Fluidized-bed reactor and regenerator for pyrolysis of heavy feedstocks.

also fed to the regenerator to fluidize the solids there, to burn the coke from the solids, and simultaneously to heat the solids. Sufficient solids are recycled to and from the regenerator and reactor to provide energy for the pyrolysis reactions. Design of a reactor system requires careful consideration of several factors.

The solid particles that are fluidized in both the reactor and the regenerator must be able to withstand rapid and large temperature fluctuations without excessive fragmentation or wear. Reactor temperatures in the BASF process vary from about 760°C for crude oil to 790°C for naphtha feeds. Perhaps slightly higher temperatures are used in the K-K process. In the regenerator, temperatures of about 900°C occur.

Erosion is always a problem in a fluidized-bed system, especially at the high temperatures employed. BASF claims the fire-resistant ceramic solids they use minimize erosion. The petroleum coke used in the K-K process is said to be rather plastic at reaction temperatures, resulting in minimal erosion.

Methods of introducing the hydrocarbons and steam to the reactor should provide uniform distribution of the hydrocarbons in the bed and uniform fluidization. In a related fluidization process of Ube Industries, Inc. (110), steam is used to atomize the liquid crude oil.

The fuel employed for heating purposes in the regenerator of the BASF process is a high-boiling hydrocarbon formed as a by-product. Design of the regenerator to realize complete combustion (with little production of carbon monoxide) and yet to use little excess air is highly desirable. The residence time of the reactants in the fluidized bed, in the disengagement space above the bed, and in the cyclonic separators used for final separation of the solids from the gas is important. The main reactions presumably occur while the gases are in the fluidized bed, whereas side reactions occur in the space above the bed and in the cyclonic separators. Hence it is desirable to minimize the time in the latter two spaces or to cool them somewhat so that side reactions are minimized.

The yields of products obtained by cracking a high-molecular-weight feedstock vary significantly with its chemical characteristics. Yields of ethylene from about 25 to 30 percent and yields of propylene from 12 to 16 percent have been reported for the K-K process.

In two related processes using a fluidized-bed reactor for the pyrolysis of crude oil (5, 110), a mixture of crude oil, steam, and oxygen is introduced to the fluidized-bed reactor. Combustion of part of hydrocarbons provides the energy required for the pyrolysis of the remaining oil. Combustion to produce carbon dioxide but not carbon monoxide is preferred in order to simplify the product-recovery steps.

The method for preparation of the ceramic solids used in the Ube process has been described (110); spherical particles are recommended. Choice of the fluidized solids, particularly in a partial-oxidation process, may require some consideration of the role the solid surfaces play in the oxidation reactions. Recent research by Menzies (111) has shown that oxygen either adsorbs or reacts on certain surfaces that subsequently are very effective for oxidizing carbon monoxide. Oxygen is generally used in these latter pyrolysis processes instead of air in order to eliminate recovery problems associated with separation of the product gases from nitrogen.

Packed-Bed Reactors: Pebble-bed furnaces were formerly employed in several plants. The gaseous hydrocarbons were passed over preheated pebbles packed in a large column. As the pyrolysis progressed, the pebbles were cooled and simultaneously carbon deposits formed on the pebbles. The hydrocarbon flow was then stopped, and air and perhaps some fuel flow was started to burn off the carbon and to reheat the pebbles. Thus pyrolysis reactions occurred over a fairly wide range of temperatures, and the conversions of the feed hydrocarbons varied significantly during the pyrolysis phase of the cycle. Such reactor systems cannot in general compete with more modern ones, such as fluidized-bed reactors.

Molten-Bath Reactors: A molten-lead bath (80) was used commercially by Monsanto Co.; the gaseous hydrocarbons were bubbled up through the melt. Such an arrangement provides excellent temperature control of the reaction mixture, but mechanical problems of building and operating such a reactor made it uneconomical. Contamination of the melt with coke, tars, sulfide compounds, etc., would be severe in most if not all cases.

The reactors for the simultaneous production of ethylene and acetylene are nontubular and will be discussed later in detail.

Energy-Recovery Operations

Much of the sensible heat of the product mixture from the pyrolysis furnaces must be recovered and reused, for economic reasons. In addition, the product mixture must be rapidly cooled to stop the reactions and to minimize side reactions that destroy ethylene, propylene, and valuable by-products. The cooling procedures used depend to a considerable extent on the feedstock for the pyrolysis units (29). When ethane, propane, or butane is used, transfer-line exchangers [called TLXs, TLEs, or by some companies (44) quench coolers] are used as the main method for cooling and recovery of energy. A combination of TLXs and oil-quenching devices is generally used when naphthas are the feedstocks. Frequently only an oil-quench system is used when a heavy gas oil is the feed, but in some cases TLXs are used to remove part of the heat.

When transfer-line exchangers are employed, more than one may be provided for a given pyrolysis furnace. As an example, two exchangers per furnace are used in a plant having furnaces with an ethylene capacity of 100 million lb/year each. The heat from the hot product gases is transferred to water to produce 400 to 2,000 lb/in^2 steam (29, 104). A steam drum is provided above each exchanger, and the steam is disengaged from the water. Approximately half the steam needed for the pyrolysis plant using ethane as a feedstock can be produced in this manner. Because of the high temperatures in these exchangers, provisions must be made for the large thermal expansions that occur in the tube, shell, bolts, etc., as the exchanger is heated from ambient conditions (60). At least three types of TLXs are in commercial use.

The Schmidtsche TLX consists of several double-pipe heat exchangers operated in parallel (134). These individual exchangers are all positioned inside the outer shell of the vertically mounted unit. The hot gases enter either at the top or the bottom of the unit. This TLX has been widely used in processes employing light feedstocks as well as heavier ones. A TLX of this design is used, for example, in a naphtha-pyrolysis process for production of 1,200 lb/in^2 steam (90).

TLXs that are basically shell-and-tube exchangers are also employed (104, 123). The hot gases flow through the tubes, and steam is generated on the shell side. Design of the inlet header is critical in order to minimize the residence time of the inlet gases in it and hence to minimize undesired further reactions. Extra steam is sometimes introduced in relatively small amounts to this header (130, 131). This steam cools the feed mixture slightly, and it can probably be introduced to at least partially blanket the header walls and minimize coke formation there. Thin tube sheets are sometimes used in these headers in order to promote rapid quenching of the inlet gases (123). This TLX is often used in a horizontal position.

Important design parameters of TLXs include the mass velocities and dew point of the product mixture and the temperatures of the walls of the tubes (44). Sufficient velocities are generally provided to obtain overall heat-transfer coefficients of about 30 to 45 Btu/(h)(ft^2)(°F). Too high a mass velocity, however, results in excessive pressure drops. Popular tubes for TLXs are reported to have a diameter of 31.0 mm (1.25 in) and a wall thickness of 4 mm (0.16 in).

The wall temperatures of the tubes in the exchangers affect the amount of coking in the tubes and header (or cone). The exit ends of the tubes have the lowest temperatures, and condensation of high-boiling hydrocarbons is most likely there. Linde (44) reports that gas temperatures from about 300 to 350°C can generally be obtained with minimum plugging problems if the tubes are clean. As the tubes slowly collect a coke film, the exit-gas temperatures rise. When the temperature reaches 400 to 450°C, the exchanger should be cleaned.

TLXs are cleaned whenever a furnace is shut down for cleaning and repairs or vice versa. To clean some exchangers, the heads are removed, and the tubes are cleaned mechanically. Horizontal exchangers are relatively easy to clean in this manner, particularly if the exchanger is on the ground level below the furnace. High-pressure water jets have also been used for cleaning exchangers.

A third type of TLX has recently been announced (104). In this unit, the hot gases are on the shell side, and the steam is on the tube side. The tubes are tied to a peripheral collector, thus eliminating the tube sheet.

For higher-molecular-weight feedstocks, the product gases cannot be cooled as much as the product streams from light paraffins because larger amounts of condensable hydrocarbons, including tars, are present. Hence, higher temperatures must be provided, and higher-pressure steam is produced; less steam is obtained, however, because of the smaller decrease in the sensible heat of the product gases in the cooler.

The gas mixtures from TLXs are sometimes cooled further, especially with heavier feeds, in order to recover more energy. A common method is to contact the partially cooled gas stream directly with a relatively nonvolatile oil. The oil serves to cool the gases and to condense some of the less volatile hydrocarbon products. The resulting hot oil can be recirculated to a heat exchanger to preheat the boiler water or feedstock. In some plants, low-pressure steam is generated in this way. Linde (44) reports that the amounts of steam generated by the hot oil vary significantly, depending on the level of coking in the quench exchanger and hence on the amount of heat added to the quenching oil system.

When an oil-quench system is used, part of the mixture of oil and dissolved hydrocarbons is sent to either a distillation or a stripping column for the recovery of valuable hydrocarbon products dissolved in the oil. The recovered hydrocarbons are then processed in the separation section of the plant. The oil stream is recycled to the oil-quench system.

Recently a novel system in which molten lead is used as a quench medium has been advertised for recovery of the sensible heat of the pyrolysis gases (117). The main advantage claimed is that coking is no longer a problem. Methods of removing contaminants that would certainly collect with time have not been reported.

Supplemental Steam Generation

Since transfer-line exchangers generate only part of the steam needed in a pyrolysis unit, additional steam must be supplied from other sources. In some cases, this steam is provided from a central steam-generating plant. In other cases, the pyrolysis furnaces are built big enough to generate sufficient additional steam in the convection sections (44, 57). Extra fuel for the pyrolysis furnace may be needed for this additional steam.

The steam system in a pyrolysis plant is large and complicated. It must be designed with sufficient flexibility to allow for start-ups and shutdowns, for changes in the recovery of the sensible heat as the quench coolers become coked, and for most problems caused by the temporary malfunctioning of any portion of the steam system.

Jaggai and Holland (90) describe the steam system in a Japanese plant. Steam at the approximate pressures of 50, 150, 400, and 1,200 lb/in^2 gage is needed. The 1,200 lb/in^2 gage steam is used to drive the turbine for the propylene refrigeration compressor, and the exhaust steam from this turbine at 400 lb/in^2 gage is resuperheated and used for driving the turbines of the product (cracked) gas compressors. Other steam arrangements are employed in other plants, but effective utilization of the steam to minimize the energy requirements of the process is a key factor in all pyrolysis plants.

Additional Cooling of Product Stream

Before the light hydrocarbons can be compressed, they must be cooled to essentially ambient temperatures, and most of the steam and some of the heavier hydrocarbon products are condensed and separated because of the cooling. A water-scrubbing tower (106) is widely employed in plants designed for ethane or propane feedstocks, as shown in Fig. 2-6. The gases from the transfer-line exchangers enter near the bottom of the tower, and water enters at the top. Several trays are provided in the top of the column, and water sprays and/or more plates are in the bottom portion of the column. The water sprays remove most of the tars and less volatile components that would plug the plates. Plates constructed from angle irons are sometimes used in the bottom sections of the column since such plates do not plug easily. The trays in the top section can be conventional trays, e.g., Glitsch ballast trays, that provide effective contact between the relatively cool water and the remaining gas stream.

Another common method for this cooling operation is to use at least two heat exchangers with air as a coolant in the first and water in the second. This method works satisfactorily after the gas stream is cooled and washed with oil since the oil removes most of the tars and less volatile hydrocarbons that cause coking or plugging problems in the tubes of the exchanger.

Water and hydrocarbon phases form during the cooling operation. The phases are normally separated in a decanter or settling drum. This drum must be large enough to provide several minutes' residence time since the separation of phases is often rather slow. In many cases, a heavy tar phase also forms at the bottom of the drum. A light organic phase, which contains considerable amounts of aromatics and is in the gasoline range, forms as the top layer, and the water phase is intermediate. Both the water and oil phases are recovered and then processed further. The oil phase is stabilized to remove volatile hydrocarbons. Part of the water phase is often air-cooled and then recycled to the water-scrubbing tower. The remainder of the water phase is sent to a distillation column to recover dissolved hydrocarbons. The resulting water, however, often still contains considerable amounts of phenols and other hydrocarbons. Such high-boiling hydrocarbons cannot be removed by distillation, and disposal of the polluted water has only been partially solved by biodegradation and other techniques. Zdonik et al. (29) report further details of the water-quenching systems.

The cooling system for the product gases is designed to cool the mixture to about 40 to 50°C (104 to 112°F) for plants using ethane-propane feedstocks. Higher temperatures are required in naphtha plants to minimize condensation of the heavier by-products. The gas pressures are normally slightly above atmospheric.

Product-gas Compressors

Centrifugal compressors for the product gases in any pyrolysis plant are large and complicated. The steam turbines used to operate these compressors require large volumes of high-pressure steam.

The feedstock of these compressors always contains a wide variety of hydrocarbons, including the uncondensed portion of the heavier hydrocarbons. Water vapor, carbon oxides, and hydrogen sulfide are also present. The gases are compressed in several stages to a final pressure of about 25 to 35 atm. Centrifugal compressors are now used exclusively in the larger, more modern plants since they are more economical than reciprocating compressors. Four- or five-stage centrifugal compressors are common, and the product stream is cooled between stages. Liquid hydrocarbons and water formed because of the compression and cooling steps are separated from the gas phase after each stage. Hence the composition of the gas stream changes somewhat during compression.

In operating the compressors, it is important not to get too high a temperature in the exhaust stream from any stage of the compressor. High temperatures promote the formation of polymer deposits in the compressors. Zdonik et al. (29) indicate that 110°C (230°F) is a maximum discharge temperature allowable when relatively large amounts of dienes and C_5 and heavier hydrocarbons are present in the gas stream being compressed. Up to 120°C is permissible, however, when ethane and propane are cracked. Generally polymer fouling is most critical in the first and second stages of the compressors.

Two techniques are sometimes used to minimize polymer buildup in the compressors. Water injection reduces the temperature of the exhaust gas from each stage. In many cases, the preferred technique is the introduction of a stable gas oil which is atomized into the suction (or inlet) line of each compressor stage. This oil "washes" the critical portions of the compressor and prevents or at least minimizes polymer deposits. Inhibitors or antifouling agents are also often mixed with the oil being used.

Compressors have been operated for periods up to 1 or even 2 years before cleaning or major maintenance was necessary. Such a performance record indicates reliable operation and has justified the use of a single large compressor unit for the product-gas stream of large plants. A three-case five-stage centrifugal compressor driven by a condensing-steam turbine is used in the huge ethylene-propylene plant of Imperial Chemicals Industries (82, 86). This plant, which produces slightly over 1 billion lb/year of ethylene, is provided with a steam turbine for the product compressors that is rated at almost 40,000 hp.

The compressor system for a plant should be designed to maintain the desired pressure and temperature levels even though the flow rate may vary. Inevitable variations are continually occurring in the flow rate, pressure, temperature, and composition of the inlet-gas stream, in the steam pressure and the amount of steam superheat, etc. Various methods of controlling the operation of the compressor and the accompanying steam turbines have been considered. Butterfly valves have been used to control the pressure in either the inlet (suction) line or the exit (discharge) line of the compressor. Another control method is to use adjustable inlet guide vanes, which affect the direction of the gas stream entering the compressor; such a technique can be used to control the exit pressure obtained in the compressor. Adjustable guide vanes tend to give highly stable operations but are somewhat more expensive. Sometimes adjustable guide vanes are installed in the first and maybe second stages of the compressor, and the suction pressure is controlled in later stages.

A compressor and turbine can sometimes be modified to produce up to 20 or even 35 percent more capacity or horsepower. In one turbine, the nozzle blocks were changed and the horsepower was increased significantly. The steam requirements, however, increased for unit horsepower obtained because of this change.

The exhaust hydrocarbon stream from each stage, as mentioned earlier, is first cooled, and the resulting liquid hydrocarbons and water are then separated from the gas phase using a separation tank. The gas stream from such a tank is introduced to the inlet (or suction) line of the next stage of the compressor. The liquid-hydrocarbon phase contains considerable amounts of valuable by-products that need to be recovered. Methods for this recovery vary in commercial plants.

The liquid hydrocarbons are often collected in a common tank maintained at the pressure following the second stage of compression. Provisions are made to recirculate any gas that flashes from the combined liquids to the third stage of the compressor; these details are not shown in Fig. 2-6.

In another recovery arrangement, the liquid hydrocarbons from the separation tank at the exit end of the previous stage are fed back into the separation tank at the end of the previous stage (90). By this method, the hydrocarbon condensate from the separation tank at the end of the first stage contains relatively few low-molecular-weight hydrocarbons. This technique is often used for naphtha and gas-oil processes, in which large amounts of hydrocarbons are condensed in both the first and second stages of compression.

The hydrocarbon liquids from the compressor system are fractionated or stripped to remove C_2 to C_3 hydrocarbons, which are recirculated to the compressors. Both C_4 and gasoline cuts are sometimes obtained. The water products obtained in the compression operation are generally combined, and the mixture is then stripped to remove valuable hydrocarbons.

Purification of Product Stream

The hydrocarbon product stream leaving the pyrolysis furnace contains relatively small amounts of acid gases (primarily carbon dioxide and hydrogen sulfide) and acetylenic hydrocarbons. In addition, the product stream initially contains considerable amounts of water vapor, most of which is removed during cooling and compressing. At least three purification steps are employed in plants producing high-purity olefins. These steps are for the removal of the acid gases, removal or elimination of the acetylenic compounds, and final drying of the hydrocarbon stream.

Removal of Acid Gases Acid gases (mainly carbon dioxide and hydrogen sulfide) are often removed after the product-gas stream has been compressed to an intermediate pressure of about 15 atm. Such a purification step frequently is provided between the third and fourth stages of the compressor, and at least two techniques are employed commercially.

In one technique, as shown in Fig. 2-6, the gases are washed first with caustic solutions (such as 3 to 15% NaOH solutions) and then with water. One or more columns are used, and they are operated with countercurrent flow of liquids and gas. Such columns are quite large and may be provided with at least 30 valve-type trays to remove essentially all acid gases. Sodium carbonate and sodium sulfide are formed in the aqueous solutions, which are recirculated several times. Plugging of the lines and the equipment with solid salts is a problem when essentially complete utilization of the caustic is attempted.

Disposal of the spent caustic is always a problem (109). Some companies, especially in the past, have transported the caustic by barge to sea, where it was dumped. Abandoned deep wells were used in other cases. More recently it has become increasingly necessary to neutralize the spent caustic to meet pollution standards.

Another technique widely used for removing the acid gases consists of first scrubbing the gas stream with a monoethanolamine solution. The resulting gases are scrubbed first with a caustic solution and then with water. The monoethanolamine solution dissolves most of the acid gases, and it is then regenerated and reused. This latter technique is preferred when the acid-gas content is relatively high since caustic use is minimized.

Other amines have been considered for removal of the acid gases including 2-(2-aminoethoxy)ethanol. When amine solutions are used, it is necessary to have a reasonably low regeneration temperature for the amines. Small amounts of hydrocarbons also adsorb in these amine solutions, and butadiene and heavier diolefins (109) polymerize at high temperatures, about 70 to 100°C. It is hoped that recent process modifications will allow even lower regeneration temperatures than were previously possible. Amines other than monoethanolamine may become competitive in the future.

Removal of Acetylenic Hydrocarbons   Acetylene is the main acetylenic compound present in the product-gas stream from a pyrolysis furnace, but others include methylacetylene and vinylacetylene. The concentration range for these compounds is generally quite low, ranging from about 2,000 to 5,000 ppm when light paraffins are feedstocks up to perhaps 0.2 or even 3 percent by weight (based on the ethylene produced) for naphtha feedstocks. When especially high temperatures, higher than usual conversion levels of cracking, or crude-oil feedstocks are used, significantly more acetylenic hydrocarbons may be present. Specifications for both ethylene and propylene necessitate that they be essentially free (less than several ppm) of acetylenic compounds. Simple distillation cannot be used for the desired separations because of the formation of azeotropes or the small differences in volatilities.

Probably the most common technique for reducing the acetylenic content in an olefin plant is to hydrogenate the triple bonds to double bonds. Unfortunately the reduction processes are not completely selective, and side reactions occur. When acetylene is hydrogenated in a C_2 stream, as much as 1 to 2 percent of the ethylene may be hydrogenated as the acetylene content is reduced to 1 to 10 ppm. Some acetylene also polymerizes to form a "green" oil.

The hydrogenation reactor in an olefin plant is generally located at one of three places (29). First, it may be located before the product stream has been separated into the various light hydrocarbons. The acetylene converter is often located after the acid gases have been removed, as shown in Fig. 2-6, and it is frequently referred to as a *front-end converter* since it is located before hydrogen is removed from the gas stream. The feed stream to a front-end converter is sometimes dried, but in other cases additional steam is added to it.

In the second method the C_2 fraction is first separated from the heavier hydrocarbons in the deethanizer, to be described later. The C_2 mixture is then sent to a *back-end* or *tail-end converter,* where the acetylene is selectively removed. The nomenclature for this converter is based on the fact that it is located near the end of the separation process. Another back-end converter is also sometimes provided for removal of methylacetylene or propadiene from the C_3 fraction.

In other processes (69), C_3 and/or C_4 hydrocarbons plus heavier components are removed from the product mixture first, leaving a mixture of light hydrocarbons that is then selectively reduced.

The catalyst used for the reduction process is normally packed in a cylindrical vessel operated adiabatically. Front-end converters frequently use a nickel, nickel-cobalt-chromium, or cobalt-molybdenum catalyst supported on alumina or silica (121). A trace of hydrogen sulfide is generally added to the feed stream to maintain the desired catalyst activity and selectivity. In this way a metal sulfide surface is provided on the catalyst. Palladium on alumina is a popular catalyst for back-end converters, but it cannot be used in front-end converters since catalyst poisons, including carbon monoxide, quickly deactivate it.

Removal of the acetylenic compounds by front-end hydrogenation has both advantages and disadvantages. Since excess amounts of hydrogen are already in the product stream, additional hydrogen need not be added. Furthermore, all acetylenic hydrocarbons are removed in front-end converters. Unfortunately significant amounts, often 30 to 60 percent, of the diolefins, including butadiene, are also reduced or polymerized in the front-end converters. Hence these converters are unsatisfactory for a plant in which butadiene or other diolefins are to be recovered, as in plants with heavier feedstocks. The polymerization of diolefins tends to coat or "coke" the catalyst surface, and higher temperatures are then normally used for the reduction steps. Higher temperatures result in less selective hydrogenations, and the catalyst has to be regenerated frequently.

Several operating variables are important in the operation of a front-end converter. Higher temperatures, higher partial pressures of hydrogen, and lower concentrations of hydrogen sulfide (or other sulfur-containing gas) all increase the rates of hydrogenation. More hydrogen sulfide, however, increases the desired selectivity of hydrogenation of acetylene compared to ethylene. About 5 to 10 percent steam is sometimes added to the hydrocarbon feed since steam increases the catalyst activity and also reduces the formation and deposition of polymers on the catalyst surface.

Often more than one catalyst bed is provided in the cylindrical reactor to maintain better temperature control and fairly uniform residence times of the gases in the column (29). The gas distributor at the top of the column is designed to provide uniform distribution of the downflowing inlet stream to minimize disarrangement of the top catalyst bed (121). Otherwise attrition of the catalyst particles may occur. To prevent runaway temperatures, part of the reactor feed may be injected partway down the column reactor.

A gas heater is usually required to heat the reactor feed stream to the desired temperatures (29, 44). Temperatures in a converter vary with time and with position in the catalyst bed. As some catalysts are used, they gradually lose activity, and higher temperatures are required. Temperatures at the top of the bed in a front-end converter may be approximately 140 to 160°C and those in the bottom from 190 to 210°C. A converter may be continuously used up to several months before regeneration. Two converters are normally provided; one is either being regenerated or being used as a standby while the other is in operation.

Regeneration of the used catalyst in a front-end converter is accomplished by using superheated steam and possibly a small amount of air at about 375 to 450°C. The organic deposits are either vaporized and vented from the bed and/or removed by oxidation reactions. Following removal of the organic hydrocarbons, the catalyst must be reactivated with hydrogen at elevated temperatures. Regeneration and reactivation of the catalyst require about 1 to 3 days, depending to some extent on the catalyst and the operating variables. Immediately after reactivation, the selectivity of hydrogenation is generally low, but it increases with increased hydrogenation; meanwhile the activity of the catalyst is decreasing.

Back-end converters, which generally use palladium or promoted palladium catalysts, are operated at about 70 to 120°C and with space velocities of 1,000 to 3,000 volumes of gas per hour per volume of catalyst bed (121); 2 to 3 mol of hydrogen is added per mole of acetylene present to reduce the acetylene content of the ethylene to less than 10 ppm. Back-end converters are significantly smaller than front-end converters because of the smaller volumes of gas involved. Unreacted hydrogen must sometimes be removed from the hydrocarbon stream leaving the back-end converters. These latter converters often have to be regenerated after several months by procedures similar to those used for front-end converters.

The green oil from the converters, which is a mixture of liquid polymers, is generally disposed of by burning in one of the furnaces.

The flowsheet for removal of acetylene presented by Zdonik et al. (29) shows an arrangement for the converters, feed heater, regeneration heater, and heat exchangers.

With an improved palladium catalyst, a much simpler process for hydrogenation of acetylenic compounds is feasible (104). The process of Farbenfabriken Bayer is described as a *cold hydrogenation process* since the feed gases do not need to be preheated (99). At the low temperatures used, relatively little thermal polymerization occurs to produce green oil and other low polymers. The catalyst packed inside the tubes of what is essentially a heat exchanger retains its activity for a long time. Only a single reactor with intercoolers is provided in some cases (104). Coolants employed to moderate the temperatures are either C_3 or C_4 hydrocarbons, such as liquid propylene or butanes at their boiling point. Trickle-bed systems are employed when liquid hydrocarbons are to be treated.

The second method for reducing the acetylene content of a gas stream in an olefin plant uses solvent absorption. Although such methods have been known for many years, increased interest and improved processes have recently resulted in wider use of this method. Effective solvents include N,N-dimethylformamide (DMF) and acetone (93, 94, 128). Several commercial units, including the Monsanto plant at Chocolate Bayou, employ DMF (136), and Linde have recently publicized (105) some of the details of their process, which uses either DMF or acetone. They claim that their process is effective for production of 99.9 percent pure acetylene from product streams containing as little as 0.2 percent acetylene. They have presented calculations indicating significant financial incentive for recovering acetylene in plants using naphtha feedstocks; the recovered acetylene has a higher sales value than ethylene.

Drying of Hydrocarbons   The hydrocarbon stream from the compressors must be dried before it is refrigerated, otherwise ice forms. Packed beds of alumina or molecular sieves are preferred desiccants. Two or more column dryers filled with desiccant are normally provided in a plant. One or more columns are on stream while another column is being regenerated. Figure 2-6 shows only a single column located after the last (fourth-stage) compressor.

The size of the bed can be designed for 12, 24, or 48 h of operation before regeneration is needed, but a 24-h cycle is common (29). In some large plants, three columns are provided, and two columns in series are on stream. The column which has been regenerated last is the guard column, i.e., the second column in the series. When the primary column (the first column in the series) is ready to be regenerated, the guard column is switched to become the primary column. A guard column ensures removal of essentially all moisture. Some modern plants are provided with motor-operated valves and a cycle time to switch the columns automatically.

The desiccant beds are reactivated by passing hot, dry residue gases over the bed for several hours. The temperature of the gas stream is slowly raised. The water and adsorbed hydrocarbons, which are primarily the less volatile ones, are stripped from the desiccant. Zdonik et al. (29) present the flowsheet of the dehydration equipment, which consists of the column dryers, dehydrator reactivation furnace, knockout drum for less volatile hydrocarbons, and heat exchanger.

Molecular sieves are currently used in several large olefin plants. The Linde Division of Union Carbide Corp., which markets molecular sieves, has reported extensive laboratory and plant data indicating that drying with molecular sieves is cheaper than with alumina or silica gel, another possible desiccant (118, 126). Molecular sieves have a much greater water capacity, often 5 or more times greater; they retain this high capacity for more cycles (drying followed by regeneration); and in addition they adsorb fewer heavy hydrocarbons. This latter factor has not yet been completely investigated but may be important since hydrocarbons that are adsorbed are not recovered during the regeneration of the bed. This preliminary evidence indicates that more by-product hydrocarbons are recovered with molecular sieves. A plant using mainly ethane as the feedstock replaces its alumina bed every 9 to 12 months, and molecular sieves probably would last much longer. The major disadvantage of molecular sieves appears to be their high purchase price.

Separation of Hydrocarbons

Ethylene, propylene, and other important by-products are recovered in pyrolysis plants mainly by low-temperature fractionation. After the heavier hydrocarbons, the acid gases, acetylene, and water have been removed from the hydrocarbon mixture, the residual mixture is refrigerated and partially liquefied before fractionation.

In most plants (see Fig. 2-6), the hydrocarbons are separated in order of decreasing volatilities. The first step in such a case involves partial condensation of the mixture followed by demethanization. Noncondensable gases, including hydrogen, nitrogen, and carbon monoxide, are first removed, and a relatively pure methane stream is recovered. Next the C_2 hydrocarbons (ethylene, ethane, and any acetylene present) are obtained as the top product from the deethanizer. The depropanizer and debutanizer are subsequently used for separating the C_3 and C_4 hydrocarbons, respectively, from the C_5 to C_8 hydrocarbons. The C_2 and C_3 fractions are separated to recover ethylene and propylene, respectively. The C_4 and heavier cuts are also sometimes further processed to recover valuable by-products such as butadiene, isoprene, and aromatics.

In some plants, the deethanizer is positioned before the demethanizer. Claims have been made that such an arrangement requires less refrigeration (44, 122), but Zasloff (138) indicates that the Lummus Company after careful study found no real advantages to the arrangement. Lummus normally positions the demethanizer before the deethanizer.

Clancy and Townsend (69) have recently discussed another separation technique used in several plants. The product mixture is first separated to obtain a light-end cut (through C_3's) and the remaining (C_4's and heavier) hydrocarbons. This arrangement is designated as a front-end depropanizer. In the flowsheet for this process (see Fig. 2-9) the depropanizer is located between the dryer and the fourth-stage compressor. The acetylenic compounds are removed from the light-end fraction, the resulting gas stream is cooled to obtain partial condensation, and the condensate is used as reflux for the depropanizer. The remaining gases are sent to the coolers provided with the demethanizer. The light ends are subsequently separated in a rather conventional manner, as shown in Fig. 2-9. The bottom product from the depropanizer is normally separated in the debutanizer to give a C_4 cut and a heavier fraction that can be used as a gasoline-blending stock. The claim is made that this separation technique is superior to the technique shown in Fig. 2-6 and less expensive to operate.

Low-Temperature Refrigeration Low-temperature refrigeration is needed to liquefy the light hydrocarbons, and cascade refrigeration is

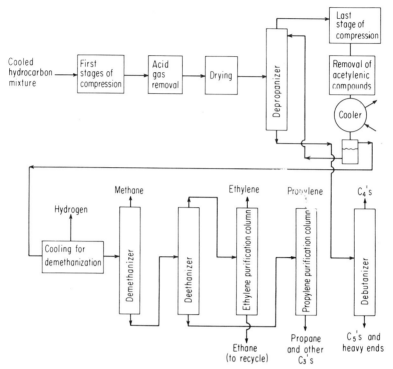

Fig. 2-9   Separation technique using front-end depropanizer.

employed.   Propylene, ethylene, and sometimes methane are the refrigerants in most cascade systems.   Tucker (129) gives many details of the temperatures and pressures used with propylene and ethylene refrigerants.   Propylene refrigerant is used for temperatures down to about −45°C (−50°F).   Ethylene provides temperatures from about −70 to −100°C (−90 to −150°F), and temperatures from −130 to −160°C (−200 to −260°F) can be obtained with methane.

The energy demands for the refrigeration compressors in an ethylene-propylene plant are large—often approximately equal to those of the product—gas compressor.   Multistage (often three-stage) centrifugal compressors are used for both propylene and the ethylene.   A positive-displacement compressor may be used if methane is to be refrigerated since the amounts of methane used are relatively small.   For a plant producing 0.5 billion lb/year of ethylene, the driver of a propylene compressor may be rated in the 15,000- to 20,000-hp range.   The ethylene compressor probably requires only about one-fourth that amount of energy.   If a methane unit is employed, energy requirements may be approximately 1,000 hp.

As increasing amounts of natural gas are transported by tanker in liquid form, the suggestion has been made that the liquefied gas could be used as a refrigerant in pyrolysis plants (98). Such a suggestion seems to have merit since the natural gas is normally distributed as a gas to industrial and home customers.

Demethanizer The demethanizer is said (29) to be the heart of the low-temperature fractionation system. When it is positioned before the deethanizer, most economical operation is at about 450 to 500 lb/in² gage (31 to 35 atm) and at temperatures low enough to ensure that most of the methane is liquefied. Liquid methane is used as a reflux in this column. A portion of the methane and residual hydrogen, nitrogen, and carbon monoxide is not condensed, however, and these gases are sent to a burner for disposal.

Several methods of operating the demethanizer are used commercially (44, 79, 133). The first consists of cooling the feed stream, first with propylene and then with ethylene refrigerant. Reflux temperatures of -95 to $-100°C$ (-135 to $-150°F$) are obtained with ethylene refrigerant.

A second method is to cool the demethanizer feed stream with propylene refrigeration to about $-40°C$ ($-40°F$). A side condenser employing ethylene refrigerant at approximately $-70°C$ ($-92°F$) is located above the feed plate. The temperatures at the top of this demethanizer are similar to those of the first method.

A third arrangement for the demethanizer is to cool the hydrocarbon mixture to about $-130°C$ ($-205°F$). Most of the C_2's and heavier hydrocarbons are condensed, and the uncondensed hydrogen and methane are separated from the condensate. The condensate is then fed to the demethanizer to separate the small amount of condensed methane.

A series of heat exchangers is often used to cool the dry feed mixture to the demethanizer. As many as 10 exchangers in series are cooled with liquid propylene and ethylene refrigerants, but for simplicity, Fig. 2-6 shows only one. As the temperature of the feed mixture is reduced, partial condensation occurs, yielding several condensates of varying composition and temperature. As a general rule, only the liquid condensates are fed to the demethanizer. Multiple feed inlets are provided so that each liquid condensate is added at the optimum plate for that liquid. To accommodate the significant changes in liquid and vapor loads throughout the column, a column with variations in diameter as a function of height is often employed.

Gas expanders, or turboexpanders, are sometimes used for obtaining temperatures lower than those which can readily be obtained with ethylene refrigeration (63, 124). In the expander, high-pressure gases are expanded, and the work energy is removed. Figure 2-6 indicates how an expander is used in the final cooling stage for the incoming hydrocarbon mixture. When high-pressure gases containing large concentrations of methane are used, some methane condenses and the condensate may be used as reflux for the demethanizer.

Separation of methane from ethylene requires only several plates in the rectifying section of the demethanizer; however, about 30 plates are needed in the stripping section of the column in order to reduce the methane content in the bottom product sufficiently. The bottom temperature is usually less than 18°C (65°F), and high-pressure refrigerant (propylene) vapor is often used as a heat source in the reboiler of the demethanizer.

Low-carbon steel is not suitable for temperatures less than −40°C (−40°F), and 3.5 percent nickel steel is used in those portions of the column below −40°C. Obviously the demethanizer must be carefully insulated to minimize heat transfer from the atmosphere.

Low-temperature equilibrium constants for methane and other light hydrocarbons (87) are recommended when designing a demethanization column.

Recovery of Pure Ethylene and Propylene Design and operational details of the remaining distillation equipment may vary significantly between commercial plants. Each plant is unique, with different hydrocarbon mixtures to separate and different specifications or requirements for the desired products. Table 2-8 summarizes typical information for fractionation towers used to produce high-purity ethylene and propylene and also C_4 and heavier cuts.

The basic separation procedure is to obtain C_2, C_3, C_4, . . . cuts. The C_2 cut contains acetylene unless a front-end converter has been used. If acetylene is present, and if it is not to be recovered, a back-end converter is employed.

The pressures and temperatures employed in the distillation columns affect the size and capital costs. Lower pressures (and hence lower temperatures) also affect the number of plates needed, since relative volatilities of the compounds generally increase with lower pressures.

The operating conditions for these columns are carefully chosen to minimize refrigeration and energy requirements. Propylene vapors, for example, are used as the heat source in the reboiler of the ethylene-purification column, in which ethylene and ethane are separated, and then a somewhat lower-pressure propylene liquid is used as refrigerant in the reflux condensers of the same column and also of the deethanizer. Low-pressure steam from several steam turbines used to operate compressor systems is available for heat-transfer purposes in several columns.

Separations of propylene-propane mixtures and of ethylene-ethane

TABLE 2-8 Typical Information for Several Distillation Steps of Pyrolysis Plants (29)

Tower	Product		Pressure, atm	Reboiler		Reflux			Plates	Reflux ratio
	Distillate	Bottom		Temp., °C	Heat-transfer agent	Temp., °C	Heat-transfer agent			
Deethanizer...........	C_2's	C_3–C_8's	27	70–75	Steam	−10	Propylene liquid		40	0.8–1.0
Ethylene purification...	99.5 to 99.8% ethylene	Ethane	20	−5	Propylene vapor	−30	Propylene liquid		100–120	4.5–5.5
Depropanizer.........	C_3's	C_4–C_8's	16	100–110	Steam	50	Water		35–45	High
Propylene purification...	99.5% propylene	Propane	19	60	Steam	45	Water		200	
Debutanizer..........	C_4's	C_5–C_8's	4	100	Steam	40	Water			

mixtures require large columns to produce high-purity olefins. In the Imperial Chemical Industries plant, which produces about 1.0 billion lb/year of 99 percent ethylene, the propylene-purification tower is 16 ft in diameter and 266 ft high (82, 86). The ethylene-purification column is 13 ft in diameter and 245 ft high. If lower-purity propylene is suitable, fewer plates and hence a smaller column are required. For ethane pyrolysis, only relatively small amounts of propylene are produced, and it may be uneconomical to separate the propane and propylene. In such a case, the entire C_3 stream may be recycled to the pyrolysis furnaces. Ethane from the bottom of the ethylene-ethane tower often contains 1 to 2 percent ethylene, and it is recycled to the pyrolysis furnaces.

Based on relative volatilities, propylene and propane are more difficult to separate and hence require more plates and a higher reflux ratio than ethylene and ethane. Selection of the pressure and other conditions at which the columns are operated requires a careful economic analysis (74). High-pressure (16 to 20 atm) separation of propylene from propane is reported to be more economical than low-pressure (8 to 12 atm) separation (108). Neither azeotropic nor extractive distillation has yet been reported for the commercial separations of these binary mixtures. Several extractive distillation agents were recently reported to be effective for separating propane-propylene mixtures (85). Possibly such agents will find commercial application in the future (112).

In late 1967, Union Carbide Corp. at their Institute olefin plant started tests with their high-flux tubing in the overhead condenser of their ethylene-purification column (73). Ethylene vapors were condensed by heat transfer with boiling propylene. The high-flux tubing promotes nucleate boiling, and overall heat-transfer coefficients of 800 Btu/(h)(ft²)(°F) have been obtained with no fouling at a temperature difference of 10°F. Such coefficients are approximately 7 times greater than those obtained in conventional tubing.

For the Union Carbide plant at Ponce, Puerto Rico, high-flux tubing was chosen for the reboiler which condenses ethane against boiling propylene and for the condenser which condenses ethylene against boiling propylene (73). This choice of tubing allowed a reduction of the heat-transfer areas in these two units by factors of approximately 6 and 7, respectively. Such a reduction resulted in much smaller heat exchangers and a 35 to 50 percent saving in their cost.

High-flux tubing will undoubtedly find extensive use in olefins plants in the future. It will be helpful for reducing the skin temperatures in reboilers and hence minimizing polymerization and thermal-decomposition reactions. Furthermore, efforts are continuously being made to minimize overall energy requirements, and high-flux tubing can be used for heat transfer between streams with only small temperature differences.

Fouling, caused primarily by polymerization, is significant in the re-boilers of the depropanizer and sometimes the deethanizer and the de-butanizer. Fouling of the depropanizer reboiler is controlled by mini-mizing temperatures of the boiling liquids to less than 125°C (29) and of the heat-transfer surfaces. Generally two reboilers are provided for the depropanizer so that one can be cleaned while the other is in service.

Fouling is normally less of a problem in the reboilers of deethanizers since both the temperatures and concentration of dienes there are lower. A company using ethane as a feedstock, however, has experienced foul-ing problems in this column and has added a spare reboiler since the plant was built. They minimize the ratio of fouling by using an anti-foulant and by recycling a C_3 stream to lower the bottom temperature. It would be interesting to know if high-flux tubing would improve their operation.

When a significant amount of isoprene and other C_5 or higher dienes are present, fouling may be a problem in the reboiler of the debutanizer. Temperatures in the rectifying section of the debutanizer are relatively low, so that butadiene polymerization is generally not a problem there. Essentially no fouling occurs in the demethanizer because of the mod-erate temperatures.

Antifouling agents are sometimes employed in small concentrations in various columns and reboiler.

Recovery of By-Products Within the last few years, an increasing number of companies have found it feasible to recover additional by-products such as butadiene, isoprene, and aromatics from C_4, C_5, C_6, and heavier cuts. Figure 2-6 does not show a separation technique for these compounds. As the molecular weight of the feedstock increases, more and more of these hydrocarbons are produced and their recovery becomes an economic necessity (135). Information obtained from Monsanto Co. (67) indicates for gas oil cracking approximately the following figures for the product stream: C_2's (mainly ethylene), 29 percent; C_3's (mainly propylene), 19 percent; C_4's, 15 percent; C_5's, 5 percent; benzene, 8 percent; and toluene, naphthalenes, and aromatic tars, 23 percent.

The C_4 fraction contains n-butane, isobutane, all C_4 olefins, butadienes (primarily 1,3-butadiene), and vinylacetylene. Conventional distillation is generally impractical since the volatilities of the compounds are too close. Extractive distillations are usually provided to separate the bu-tane and butenes from the mixture and then to separate butadiene from vinylacetylene. Extractive-distillation solvents include an aqueous solution of N,N-dimethylacetamide (59, 64, 70, 71, 83), N-methylpyrrolidone (96, 106), furfural (116), dichloroethyl ether (64, 83), and β-methoxy-propionitrile (66). This latter solvent is used by Monsanto Co. at their Chocolate Bayou plant; they found that a much lower solvent-to-C_4 feed ratio was feasible with it. Such a reduced ratio results in lower operating costs.

Care should be taken in operating the butadiene-recovery system to prevent formation of gas mixtures containing high concentrations of vinylacetylene because they are highly explosive. A distillation column was destroyed by an explosion when a high concentration was formed during start-up (64, 81, 83, 92).

When C_5 and heavier cuts are present in only small amounts, they are often used as gasoline-blending stocks since they have relatively high octane numbers. When substantial amounts of these cuts can be obtained, it is economical to recover the isoprene and aromatics (72).

Economic Aspects of Distillation System   The distillation train must be designed for considerable flexibility of operation. Feed rates to each column may normally vary by perhaps 20 to 25 percent. Such variations occur whenever a pyrolysis furnace is down for cleaning or repairs. Even when all pyrolysis furnaces are being operated, gradual changes in the composition and amounts of various hydrocarbons occur as a furnace gradually becomes coked.

The capital and operating costs for the distillation section of an ethylene-propylene plant are high, accounting for significant fractions of total plant costs. Hence there is a considerable incentive to optimize the designs of these columns. The types of trays used are important for plate efficiencies, pressure drops in column, column dimensions and operation, and costs (77, 78). Although bubble-cap trays allow considerable flexibility of operation, they have relatively poor plate efficiencies compared with other types and they also result in fairly large pressure drops. Sieve (or perforated) plates have higher plate efficiencies, have lower pressure drops per tray, allow higher vapor loadings in the column, and are cheaper than bubble-cap trays. Sieve trays are less adaptable, however, to changes in the feed rates to the tower. Sieve trays and valve-type trays are used in some units of relatively recent design. Multiple-downcomer trays, which use multiple downcomers for each perforated plate, have been proposed for many of the distillation columns in olefin plants, including the ethylene- and propylene-purification columns. Delnicki and Wagner (75) have compared propylene-purification columns using multiple-downcomer trays with conventional (multipass sieve or valve) trays. Much smaller columns, lower reflux ratios, and/or better separations are possible with the multiple-downcomer trays. Cities Service Oil Co. was able to increase by 50 percent the propylene output of one of their columns when they switched to the multiple-downcomer tray (91).

Processes for Production of Both Ethylene and Acetylene

Several processes (22, 58, 62, 68, 89, 93, 94, 120, 128, 137) are used for the simultaneous production of ethylene and acetylene. The reactors in these processes are operated at temperatures significantly higher than in tubular reactors since acetylene production is favored by higher temperatures. Ethylene, propylene, and butenes are the predominant unsaturated products formed at temperatures up to about 900°C (58, 93). At higher temperatures, ethylene and acetylene become the preferred compounds. From thermodynamic considerations, acetylene is the preferred product above 1130°C at essentially atmospheric pressure. The ratio of acetylene to ethylene increases in the product as the temperature for pyrolysis is increased.

Since suitable materials of construction are not available to build tubular reactors for temperatures significantly above 900°C, other types of reactors are needed.

Wulff Process The Wulff process was first proposed in 1925, but several years elapsed before the process was commercialized. Several Wulff units have been built within the last few years. Feedstocks that can be used range from ethane to heavy gas oils, and flowsheets have been discussed in the literature (22, 88).

Regenerative furnaces are employed in the Wulff process. These furnaces are filled with lattices of refractive checkers with numerous parallel-flow channels for the hydrocarbons. The furnaces are alternately on stream and then off for about 60 s each. While off stream, hot combustion gases flow through the furnace to heat the refractory checkers to high temperatures and to burn off the tars and coke produced during pyrolysis. A fuel such as natural gas is supplied for the combustion phase. Upon completion of the heating phase, a preheated mixture of hydrocarbons and steam is fed to the furnace for the pyrolysis phase.

The reaction mixture is maintained at essentially a total pressure of 0.5 atm. Because of steam dilution, the hydrocarbons are at a partial pressure significantly less than this. Weaver (137) indicates that the yields of acetylene increase rapidly as the pressure is decreased. Higher acetylene yields could be obtained by operating at higher temperatures, but operating problems and costs increase in such a case. The temperature and pressure are then chosen to minimize the production cost of acetylene. At the conditions chosen, the kinetics of pyrolysis is very fast, and residence times of the gases in the hot furnace are very low, about several hundredths of a second.

Several regenerative furnaces are used in a Wulff plant, half being on stream at any given time. The plant built by Union Carbide Corp. in Brazil has 12 furnaces, but more are planned for the future (68). The plant was designed so that the ratio of ethylene to acetylene could be varied from 8:1 to 2:1. The planned capacity of the plant at start-up was for 281 and 35 million lb/year of ethylene and acetylene, respectively. A British plant is reported to be operating with a 0.5:1 ratio of ethylene to acetylene.

Factors that must be considered in operating the furnaces include the selection of cycle times. During the pyrolysis phase of the cycle, the temperatures in the furnace drop substantially, and higher temperature drops are experienced with longer cycles. When the pyrolysis phase of the cycle is about 60 s, temperature variations are in the range of 100°C. Hence significant variations in the yields of ethylene and acetylene occur during the cycle.

The type of refractory checker used is of importance not only to provide long and dependable furnace life but also to provide uniform temperature patterns for the hydrocarbon gases. The checkers must withstand the large temperature changes that are experienced during start-up or shutdown and also during the repeated pyrolysis and heating phases. The checkers are obviously subjected to mechanical stresses and to steam, which attacks ceramics at high temperatures (94). Some difficulties are experienced in maintaining uniform flow channels and temperature patterns for the gases in the furnace.

Consideration must also be given to flushing out the combustion gases and excess oxygen of the heating step of the cycle before the hydrocarbon phase to be cracked is added. Steam would be effective in removing most of the combustion gases. Some oxygen, however, may react or be adsorbed at the walls and so activate them. Such oxygen would certainly modify the subsequent pyrolysis reactions.

Wulff furnaces are not easy to operate, and difficulties have been experienced in obtaining the design capacities. In general, units designed to produce lower ratios of ethylene to acetylene have had more difficulties since they operate at higher temperatures. Modifications in operating procedures are planned in the future, including more frequent cyclical decoking of the unit, since coke and tar accumulations occur when liquid hydrocarbons are used as feedstocks.

Numerous heat exchangers are provided in Wulff plants for effective utilization of the heat of combustion. The air, fuel, and hydrocarbon feedstock are all preheated before entering the furnace.

Flame-Pyrolysis Processes   In flame-pyrolysis processes (58, 93, 94, 128), the hydrocarbons to be cracked are introduced directly in a hot mixture of combustion gases and steam, as shown in Fig. 2-10 (93,

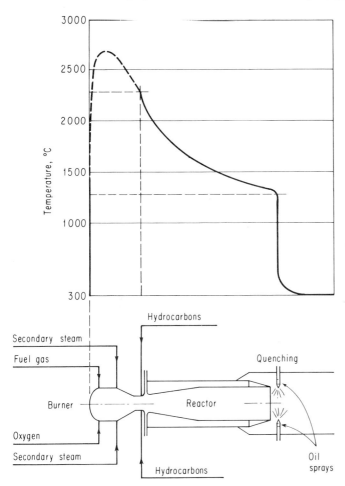

Fig. 2-10 Temperature profiles in reactor for flame pyrolysis.
[*From* (93) *with permission of Chemical Engineering.*]

94). The combined mixture is at a sufficiently high temperature for the desired cracking reactions to occur very rapidly, sometimes within 0.002 to 0.003 s. Figure 2-10 also shows the temperatures of the gases in the furnace for the Hoechst process.

In the burner section of the Hoechst reactor, the gas temperatures are high, with a maximum of about 2600°C and an exit temperature of approximately 2300°C. The secondary steam, so designated to distinguish it from the steam formed by combustion of the fuel, is added to decrease the gas temperature and also to lower the concentration of free radicals in the gas to a more desirable level. Prevaporized naphtha or other hydrocarbons to be reacted are then injected into the reactor. The method of mixing these hydrocarbons with the combustion gases and secondary steam is undoubtedly critical. Rapid and thorough mixing is desired, requiring an effective injection system. In the reactor proper, the cracking occurs adiabatically, and the temperatures drop from about 2300 to 1250°C. The hot gases leaving the reactor are quenched with oil or water to a relatively low temperature. A flowsheet for a flame-pyrolysis process is described by Kamptner et al. (93).

The ratio of ethylene to acetylene decreases significantly as the reaction temperature is increased. The process can be employed with light to heavy hydrocarbons. Acetylene can also be produced from methane when temperatures above 1260°C are provided.

Akin et al. (58) discuss the relative advantages of using oxygen or air for the combustion step. Higher concentrations of ethylene and acetylene are realized when oxygen is used because little nitrogen is present as a diluent. Hence recovery of the desired products is cheaper. More steam is usually mixed with the hydrocarbons when oxygen is provided to reduce the partial pressure of the hydrocarbons sufficiently. The steam is much easier to separate from the hydrocarbons than nitrogen. Choosing between air and oxygen is primarily on economic grounds. Oxygen is employed in the process developed by Farbwerke Hoechst A.G. of Germany (93, 94), but the process of Kobe Steel, Ltd., of Japan, uses air (128). Temperatures in the latter process are lower than those of the Hoechst process.

Kamptner et al. (94) have discussed the design of the reaction system, which must have the following features: great thermal resistance, reasonable mechanical stability, relatively low heat losses to the surroundings, and complete combustion of the fuel gas before addition of the feedstock to be cracked. Ceramic materials, although reasonably resistant to the high temperatures, are badly degraded by high-temperature hydrolysis reactions in the presence of steam. A water-cooled metal combustion chamber has been adopted for the process, and energy release rates of 110 million btu/(ft³)(h) are obtained. The energy losses by heat transfer to the walls of the reactor are negligibly small compared to the total energy input to the furnace.

An oil-quench system is used by Farbwerke Hoechst at the exit end of their furnace. A significant fraction of the sensible heat is recovered and used to generate low-pressure steam.

In the flame-pyrolysis process, free radicals produced in the flame are available to start the free-radical chains for the pyrolysis steps. Such a factor is of importance in obtaining rapid pyrolysis reactions.

Considerable amounts of carbon dioxide and some carbon monoxide are in the exit-product stream from the flame-pyrolysis reactor. In one plant, the concentration of carbon dioxide is about 2 percent (93). Hoechst has chosen a potassium methyltaurinate solution to reduce the carbon dioxide concentration to 50 ppm or less. A mild caustic soda scrubbing is used to remove the residual carbon dioxide.

Kamptner et al. (94) claim that relatively few by-products are formed in the Hoechst process. The residual-gas stream consists primarily of hydrogen, carbon monoxide, and methane, and this stream is used to supplement the fuel gas. They claim that ethylene and acetylene can be produced at total manufacturing costs of 3.1 and 4.9 cents/lb, respectively, when naphtha is used as a feedstock for a plant making 187 million lb of total C_2's. Such a plant is small relative to modern plants using conventional thermal cracking.

Product Recovery from Ethylene-Acetylene Plants   The product mixture of all ethylene-acetylene plants contains appreciable quantities of C_3 and heavier acetylenes and dienes (58). Separation of such mixtures necessitates techniques that are quite different from those used in pyrolysis plants used to produce ethylene and propylene primarily. Care must be taken to ensure that high concentrations of heavier acetylenes are never obtained in the separation process. Hoechst (93) uses the naphtha feedstock as an absorbent to remove the heavier acetylene and dienes. The resulting naphtha stream is fed to the flame-pyrolysis units.

The exact techniques employed to separate the product mixture of course depend on the composition of the product-gas stream. After the carbon dioxide has been separated, several separation steps are normally employed (22, 93, 128, 137), including conventional distillation and extraction or extractive distillation using acetone or dimethylformamide as the extractive agent.

Economic Considerations

Choosing a process for production of ethylene and propylene involves considerations of many technical and economic factors. The choice of the feedstock is probably the most important. When ethane is the feedstock, ethylene is primarily the only hydrocarbon product. Propane, however, also produces significant quantities of propylene. As the molecular weight of the feedstock increases, the plant becomes relatively less an ethylene or propylene process and becomes primarily a heavy-hydrocarbon plant.

Modern plants to produce ethylene and propylene plus by-products are large, complex, and expensive. The capital cost of a plant of course depends on many factors including feedstock, overall plant capacity, product distribution desired, location, and design details. Zasloff (138) in 1970 indicated that a plant to produce 440,000 metric tons/year (970 million lb) of ethylene costed about $37 million. This plant for a feedstock mixture of ethane and propane was specified as being on the Gulf Coast of the United States, and the cost per ton of annual capacity was $85. The breakdown of the capital costs of the equipment for various steps of the process is reported below as the percent of the total costs:

Percent

Pyrolysis furnaces, quench tower, and waste-heat-recovery equipment...........	29
Product compressors and equipment for acid-gas removal, condensate stripping, and drying............................	17
Fractionation equipment and related heat exchangers............................	34
Refrigeration equipment.................	20

The capital costs of plants for heavier feedstocks are more expensive than those for lighter feedstocks. The distillation and recovery of propylene, butadiene, isoprene, aromatics, etc., in such a plant are complicated and expensive. Likewise waste-heat recovery is more difficult and in general less efficient because of the increased problems of coking or fouling of equipment.

Capital-cost information for olefin plants using various feedstocks has been presented by Stobaugh (127).

Capital costs for ethylene-acetylene plants are also high. The Wulff plant of Union Carbide Corp. in Brazil to produce about 400 million lb/year of ethylene, acetylene, and benzene had an original cost of $20 million (89). The plant has experienced start-up problems, and a sizable additional investment is being required to modify the gas-processing equipment.

An olefin plant should always be designed for some flexibility of operation because several factors may change during the life of a plant, e.g., the availability and cost of various hydrocarbon feedstocks. In addition, the demand and selling prices for ethylene, propylene, and various by-products may also vary with time (84). Hence some modifications in the plant operation are desirable from time to time in order to maximize profits. Data for the pyrolysis of light naphthas (107) indicate (Table 2-9) how the yields of ethylene and propylene can be varied to some extent independently of each other. Shorter residence times caused by higher temperatures favor ethylene production.

The capital and operating costs for a plant that can be operated in a highly flexible manner are greater than comparable costs for a less flexible plant. To provide flexibility, some portions of the plant must be overdesigned relative to what is actually required for the planned "normal" operating conditions. When a plant is being designed, impor-

**TABLE 2-9 Yields of Ethylene and Propylene from
Light-Naphtha Pyrolysis (107)**

Residence time,	Yield, wt %	
	Ethylene	Propylene
0.35	30	18.0
0.35	35	18.5
0.35	40	18.0
0.70	20	17.5
0.70	30	20.0
0.70	40	17.5
1.20	20	20.5
1.20	25	21.0
1.20	30	19.0

tant questions include the degree of flexibility to be provided and whether plans should be made for proposed changes in the future.

In a plant as complex and integrated as an olefin plant, certain pieces of equipment are the bottlenecks. Quester (119) has discussed how the overall plant operation should be analyzed to determine the bottlenecks and suggests procedures for removing them. Frequently a plant can be modified as desired by adding more pyrolysis furnaces, increasing the compressor capacities, making certain modifications of the distillation equipment, or replacing the equipment that is the bottleneck with improved or bigger equipment.

Energy requirements for olefin plants are always high and depend on the feedstock being used, as shown in Table 2-10. Energy costs are a significant share of the total operating cost. Many heat exchangers are used in olefin plants in order to minimize energy requirements. An optimum number of exchangers is determined by balancing the costs of additional exchangers against the savings resulting by reduced energy requirements.

**TABLE 2-10 Approximate Energy Demands of Plant to
Produce Ethylene at a Rate of 1 Billion lb/yr**

Feed-stock	Energy demand, million kcal/h
Ethane	340
Propane	425
Naphtha	470
Gas oil	630

Improved methods of controlling and optimizing plant operation have been developed in the last few years. Computer controls are now available for at least portions of the overall process, e.g., the pyrolysis furnaces. Such a control system has helped immensely in maintaining optimum conditions at all times. Without a computer, it was not possible to evaluate as quickly and completely how a change in the operation of one portion of the process would affect the remainder of the plant. In addition, as discussed earlier, operating changes occur routinely with time in several portions of the plant, e.g., the pyrolysis reactors, the transfer-line exchangers, and the reboiler of the depropanizer.

Efficient plant operation requires highly dependable equipment throughout the entire plant (104, 132). Faulty operation of the equipment at any of several key points may necessitate shutdown of the entire plant. Most mechanical failures resulting in complete plant shutdown are caused by mechanical problems in the compressors, their drivers, and various auxiliary equipment. A few companies reportedly use parallel or spare compressors for the product stream and/or the propylene. Most companies do not believe that the additional cost of the spare compressors can be justified. Preventive-maintenance techniques should be practiced to minimize compressor difficulties. A stock of key spare parts is necessary at all times to expedite repairs when required.

Other portions of olefin plants that sometimes experience trouble have been reviewed by Tucker and Cline (132). Trouble spots include the systems for providing steam, electric power, cooling water, and fuel. They also reviewed potential hazards and the resulting safety provisions.

Plant location requires consideration of the methods for transporting the products to the customer. Ethylene, being the most volatile product, is the most difficult to transport. In the Gulf Coast area, pipelines are the preferred and cheapest method (127). Truck-and-trailer combinations have been used to transport liquefied ethylene at $-100°C$ from the Gulf Coast to as far as California at a cost of about 0.8 cents/lb. Other methods currently being seriously considered or under development include railroad tank cars, barges, and ocean tankers.

Storage facilities for the light-hydrocarbon feedstocks and ethylene are often important considerations in plant location. Several ethylene plants are located in areas where the ethylene and ethane can be stored in depleted gas wells or other suitable underground domes.

Plant layout and location require careful planning relative to safety and methods of minimizing and controlling fires and explosions. Imperial Chemical Industries, Ltd. at their large olefins plant at Wilton, Tesside, have installed firewall steam curtains around potentially hazardous areas. The curtains, constructed with prefabricated concrete, have perforated steampipes mounted on their top. The result, the builders believe, is that a more compact plant can be safely built than would otherwise be possible.

Summary

Major changes have evolved within the last 10 years in the commercial production of ethylene, propylene, and important by-products, and even further changes can be expected in the future. The commercial units are now much bigger and more complicated and have increased recovery of the by-products. The production capacity of ethylene and propylene can be expected to grow rapidly in the next few years, and additional problems in obtaining suitable and cheap feedstocks for the processes can be anticipated.

Literature Cited

1. Anderson, E. V.: Petrochemical Feedstock, *Chem. Eng. News*, June 15, 1970, p. 18.
2. Burke, D. P., and R. Miller: Ethylene, *Chem. Week*, Oct. 23, 1965, p. 64; Nov. 13, 1965, p. 69.
3. Catry, P. P., M. Lemee, and M. Vigner: Europe: Ethylene Demand Stronger, *Hydrocarbon Process.*, March 1967, p. 137.
4. Petrochemicals 1970–1980: Plugging Worldwide Gaps, *Chem. Eng.*, July 13, 1970, p. 34.
5. Japan Aims for Ethylene-from-Crude Process, *Chem. Eng. News*, Jan. 15, 1968, p. 32.
6. Propylene Demand: Up 50% by 1973, *Chem. Eng. News*, Aug. 5, 1968, p. 26.
7. Propylene: Tight Supply through 1975, *Chem. Eng. News*, Apr. 21, 1969, p. 28.
8. Forecast Cloudy for Foreign Propylene, *Chem. Eng. News*, Sept. 29, 1969, p. 24.
9. European Ethylene Capacity Keeps Growing, *Chem. Eng. News*, Jan. 26, 1970, p. 28.
10. Olefins Demand Forcing Feedstock Change, *Chem. Eng. News*, May 25, 1970, p. 21.
11. Ethylene Makers May Face Feedstock Change, *Chem. Eng. News*, June 8, 1970, p. 18.
12. Olefins, Aromatics Face Steady Growth, *Chem. Eng. News*, June 8, 1970, p. 24.
13. Chopey, N. P.: New Ethylene Processes Cater to Heavy Feedstock, *Chem. Eng.*, Sept. 2, 1963, p. 34.
14. Coldiron, D. C.: "The Ethylene Industry in the United States," M.S. thesis, Alfred P. Sloan School of Management, Massachusetts Institute of Technology, Cambridge, Mass., 1966.
15. Collingswood, P.: World Developments in Olefin Supplies, *Amer. Chem. Soc. Chem. Inst. Can. Meet., Toronto, May* 1970.
16. Freiling, J. G., B. L. Huson, and R. N. Summerville: Which Feedstock for Ethylene, *Hydrocarbon Process.*, November 1968, pp. 145–152.

17. Freiling, J. G., C. C. King, and J. Newman: Ethylene Raw Materials and Production Economics, *Am. Chem. Soc. Chem. Inst. Can. Meet., Toronto,* 1970.
18. Greek, B. F.: Ethylene, *Chem. Eng. News,* Feb. 22, 1971, pp. 16–19.
19. Kamptner, H. K., W. R. Krause, and H. P. Schilken: HTP: After Five Years, *Hydrocarbon Process.,* April 1966, p. 187.
20. Miller, R. L.: Butadiene's Technical Shift, *Chem. Eng. News,* Jan. 24, 1972, pp. 52–53.
21. Phillips, R. F.: Petrochemicals, *Chem. Eng.,* May 22, 1967, pp. 153–175.
22. Rosenzweig, M. D.: Wulff Route Stars in Brazilian Ethylene Plant, *Chem. Eng.,* Nov. 3, 1969, p. 64.
23. Spencer, P.: Petrochemical Feedstocks: Economic Potential of LNG, *Chem. Eng.,* Mar. 9, 1970, p. 160.
24. Reis, T.: Economic Forecast for Europe in Ethylene and By-Product Formation, *Ind. Eng. Chem.,* **62**(7): 44 (1970).
25. Rothman, S. N.: Ethylene Plant Optimization, *Chem. Eng. Prog.,* **66**(6): 37 (1970).
26. Stretzloff, S. Z.: Ethylene and Propylene: Booming Building Blocks, *Chem. Eng.,* Aug. 24, 1970, p. 75.
27. Tewksbury, J. F.: Ethylene and Derivatives Trade, *Chem. Eng. News,* July 28, 1969, p. 16.
28. Tucker, W., and W. E. Cline: Large Plant Reliability, *Chem. Eng. Prog.,* **67**(1): 37 (1971).
29. Zdonik, S. B., E. J. Green, and L. P. Hallee: "Manufacturing Ethylene" (articles published in *Oil Gas J.,* 1966–1970), Petroleum Publishing Co., Tulsa, Oklahoma, 1970.
30. Begley, J. W.: Find Butane Dehydrogenation Equilibria, *Hydrocarbon Process.,* July 1965, p. 149.
31. Blakemore, J. E., and W. H. Corcoran: Validity of the Steady-State Approximation Applied to the Pyrolysis of *n*-Butane, *Ind. Eng. Chem. Process Des. Dev.,* **8**: 206 (1969).
32. Buekens, A. G., and G. F. Forment: Thermal Cracking of Propane: Kinetics and Product Distribution, *Ind. Eng. Chem. Process Des. Dev.,* **7**: 435 (1968).
33. Buekens, A. G., and G. F. Forment: Thermal Cracking of Isobutane, *Ind. Eng. Chem. Process Des. Dev.,* **10**: 309 (1971).
34. Crynes, B. L., and L. F. Albright: Pyrolysis of Propane in Tubular Flow Reactors: Kinetics and Surface Effects, *Ind. Eng. Chem. Process Des. Dev.,* **8**: 25 (1969).
35. Davis, H. G., and K. D. Williamson: The Pyrolysis of Ethane and Related Hydrocarbons, *World Pet. Congr.,* Proc., 5th New York, Sec. 4, 37–54 (1959).
36. Fabuss, B. M., J. O. Smith, R. I. Lait, A. S. Borsanyi, and C. N. Satterfield: Rapid Thermal Cracking on *n*-Hexadecane at Elevated Pressures, *Ind. Eng. Chem. Process Des. Dev.,* **1**: 293 (1962).
37. Herriott, G. E., R. E. Eckert, and L. F. Albright: Kinetics of Propane Pyrolysis, *AIChE J.,* **18**: 84 (1972).
38. Kershenbaum, L. S., and J. J. Martin: Kinetics of Nonisothermal Pyrolysis of Propane, *AIChE J.,* **13**: 148 (1967).
39. Kunugi, T., T. Sakai, K. Soma, and Y. Sasaki: Kinetics and Mechanism of the Thermal Reaction of Ethylene, *Ind. Eng. Chem. Fundam.,* **8**: 374 (1969).
40. Kunugi, T., T. Sakai, K. Soma, and Y. Sasaki: Thermal Reaction of Propylene: Kinetics, *Ind. Eng. Chem. Fundam.,* **9**: 314 (1970).

41. Kunugi, T., K. Soma, and T. Sakai: Thermal Reaction of Propylene: Mechanism, *Ind. Eng. Chem. Fundam.*, **9**: 319 (1970).
42. Laidler, K. J., N. H. Sagert, and B. W. Wojciechowski: Kinetics and Mechanisms of the Thermal Decomposition of Propane, *Proc. R. Soc.*, **270A**: 242, 254 (1962).
43. Laidler, K. J., and B. W. Wojciechowski: Kinetics and Mechanisms of Thermal Decomposition of Ethane: The Uninhibited Reaction, *Proc. R. Soc.*, **206A**: 91, 103 (1961).
44. Linde Aktiengesellschaft, Wiesbaden: "Special Petroleum Chemicals Issue," Linde Reports on Science and Technology, Wiesbaden, Germany, 1967.
45. Longwell, P. A., and B. H. Sage: Thermal Decomposition of *n*-Hexane at High Pressures, *J. Chem. Eng. Data*, **5**: 322 (1960).
46. Minkoff, G. J., and C. F. H. Tipper: "Chemistry of Combustion Reactions," pp. 237–257, Butterworth, London, 1962.
47. Pease, R. N.: The Thermal Dissociation of Ethane, Propane, *n*-Butane, Isobutane Preliminary Study, *J. Am. Chem. Soc.*, **50**: 1770 (1928).
48. Poltorak, V. A., and V. V. Voevodskii: Kinetics of the Cracking of Propane in the Presence of Oxygen, *Russ. J. Phys. Chem.*, **35**: 82 (1961).
49. Purnell, J. H., and C. P. Quinn: The Pyrolysis of *n*-Butane, *Proc. R. Soc.*, **270A**: 267 (1962).
50. Semenov, N. N.: "Some Problems in Chemical Kinetics and Reactivity," vols. I and II, trans. by M. Boudart, Princeton University Press, Princeton, N.J., 1959.
51. Snow, R. H.: A Chemical Kinetics Program for Homogeneous and Free-Radical Systems of Reactions, *J. Phys. Chem.*, **70**: 2780 (1966).
52. Snow, R. H., R. E. Peck, and C. G. von Fredersdorff: A Computer Study of a Free-Radical Mechanism of Ethane Pyrolysis, *AIChE J.*, **5**: 304 (1959).
53. Vanderkooi, W. N., and R. A. Mock (to Dow Chemical Co.): Method for Cracking Aliphatic Hydrocarbons, U.S. Pat. 3,244,760 (Apr. 5, 1966).
54. Voevodsky, V. V.: The Thermal Decomposition of Paraffin Hydrocarbons, *Trans. Faraday Soc.*, **55**: 65 (1959).
55. Wang, Y. L., R. G. Rinker, and W. H. Corcoran: Kinetics and Mechanism of the Thermal Decomposition of *n*-Butane, *Ind. Eng. Chem. Fundam.*, **2**: 161 (1963).
56. Wojciechowski, B. W., and K. J. Laidler: Free Radical Mechanisms for Inhibited Organic Decompositions, *Can. J. Chem.*, **38**: 1027 (1960).
57. Aalund, L.: Houston's Big Ethylene Plant Marks New Era in Olefins, *Oil Gas J.*, Nov. 20, 1969, p. 158.
58. Akin, G. A., T. F. Reid, and R. J. Schrader: Eastman Process for Cracking Light Hydrocarbons to Acetylene and Ethylene, *Chem. Eng. Prog.*, **54**(1): 41 (1958).
59. Bannister, R. R., and E. Buck: Butadiene Recovery via Extractive Distillation, *Chem. Eng. Prog.*, **65**(9): 65 (1939).
60. Bergman, D. J.: High Temperature Exchanger Problems, *Hydrocarbon Process.*, October 1966, p. 158.
61. Ethylene from Crude Oil, *Br. Chem. Process Suppl.*, November 1967, p. 97.
62. Acetylene and Ethylene by Submerged Flame Process (BASF), *Br. Chem. Process Suppl.*, November 1967, p. 99.
63. Bucklin, R. W. (to Coastal States Gas Producing Co.): Method and Equipment for Treating Hydrocarbon Gases for Pressure Reduction and Condensate Recovery, U.S. Pat. 3,292,380 (Dec. 20, 1966).

64. Buehler, J. H., R. H. Freeman, R. G. Keister, M. P. McCready, B. I. Pesetsky, and D. T. Watters: Report on Explosion at Union Carbide's Texas City Butadiene Refining Unit, *Chem. Eng.*, Sept. 7, 1970, pp. 77–86.

65. Ethylene Expansion Boosts Furnace Making, *Chem. Eng. News*, June 26, 1967, pp. 40–42.

66. Solvent Shift Doubles Butadiene Output, *Chem. Eng. News*, Sept. 22, 1969, p. 68.

67. Cracking Coproducts Determine Profits, *Chem. Eng. News*, Nov. 2, 1970, p. 31.

68. Brazil Enters Petrochemical Age, *Chem. Eng. Prog.*, July 1970, p. 47.

69. Clancy, G. M., and R. W. Townsend: Ethylene Plant Fractionation, *Chem. Eng. Prog.*, 67(2): 41, 1971.

70. Coogler, W. W.: Butadiene Recovery Process Employs New Solvent System, *Chem. Eng.*, July 31, 1967, pp. 70–72.

71. Coogler, W. W.: Butadiene from Special Distillation, *Hydrocarbon Process.*, May 1967, pp. 166–168.

72. Craig, R. G.: Benzene Bonanza Beckons via Dripolene Conversion, *Chem. Eng.*, Oct. 6, 1969, p. 142.

73. Creighton, R. L., Linde Division, Union Carbide Corp., Tonawanda, N.Y.: personal communication, June 1971.

74. Davison, J. W., and G. E. Hays: Process Design for Ethylene-Ethane Fractionator, *Chem. Eng. Prog.*, 54(12): 52 (1958).

75. Delnicki, W. V., and J. L. Wagner: Performance of Multiple Downcomer Trays, *Chem. Eng. Prog.*, 66(3): 50 (1970).

76. Demarest, K. D.: Current Status of Reactor Furnaces, *Chem. Eng. Prog.*, 67(1): 57 (1971).

77. Fair, J. R.: Selecting Fractionating-Column Internals, *Chem. Eng.*, July 5, 1965, p. 107.

78. Fair, J. R., and W. L. Bolles: Modern Design of Distillation Columns, *Chem. Eng.*, Apr. 22, 1968, pp. 156–178.

79. Fair, J. R., W. L. Bolles, and W. R. Nisbet: Ethylene Purification: Demethanization, *Chem. Eng. Prog.*, 54(12): 39 (1958).

80. Fair, J. R., J. W. Mayer, and W. H. Lane: Commercial Ethylene Production by Propane Pyrolysis in a Molten Lead Bath, *Chem. Eng. Prog.*, 53(9): 433 (1957).

81. Freeman, R. H., and M. P. McCready: Butadiene Explosion at Texas City, *Chem. Eng. Prog.*, 67(6): 45 (1971).

82. Friend, W. L., and J. J. Crawford: ICI Getting World's Largest Single-Line Ethylene Unit, *Oil Gas Int.*, 7(1): 78 (1967).

83. Griffith, S., and R. G. Keister: This Butadiene Unit Exploded, *Hydrocarbon Process.*, September 1970, pp. 323–327.

84. Giudici, F., and B. Arcelli: Propylene in Europe, *68th Natl. Meet. Am. Inst. Chem. Eng., Houston, Texas, March 1971.*

85. Hafslund, E. R.: Propylene-Propane Extractive Distillation, *Chem. Eng. Prog.*, 65(9): 58 (1969).

86. Harvey, R. D.: Huge British Ethylene Project Sets New Standards, *Oil Gas J.*, Dec. 22, 1969.

87. Hengstebeck, R. J., and R. E. Bosanac: Equilibrium Relationships for Designing Demethanization Facilities for Ethylene Plants, *69th Natl. Meet. Am. Inst. Chem. Eng., Cincinnati, Ohio, May 1971.*

88. Ethylene and Propylene Flowsheets, *Hydrocarbon Process.*, November 1965, pp. 202–208; November 1967, pp. 139, 170–178; November 1969, pp. 176–177, 195.

89. Flexibility Is Key to New Olefin Plant, *Hydrocarbon Process.*, July 1970, p. 11.

90. Jaggai, F. L., and R. F. Holland: Japan: Ethylene Byproducts Down, *Hydrocarbon Process.*, March 1967, p. 131.

91. Jamison, W. D.: Propylene Output Leaps 50%, *Chem. Process.*, **32**: 38 (November 1969).

92. Jarvis, H. C.: Butadiene Explosion at Texas City, *Chem. Eng. Prog.*, **67**(6): 41 (1971).

93. Kamptner, H. K., W. R. Krause, and H. P. Schilken: Acetylene from Naphtha Pyrolysis, *Chem. Eng.*, Feb. 28, 1966, pp. 80–82, 93–98.

94. Kamptner, H. K., W. R. Krause, and H. P. Schilken: HTP: After Five Years, *Hydrocarbon Process.*, April 1966, p. 187.

95. Kitzen, M. R., F. M. Wall, and H. J. Cijfer: Gas Oil Pyrolysis in Tubular Reactors, *Chem. Eng. Prog.*, **65**(7): 71 (1969).

96. Klein, H., and H. M. Weitz: Extract Butadiene with NMP, *Hydrocarbon Process.*, November 1968, pp. 135–137.

97. Knapp, W. J.: New Midwest Ethylene Plant Onstream, *Hydrocarbon Process.*, October 1968, p. 111.

98. Kniel, L.: Utilization of the LNG Cold Potential with Particular Reference to the Manufacture of Ethylene, *Int. Conf. Cryog. Fuel Power Technol, Int. Inst. Refrig., Lond., Mar. 28, 1969.*

99. Kronig, W.: Cold Hydrogen Treat Pyrolysis Cuts, *Hydrocarbon Process.*, March 1970, p. 121.

100. Kroper, H., H. M. Weitz, and U. Wagner: New Butadiene Recovery Process, *Hydrocarbon Process.*, November 1962, pp. 191–196.

101. Lee, F. A., P. von Weisenthal, and K. R. Wagner (to Foster Wheeler Corp.): Fired Heater, U.S. Pat. 3,182,638 (May 11, 1965).

102. Lenoir, J. M.: Furnace Tubes: How Hot, *Hydrocarbon Process.*, October 1969, p. 97.

103. Leonard, E. C.: "Vinyl and Diene Monomers," pp. 577–689, Wiley-Interscience, New York, 1971.

104. Loftus, J.: Reliability in Ethylene Plants, *Chem. Eng. Prog.*, **66**(12): 53 (1970).

105. Lorber, U., H. Reimann, and F. Rottmayr: Acetylene Recovered from Ethylene Feedstock, *Chem. Eng.*, July 26, 1971, pp. 83–85.

106. Lummus Company: "Conoco: Lake Charles," New York, 1968.

107. Maddock, M. J., and R. J. Dorn: Ethylene Processes Profit from SRT Heaters, *Oil Gas J.*, Feb. 1, 1971, pp. 65–69.

108. Marik, J., V. Miller, and J. Rosak: Propylene Isolation in Separation of Pyrolysis Gases, *Tech. Chem. (Prague)*, **1969**(3): 25–28; *Chem. Abstr.* **71**: 40928u.

109. Marshall, L., and H. B. Zasloff: Gas Oil Feedstock for Ethylene Production, *Chem. Eng. Prog.*, **65**(10): 65 (1969).

110. Matsunami, T., Y. Suxukawa, G. Nawata, and H. Kono: Crack Crude for Olefins, *Hydrocarbon Process.*, November 1970, pp. 121–126.

111. Menzies, W. R.: "Surface Effects on the Partial Oxidation of Methane in Tubular Reactors," M.S. thesis, Purdue University, Lafayette, Ind., 1971.

112. Null, H. R.: Azeotropic and Extractive Distillation, *Chem. Eng. Prog.*, **65**(9): 47 (1969).

113. O'Sullivan, T. F., and L. L. Hofstein: SRT Heater: A Break from Tradition, *Oil Gas J.*, Feb. 1, 1971, pp. 70–73.

114. Palchik, E. H., T. F. O'Sullivan, and W. Tucker (to the Lummus Co.): Vertical Tube Fluid Heater, U.S. Pat. 3,274,978 (Sept. 27, 1966).

115. Paushkin, Y. M., and T. P. Vishnyakova: "The Production of Olefin-Containing and Fuel Gases," trans. by D. Finch and B. P. Mullins, Pergamon, New York, 1964.

116. Peters, W. D., and R. S. Rogers: Improved Furfural Extraction Process, *Hydrocarbon Process.*, November 1968, pp. 131–134.

117. Selas to Sell System Using Lead Quench for Ethylene, *Petro-Chem. Eng.*, January 1970, p. 1.

118. Pierce, J. E., and D. L. Stieghan: Dry Cracked-Gas with Molecular Sieves, *Hydrocarbon Process.*, March 1966, pp. 170–172.

119. Quester, F. J.: Revamp for New Ethylene Capacity, *Petro-Chem Eng.*, February 1970, p. 22.

120. Reid, J. M., and H. R. Linden: Acetylene/Ethylene via Thermal Cracking, *Chem. Eng. Progr.*, 56(1): 47 (1960).

121. Reitmeier, R. E., and H. W. Fleming: Acetylene Removal from Polyethylene Grade Ethylene, *Chem. Eng. Prog.*, 54(12): 48 (1958).

122. Ruhemann, M., and P. L. Charlesworth: Thermodynamic Efficiency of Gas Separation Plants, *Br. Chem. Eng.*, 11: 839 (August 1966).

123. Sanders, H., and F. Baumgart: Festigkeitberechnung eines speziellen Rohrbundelwarmeaustauchers, *Chem. Ing. Tech.*, 41: 27 (1969).

124. Scheel, L. F.: What You Need to Know about Gas Expanders, *Hydrocarbon Process.*, February 1970, p. 105.

125. Sheldrick, M. G.: Ethylene from Heavy Feedstocks, *Chem. Eng.*, Apr. 15, 1966, pp. 122–124.

126. Silbernagel, D. R.: New Uses of Molecular Sieves in Olefin Plants, *Chem. Eng. Prog.*, 63(4): 99 (1967).

127. Stobough, R. B.: Ethylene: How, Where, Who—Future, *Hydrocarbon Process.*, October 1966, p. 143.

128. Sugai, K.: Low Cost, High Yield for Forming Ethylene, *Chem. Eng.*, June 15, 1970, p. 126.

129. Tucker, J. F.: Ethylene Liquefaction: A Case History, *Hydrocarbon Process.*, February 1969, p. 99.

130. Tucker, W. (to the Lummus Co.): Inlet Cone Device and Method, U.S. Pat, 3,374,832 (Mar. 26, 1968).

131. Tucker, W. (to the Lummus Co.): Inlet Cone Device and Method, U.S. Pat. 3,477,495 (Nov. 11, 1969).

132. Tucker, W., and W. E. Cline: Plant Reliability and Its Impact on Large Plant Design and Economics, *Am. Inst. Chem. Eng., Meet. San Juan, P.R., May 1970.*

133. Tucker, W., and H. B. Zasloff (to the Lummus Co.): Demethanization in Ethylene Recovery with Condensed Methane Used as Reflex and Heat Exchange Medium, U.S. Pat. 3,320,754 (May 23, 1967).

134. Vollhardt, F. (to Firma Schmidtsche Heissdampf): Heat Exchanger for the Cooling of Freshly Cracked Gases or the Like, U.S. Pat. 3,144,080 (Aug. 11, 1964).

135. Walker, H. M.: Product Optimization in the Petrochemical Refinery, *Am. Chem. Soc. Meet., New York, Sept. 11, 1969.*

136. Walker, H. M.: Advances in Heavy Liquid Cracking, *Chem. Eng. Prog.*, **65**(5): 53 (1969).
137. Weaver, T.: Economics of Acetylene by Wulff Process, *Chem. Eng. Prog.*, **49**(1): 35 (1953).
138. Zasloff, H. B., The Lummus Co., Bloomfield, N.J.: personal communications, 1970.
139. Zeis, L. A., and E. Heinz: Catalyst Tubes in Primary Reforming Furnaces, *Chem. Eng. Prog.*, **66**(7): 68 (1970).

CHAPTER THREE

Production of Polyethylene

I. POLYMER STRUCTURE AND PROCESS INTRODUCTION

For some time polyethylene has been the major plastic in the United States on a weight basis. In 1966, it accounted for about 27 percent of the total plastics produced (1), but since 1970 has increased to about 30 percent (7). Within the last few years, significant improvements have been made in the technology and operation of commercial polyethylene plants.

The approximate production of all types of polyethylenes in the United States was 1 billion lb in 1959 and 5 billion lb in 1970 and is predicted to be over 13 billion lb in 1980 (7). Large and increasing amounts are being produced (5) throughout the world as ethylene and then polyethylene plants are being built or expanded. By about 1970 production in western Europe surpassed that of the United States, and Japanese production was perhaps half of that of the United States. American production is probably somewhat over 30 percent of the world's total.

Significantly greater amounts of polyethylenes will be produced in the future for the following reasons:

1. Polyethylene finds numerous applications because of its desirable physical and chemical properties. Improved optimization of properties for specific end uses of the plastic seems probable as a better understanding is obtained of how the structure, or architecture, of the polyethylene molecule affects both the physical and chemical properties and as methods of tailoring the structure are determined (6, 8, 9, 13, 14, 20, 21, 26). Polyethylene is actually the name for a large family of plastics that have a wide range of physical properties and, to a lesser extent, chemical properties.

2. Low cost of polyethylene. Since this polymer was initially introduced in the late 1940s, its selling price has decreased. By 1970, list prices for certain polyethylenes were as low as 10 cents/lb in hopper-car quantities, but many varieties have much higher list prices. Actual prices are often lower than list prices. Further reductions in prices are less likely in the future, however, since the manufacturing costs are not much lower in some cases. The ratio of the selling prices to the manufacturing costs is reported to be as low as 1.2 for some types of polyethylene (16).

3. Numerous producers assure availability of polyethylene at all times.

4. Technical information and know-how for the forming operations needed to produce the final products can frequently be obtained from the seller. This feature is of considerable importance to the plastics processor. Many major producers of polyethylene have spent time, effort, and money in testing and developing polyethylenes for specific uses.

Although the following simplified equation is frequently used to indicate the polymerization of ethylene (2), the actual reactions and structure of the polyethylene molecules are complicated and will be considered later:

$$nCH_2{=}CH_2 \rightarrow -(CH_2-CH_2)-_n$$

From 30 to 60 types of polyethylenes are often produced by a single company. Figures 3-1 and 3-2 give the approximate range of molecular weights, densities, and crystallinities of commercial polyethylenes (20, 24). Although there are numerous processes for producing polyethylene, they are frequently divided into two groups (2, 23):

1. Low-density polyethylenes (LDPE) are normally produced by high-pressure processes operated at 15,000 to 50,000 lb/in^2 (1,000 to 3,000 atm). Polymer densities vary from about 0.91 to 0.935 g/cm^3 or perhaps slightly higher in some cases. The degree of crystallinity is relatively low, especially for lower-density polymers.

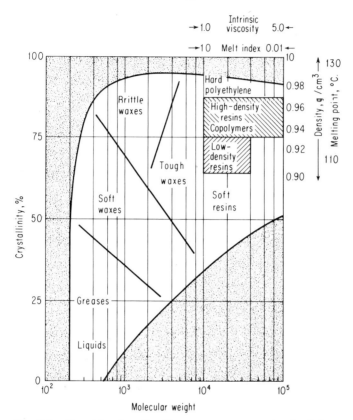

Fig. 3-1   Transition from liquids, greases, waxes, and polyethylene as a function of molecular weight and crystallinity.   [*From (1) with permission of Chemical Engineering.*]

2. High-density polyethylenes (HDPE) are manufactured using low-pressure processes operated at 15 to 3,000 lb/in^2 (1 to 200 atm)*. When pure ethylene is polymerized, the densities usually vary from about 0.955 to 0.97 g/cm^3; over 75 percent crystallinity is obtained. Copolymerization of ethylene with relatively small amounts of other olefins such as 1-butene, 1-hexene, and propylene produces copolymers with lower densities; densities as low as 0.94 g/cm^3 are common. Low-pressure processes have been commercialized since 1955, and 30 percent of the polyethylene is now produced by these methods.

The density of polyethylene has a significant effect on many physical properties, as indicated by Table 3-1.   In general, increased densities

* Some authors, including Raff and Doak (21), divide processes of this type into low- and medium-pressure.

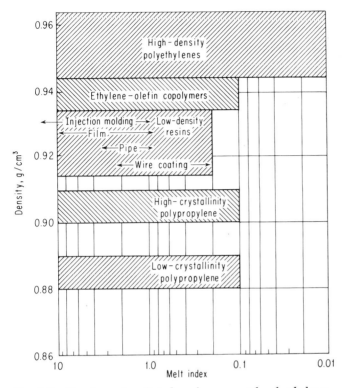

Fig. 3-2 Density and melt index of commercial polyethylenes.
[*From (1) with permission of Chemical Engineering.*]

**TABLE 3-1 Relationship of Several Physical Properties to
the Density of Polyethylenes (15)**

Property	Type of polyethylene		
	Low-density	Medium-density	High-density
Specific gravity............	0.910–0.925	0.926–0.940	0.941–0.965
Tensile strength, lb/in²......	1,000–2,300	1,200–3,500	3,100–5,500
Elongation, %.............	90–800	50–600	15–100
Impact strength (Izod test),			
ft-lb/in.................	No break	0.5–>16	1.5–20
Flexural modulus, lb/in².....	8,000–60,000	60,000–115,000	100,000–260,000
Hardness, Shore D.........	41–46	50–60	60–70
Resistance to heat (con-			
tinuous), °F..............	180–212	220–250	250

result in increased tensile strength, stiffness, hardness, and resistance to heat. Elongation (before failure) of polyethylene as it is stressed decreases as the density increases; in addition, gas and moisture permeabilities decrease.

Major uses for polyethylene include film, sheeting, pipe, coatings (for paper, cellophane, metal foil, cloth, glass fiber, etc.), coatings for wire and cables, toys, pipe fittings, garbage cans, wastebaskets, cans, bottles, and cartons. Low-molecular-weight polyethylenes are related to paraffin waxes, and some have replaced waxes in coating paper, cardboard, etc. (27).

Both the process used for polymerizing ethylene and the operating conditions have a significant effect on the structure of the polyethylene molecules and an indirect effect on the final properties of the plastic (8, 25).

Structural Characteristics of Polyethylene Molecules

Since the method of polymerization affects the structural features (or architecture) of the polyethylene molecules and hence the properties of the final plastic (6, 8, 13, 14, 20, 21, 26), these features will be reviewed briefly.

Chain Branching Both short and long branches have a significant effect on the physical properties of polyethylenes. The short branches are primarily ethyl and n-butyl groups in polyethylenes made by high-pressure processes. Up to 15 to 30 such side groups occur in some polyethylenes per 1,000 carbon atoms in the chain. This value can be varied by the method of polymerization. Short branches are particularly effective in reducing the symmetry of the polymer chain and the packing ability of the chains. As a result, short chains decrease the density and crystallinity of the polyethylene. A high-density polyethylene is often characterized as being highly crystalline, perhaps 90 percent or greater, whereas low-density polyethylenes are as low as 50 percent crystalline (see Fig. 3-1).

The degree of crystallinity is generally determined by x-ray diffraction patterns or infrared analysis (2, 21). In interpreting these experimental measurements, account must be taken of the probable physical conformation of the molecules in the crystalline and amorphous regions of the polymer. Relatively good agreement is obtained for the values determined by the two methods, and a good correlation is found between the crystallinity and the density.

Short branches are sometimes deliberately added to polyethylenes produced in low-pressure processes by means of comonomers. When used as comonomers, propylene, 1-butene, and 1-hexene produce methyl, ethyl, and butyl side groups, respectively.

Long-chain branches occur primarily in polyethylenes produced by high-pressure processes, but these branches are always much less frequent than short ones. Several carbon atoms away from the branch point, it is impossible to distinguish between the main polymer chain and the long side chain. The longer branches have important effects on polymer processability, clarity of polyethylene film, and drawdown of coating resins.

Molecular-Weight Considerations   The average molecular weight and the range of molecular weights of the molecules in the polymer are important (18) in regard to the physical properties of polyethylene. Number-average molecular-weight values M_n generally vary from about 5,000 to 40,000, but weight-average molecular-weight values, M_w, which are always greater, vary from 50,000 to 800,000.

Fractionation of polyethylenes by selective solubility and/or crystallization does not give a sharp separation. A much better separation is obtained, however, with a gel-phase permeation column, a chromatographic unit. Figure 3-3 shows the curves for typical low-density polyethylenes used for injection-molding, high-clarity, extrusion-coating, and low-clarity (liner-grade) resins. A problem with low-density polyethylenes is how to interpret the chromatograms since the branches influence the effective size of the molecules. Figure 3-3 clearly indi-

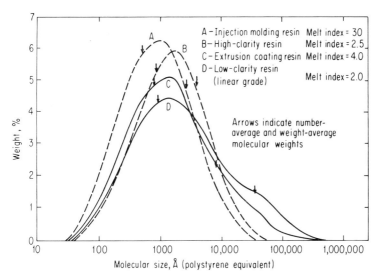

Fig. 3-3   Chromatograms for low-density polyethylenes. (*Cities Service Co.*)

cates, however, that various polyethylene resins have different molecular-weight distributions.

Chromatograms for several other types of polyethylenes have also been published (28). Standard reference polyethylenes can be obtained from the National Bureau of Standards in Washington, D.C., or from Waters Associates, Framingham, Massachusetts. These materials are useful for calibration of instruments employed with polyethylenes.

Lower-molecular-weight polymers act as plasticizers for high-weight materials. As the range of molecular weights for various molecules in the polymer increases, the ratio of M_w/M_n increases. A large ratio of 10:1 to 20:1 means a broad molecular distribution, whereas a ratio of 3:1 to 5:1 indicates a much narrower distribution. The latter products are especially suited for films (18). To increase the molecular-weight spread and to improve processability, several companies mechanically blend polyethylenes of relatively different molecular weights.

Polyethylenes with number-average molecular weights from 1,000 to 10,000 are classified as low-molecular-weight polyethylenes (27); they can be produced in an emulsifiable or nonemulsifiable form. Recently there has been increased commercial interest in producing ultrahigh-molecular-weight polyethylenes that have weight-average molecular weights greater than 2 million (3, 13, 14).

Molecular weight has some effect on the density (or crystallinity) since higher-molecular-weight chains result in more chain entanglement and less crystallization (14). Higher-molecular-weight polyethylenes in general have higher impact strength, lower brittleness temperatures, more abrasion resistance, and greater elongation at break. The importance of molecular weight on the physical properties is discussed in detail by several authors (3, 6, 18, 20, 27).

Cross-linked Polyethylenes Polyethylenes can be cross-linked by using radiation or chemical initiators (17, 21, 23). Commercial irradiation is with high-energy electron beams. Gamma radiation or x-rays are also effective for such cross-linking. In chemical methods, an initiator such as a peroxide is mechanically mixed with polyethylene. The mixture is heated to perhaps 180°C, whereupon the peroxide decomposes to form free radicals and cross-linking reactions occur. The softening temperatures and the tensile strengths of cross-linked polyethylenes are higher than those of non-cross-linked ones.

Chemical Modifications

In addition to cross-linking procedures, polyethylene is commercially modified in several ways (4), including copolymerization with various comonomers that are added in relatively small amounts.

PROPYLENE: Both random and block copolymers are produced by low-pressure processes. Random copolymers have a decreased symmetry in the polymer chain, which causes poorer packing and less crystallinity. When appreciable quantities of propylene are added, the copolymer has no crystallinity and it is an important elastomer. Block copolymers are produced by first polymerizing ethylene to form medium-length polyethylene; then propylene is added, and polymerization continues to increase the chain length. These copolymers are quite crystalline and have properties very different than those of the random copolymers. Tennessee Eastman Co. designates its block polymers as *polyallomers*.

1-BUTENE AND 1-HEXENE: These olefins are copolymerized in small quantities, often 2 to 5 percent, with ethylene in low-pressure processes to improve both environmental stress-crack resistances and impact strength. Ethyl and *n*-butyl side chains form, respectively, and the low-temperature properties of high-density polyethylenes are improved as the density and stiffness are reduced.

VINYL ACETATE: Low-density copolymers containing 2 to 5 percent vinyl acetate are produced by high-pressure processes. The copolymers find important uses in film, wire or cable coating, and molding applications. Improved optical and strength properties are realized. Stress cracking and low-temperature brittleness are both decreased.

ACRYLIC ACID: Copolymers of ethylene and acrylic acid are treated with compounds of sodium, potassium, zinc, etc., to form salts attached to the copolymer chain. The resulting polymeric salts are called *ionomers* by Du Pont, which markets them under the tradename of Surlyn. A patent issued to Rees (22) is probably a description of these polymers, which are highly transparent. Uses of the copolymers include goggles, face masks, bottles, tubing, toys, and electric power cables.

In a sense, copolymers are no longer polyethylene; but if the amount of comonomer is small, they are certainly closely related.

Chlorinated polyethylenes are produced commercially. The amount of chlorine and the type of polyethylene employed affect the properties of the final chlorinated product. Significantly different products are obtained from high- and low-density polymers. Chlorinated low-density polyethylenes are waxy and gummy products, whereas chlorinated high-density polyethylenes, containing 35 to 40 percent by weight chlorine, are extruded to produce weather flashings for buildings. These chlorinated materials have good weatherability and are excellent plasticizers for polyvinyl chloride.

Research is now in progress to find additional methods for the chemical modification of polyethylenes.

Physical Conformation of Molecules

Physical conformation of molecules in polyethylene is important (9). Since polymerization techniques may influence conformation, the most recent concepts are reviewed here briefly.

Crystallization of polyethylene is of considerable importance in terms of the physical properties. Above the first-order transition temperature, which increases from about 221 to 269°F (or 105 to 132°C) as branching of the polyethylene decreases (21), the crystals melt to form an amorphous material. As the polyethylene is cooled below this temperature, crystallization begins. The rate of cooling and the degree of mechanical working affect the rate and type of crystallization.

Until several years ago it was believed that a single polymer molecule passed through several crystallite regions surrounded by amorphous regions, i.e., the fringed-micelle model. Within the last 10 years, the folded-chain model has been suggested as the common form of the crystallite (10, 11, 12). Such crystallites were initially detected when polymers, including polyethylene, were precipitated from dilute solutions. The lamellae (or single crystals) are thin flat platelets of about 100 Å thickness. Electron-diffraction measurements indicate that the polymer chains are essentially normal to the surface of the lamellae and that the chains are folded essentially 180° at the surface. These lamellae often contain pleats or corrugations and frequently take the shape of a hollow pyramid. The lamellae pack or arrange themselves in groups called spherulites.

Crystallization in commercial polymers occurs as the melt is cooled, rather than from dilute solutions. There is less substantial evidence that all crystallites from melts are of the folded-chain variety. Evidence has been reported that polyethylenes of extremely high molecular weight crystallize only slowly, and then with great difficulty, even though they have only a few (or no) side chains. Such polyethylene molecules may be incorporated in more than one crystallite. Perhaps the type of crystallization is quite different with this type of polyethylene.

Mechanical working of the polyethylene as it is molded, extruded, etc., is also important because the arrangement or orientation of the crystallites is affected. As a result, polyethylenes often exhibit degrees of anisotropy, i.e., different physical properties, depending on the direction of working. Southern and Porter (19) have reported a great increase in tensile strength when high-density polyethylenes were extruded under high pressures at about 138°C. A high degree of crystal perfection and orientation was achieved. This approach certainly deserves more experimentation.

Compounding of Polyethylenes

Compounding or mechanically mixing polyethylene with various materials often has a significant effect on the properties. In most cases, an intimate degree of mixing is desired. One obvious reason for compounding is to produce a colored product, and numerous pigments are used for this purpose. In other cases, the chemical and physical properties of the polyethylene are changed. Carbon black, in addition to imparting a black color to the polymer, increases the stress at the draw point of low-density polyethylenes from about 1,330 to 2,000 lb/in^2 when 25 percent by weight of carbon black is added (29). Weatherability of polyethylene is significantly increased by the addition of 2.0 to 2.5 percent carbon black.

Quality-Control Tests

Correlations have been developed of the relationship between the chemical structure and physical properties of polyethylene (8, 18, 21, 23, 26). However, the chemical structure cannot be determined in routine quality-control tests for day-to-day plant operation.

A series of routine tests is used for quality-control purposes (25). Several include measurements of tensile strength, impact strength, stiffness, hardness, elongation, and abrasion resistance. Such tests are primarily important for the properties of the final product. Another series of tests relates more specifically to the properties of the polyethylene during molding, extrusion, or other forming operations to produce the final plastic material. These latter tests (2, 18, 25) often include the following:

1. Melt Index. This test measures the ease of flow of polymer at high temperatures like those used in forming operations. It is inversely related to the average molecular weight of the polymer. In general, a decrease in the melt index causes a higher ultimate tensile strength, higher elongation before rupture, and material with a higher load-bearing capacity.

2. Melt Viscosity. Approximately inversely proportional to the melt index at a specific shear stress, melt viscosity is a function of the molecular weight, the molecular-weight distribution, long-chain branching, and the rate of shear. Polymers with very broad molecular-weight distributions exhibit the greatest reduction in melt viscosity with increasing shear rates. The viscosity of the hot polymers often increases with

time if the measurements are made in air since oxidation causes some cross-linking of the polymer chains.

3. Melt Elasticity. The elasticity of the hot polymer is important to blow molding, sheet extrusion, laminating, and extrusion coating of wires and cables. It is also of some importance in injection molding.

4. Melt Extensibility. This property is of importance when a uniformly drawn or blown specimen of polyethylene is required, as in blow molding, vacuum forming, or paper coating. Various producers have devised specific tests to measure this property.

5. Melt Fracture. This fracture refers to visible defects either in the melt or the solid stage of the extruded or molded product.

These quality control tests are empirical and require judgment in interpreting the results. For example, since polyethylenes (and many other plastics) are subject to creep under stress; the tensile strength varies as the rate of strain is changed. In general, the tensile strengths reported cannot be used for designing a structural item which will be subjected to large stresses. Research is now in progress to obtain more meaningful tests for characterizing the polyethylene products.

II. HIGH-PRESSURE PROCESSES

High-pressure processes were the first processes developed for the commercial production of polyethylenes, and they still account for the major amounts—slightly over 70 percent in the United States (50). These processes operate at pressures from about 15,000 to 50,000 lb/in^2 (1,000 to 3,400 atm). The resulting polyethylenes are of the low-density variety, with a range of about 0.91 to 0.935 g/cm^3. Two types of reactors are used commercially (20, 21, 25, 30, 36, 41, 51, 56): a continuous-flow stirred autoclave, having several variations, and a tubular reactor, also used in a large number of installations.

Pure ethylene is normally added to the reactor along with small quantities of initiators and various organic compounds, and the polymerization can be classified as bulk polymerization. Water and/or solvents have also been suggested as additives in some modified processes.

The operating conditions for polymerization have a large effect on the chemical structure and on the properties of the polyethylene produced. The chemical steps and the engineering features of the reaction will be reviewed before the commercial processes are discussed.

Chemistry of High-Pressure Polymerization

The free-radical polymerization that occurs during high-pressure processes is similar to that with other olefins, but several complications arise for ethylene polymerization:

1. Numerous side chains form on polyethylene molecules, whereas in many other polymerizations little or no branching occurs.

2. Heat of polymerization for ethylene is higher on a weight basis than that of other common olefins.

Many chemical steps occur during polymerization, and digital and analog computers have been employed by several investigators to simulate the complicated reaction scheme. Most models involve some simplifying assumptions. Some models have not been published for proprietary reasons, but others have been (48, 55).

At least four basic steps occur during polymerization: initiation, propagation, termination, and chain transfer (often resulting in branching) (2, 21, 31, 47, 49, 55).

Initiation of the chain reaction requires an initiator, frequently called a catalyst even though it is not one according to strict definition. Peroxides (58), azo compounds, and oxygen are all used commercially for initiation. Choice and quantity of initiator are important from both cost and performance standpoints. A sufficient concentration of free radicals must be obtained during the entire course of polymerization. Initiators have been compared on the basis of the temperature required to obtain a 1-min half-life (21). The temperature for polymerization dictates to some extent the initiator(s) required for polymerization, or vice versa (57). Oxygen, which is a high-temperature initiator, is employed by several major producers. It probably acts to form peroxide compounds in situ.

Increasing the amount of initiator increases the overall rate of polymerization, but the additional amount results in lower-molecular-weight polymers. The amount of peroxide initiators required varies from about 10 to 100 ppm. Often the initiators are dissolved in a solvent or oil to facilitate a uniform rate of addition and to minimize high local concentrations of initiator in the reactor.

Carbon dioxide and carbon monoxide are by-products of peroxides. Carbon monoxide reacts during polymerization to form carbonyl groups in the polymer. These groups are undesirable since they sometimes affect the quality of the polyethylene formed adversely. Hence azo initiators, which do not form carbon monoxide, are often preferred. Infrared analyses are frequently able to identify the fragments of the initiator in polyethylene samples.

Gamma radiation of ethylene is effective for the initiation step (20, 59) but is probably not used commercially.

Propagation of the chain reaction involves the continued reaction of ethylene with the free radicals formed during initiation:

$$R-(CH_2-CH_2)_n-CH_2-CH_2 \cdot + C_2H_4 \rightarrow$$
$$R-(CH_2-CH_2)_{n+1}-CH_2-CH_2 \cdot$$

where n can vary from zero to a very large number and R is the initiator fragment.

Termination of the free radicals (and the production of polymeric chains) occurs by the combination (coupling) of two free-radical polymeric groups or by a disproportionation reaction. In the latter reaction, two long-chain free radicals react, with the transfer of a hydrogen radical from the end of one chain to that of the other. The result is a saturated group on the end of one polymer molecule and an unsaturated group, $-CH=CH_2$, on the end of the other.

Chain transfer occurs both intramolecularly and intermolecularly. The former occurs during propagation by a "backbiting" mechanism that produces short branches. The end of the chain coils backward to extract a hydrogen radical from the fifth carbon atom back in the chain, and chain growth then starts there. More details of the generally accepted mechanism for forming the butyl and ethyl side groups are presented by Billmeyer (2).

Intermolecular chain transfer is employed to adjust the molecular weight of the polymer molecules to desired values. Chain-transfer compounds (often called modifiers or telogens) are frequently light paraffins, hydrogen, or propylene. They are used in concentrations up to several percent in the reactor. These compounds act to terminate one polymeric chain and start another, as indicated by the following example with propane:

$$R-(CH_2CH_2)_n-CH_2CH_2\cdot\ +\ CH_3CH_2CH_3 \rightarrow$$
$$R-(CH_2CH_2)_n-CH_2CH_3\ +\ CH_3\dot{C}HCH_3$$

Propagation then starts on the isopropyl free radical.

Ethane, which is relatively cheap, is sometimes used as a transfer agent, but it is not very effective because it contains only primary C—H bonds. Methane is even more ineffective. Propane and n-butane are more effective because they have secondary C—H bonds, and the hydrogen at these bonds is much easier to extract. Branched hydrocarbons, e.g., isobutane, with tertiary C—H bonds are even more reactive. Propylene is relatively unreactive to propagation; i.e., few polymerization steps occur, but the allylic hydrogen is effective in chain-transfer steps. The isopropyl, *tert*-butyl, and allyl groups formed by chain-transfer steps with propane, isobutane, and propylene are the end groups on the polyethylene molecule then produced.

Chain transfer also occurs with polymer molecules. Long-chain branching results from such a transfer, since propagation now occurs as a branch on the polymer chain starting at the carbon atom from which the hydrogen was removed. Several other materials that tend to act as chain-transfer agents include ethylene, the initiator, or other compounds added deliberately or accidentally to the reactor.

Long-chain branching also occurs when an unsaturated group of a polyethylene molecule reacts during propagation and is copolymerized with the ethylene.

Other features of polyethylene molecules are some internal unsaturated groups, $-CH=CH-$, and vinylidene groups (20). Infrared analyses are useful for identifying these and other features of the polyethylene molecules.

Polymerization Conditions Used Commercially

The ranges for the most important polymerization conditions and their effect on the final polymer are as follows.

Temperatures vary from about 150 to 300°C (302 to 572°F), although ranges from 200 to 250°C (392 to 482°F) are more common. Temperature often affects the overall polymerization reactions in at least three ways.

1. When only one phase occurs in the reactor, i.e., when the polyethylene dissolves in the ethylene phase, lower temperatures decrease the rate of polymerization, increase the average molecular weight, decrease the number of short- and long-chain branches, and decrease the number of terminal vinyl groups. Higher temperature increases the rates of the chain-transfer steps to a greater extent than the rates for propagation.

2. Two "fluid" phases sometimes occur, an ethylene and a polyethylene phase (discussed in more detail later). The degree of miscibility decreases as temperature is lowered. Presumably the ethylene phase is always the continuous one since the conversion levels are relatively low. Polymerization occurs in each phase (25), and the polymer produced in the polymer phase is thought to be highly branched at least with long chains and probably also with short chains. Diffusion of ethylene molecules to the free-radical end of the growing chain in the polymer phase is probably the controlling step. The relative amounts of polymers formed in each phase, if known, are not reported.

3. Resin buildup on the reactor walls increases with lower temperatures for the reaction mixture and for the reactor walls (which are sometimes cooled). For portions of a reactor where higher levels of polyethylene are present, wall temperatures below 210°C are not desirable (52).

Pressures in the reactor vary from about 15,000 to 50,000 lb/in² (1,000 to 3,400 atm). For most polyethylenes, the pressures are 20,000 to 35,000 lb/in². Little or no polymerization occurs below 9,000 lb/in², and the overall rate of polymerization increases linearly as the pressure is increased, at least in the range 9,000 to 22,000 lb/in² (39). As the pressure increases, the rate of the propagation steps increases more rapidly than the rates for the termination and backbiting steps. Hence higher pressures result in higher densities and less branching, higher molecular weights, and fewer vinyl groups.

Initiator concentrations, as previously discussed, are low and vary with the type of initiator.

Modifier concentrations are directly related to the average molecular weight of the polyethylene desired. As will be discussed later, these concentrations may have more than a chemical effect on the reaction.

Polymer concentrations in the reactor, i.e., the conversion level of ethylene to polyethylene, depend on the type of reactor used. In a tubular reactor, the weight concentration varies from zero up to perhaps 20 percent. In an autoclave reactor, the range is from approximately 10 to 20 percent. Long-chain branching increases in a complicated manner as the concentration of the polymer increases.

Engineering Considerations

Engineering factors of importance for the polymerization are not well covered in the literature, but some have been discussed (25, 30, 52). At the conditions used, ethylene is in the fluid phase since it is above its critical temperature. Because of the pressures involved, ethylene density varies from about 0.4 to 0.6 g/cm³, which is in the range for most hydrocarbon liquids. Figure 3-4 shows the plot of the density of pure ethylene vs. temperature with pressure as a parameter. This

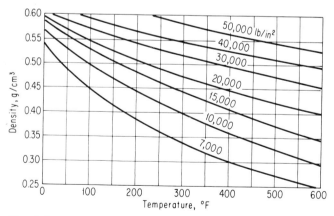

Fig. 3-4 Density of ethylene at high pressures. [*From* (30) *with permission of Chemical Engineering.*]

graph was prepared by using generalized compressibility-factor plots (44, 45).

Two phases, ethylene and polyethylene phases, sometimes occur in the reactor, especially at lower temperatures, at higher conversions of ethylene (when more polyethylene is present), and at lower pressures when the density of the ethylene phase is less. Complete miscibility occurs at high pressures. Other variables affecting the relative solubility of the two phases in each other include the type of polyethylene being produced and the specific modifier used. Higher-molecular-weight polyethylenes are less soluble, and the extent and type of branching also affect solubility.

Concentrations of several percent for the modifiers may affect the solubility of polyethylene in the ethylene phase. Since isobutane is a more effective modifier than propane or (especially) ethane, it is generally used in smaller molar concentrations. The relative solubilities of the two phases may therefore differ somewhat when isobutane is used.

When two phases occur in the polymerization vessel, the polyethylene dissolved in the ethylene phase undoubtedly has a lower molecular weight than that in the polyethylene phase. At least several companies or groups have obtained solubility information, but it is considered proprietary.

High (or at least reasonably high) turbulence seems to be used in all commercial reactors, and when two phases are present, they are emulsified. The ethylene phase, because it is present in so much larger amounts, is undoubtedly the continuous phase, while the polyethylene phase is the dispersed one. Mass transfer is of special importance with two phases because polymerization reactions occur in both (25). Heat transfer and temperature control are also always important regardless of the number of phases.

Viscosity considerations are complicated by the presence of two phases, variations in the concentrations of polyethylene in the system, i.e., conversion levels of ethylene, and the nonnewtonian character of polyethylene. Relatively little has been reported about the design of the impellers used in commercial reactors.

Safety considerations are obviously important in these reaction systems because of the very high pressures involved and because of the exothermic reactions. Safety barricades are always provided for the reactors. In addition, explosions of the ethylene occasionally occur.* At the pressures and temperatures involved, ethylene is thermodynamically unstable and may explosively decompose into carbon, hydrogen, and

* These are actually decompositions, commonly called *decomps.*

methane. Hot spots or high local concentrations of initiator may initiate a decomp. Possibly surface reactions or improper design of the moving parts of an agitator are also responsible. Safety disks are always provided on a reactor to relieve the sudden pressure rise during a decomp. Higher temperatures, especially above 280°C, higher pressures, and higher concentrations of initiators accentuate the possibility of a decomp.

The safety disk and its vent line must be designed so that the exhausting fluids can be safely discharged to the atmosphere. A reactor making about 30 to 50 million lb of polyethylene per year was provided with a 10-in vent line extending above the top of its barricade. Following a decomp, the exhausting gases caught on fire and burned for several minutes. A larger vent line was later installed to minimize the possibility of a fire following a decomp. A reactor is usually not damaged because of a decomp, and after the safety disk is replaced, it can generally be returned to service within a day. A well-designed and well-operated reactor will normally last at least 5 to 10 years. Damaging explosions have occurred, however, in commercial reactors.

Important features from both an operating and a safety standpoint include the method for introducing the ethylene feed and the initiator in the reactors. Good mixing is necessary in order to minimize high local concentrations of initiator in the reactor. If a mechanical agitator is used, the packing gland must be carefully designed and maintained. The method of introducing the feed often has an important effect on the type of polyethylene produced.*

Process Details

Most producers use highly purified ethylene, at least 99.9 percent pure, containing trace amounts of ethane and methane. Impurities often act as chain-transfer agents. Acetylene must be removed from the ethylene, and this is usually done by a selective hydrogenation technique in the ethylene plant. Figure 3-5 indicates a typical flowsheet for a process using tubular reactors or autoclave reactors.

Compressors Compressing ethylene to the high pressures needed for polymerization is a major step of the overall process. Large positive-displacement (piston) compressors (sometimes designated as pumps) are used, and a two-step compression system is normal. In the first step, ethylene at a relatively low pressure is compressed to about 250 atm (3,500 lb/in²) by the primary pump (Fig. 3-5). The

---

* Additional details of safety precautions often followed in high-pressure polyethylene plants were presented at a symposium during the 1972 meeting of the AIChE held in St. Louis; papers of this symposium were published in *Chem. Eng. Prog.*, November 1972.

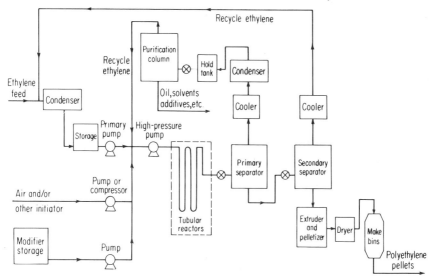

Fig. 3-5 Flowsheet for high-pressure polymerization of ethylene using tubular reactors. [*From (30) with permission of Chemical Engineering.*]

exact pressure depends on that employed in the separation phase of the process to be discussed later. In the second step of compression, the ethylene feed is combined with recycle ethylene, and the mixture is compressed to the reaction pressure by the high-pressure compressor (or pump).

For polymerization of 100 million lb (45,000 metric tons) of polyethylene per year at a polymerization level of 15 percent per pass, flows in the compressor steps are approximately as follows (53): ethylene feed to the first-step compressor, 13,000 lb/h (6 tons); combined ethylene feed to the second-step compressor, 88,000 lb/h (40 tons). Energy requirements for the compression steps total approximately 10,700 hp (8,000 kW) when the reaction pressure is 45,000 lb/in² (3,000 atm). The cost of compression is large, sometimes being at least one-third the total operating cost of the plant.

Thermodynamic graphs for ethylene have been prepared by several companies, and one such graph was published by Benzler and Koch (32) for pressures up to 140,000 lb/in² (over 9,000 atm). The enthalpy for a given temperature is a minimum at a pressure of about 4,000 to 10,000 lb/in² in the temperature range from 10 to 315°C (50 to 600°F). There are still uncertainties in the values of the thermodynamic properties at the higher pressures.

The initiator is sometimes added to the ethylene before the second compression step (the high-pressure step). This step, as might be ex-

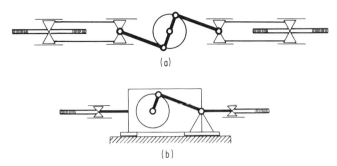

(a)

(b)

Fig. 3-6 Designs for driving pistons in a high-pressure compressor: two opposing pistons driven by (a) a conventional gearbox and (b) by well-guided frames.

pected, is the more difficult one. The temperature of the ethylene must never exceed 100°C (preferably 80°C) as the ethylene is compressed. Otherwise some polymerization might occur. Normally only two stages are required for the high-pressure compression, and the ethylene is cooled between stages.

Only a relatively small displacement volume is allowed in the piston-cylinder arrangement of the compressor, which must be ruggedly constructed because of the high pressures involved and which must provide long and dependable service. Design features of these compressors include methods for centering and guiding the piston; the pistons are operated at rather limited speeds. Accessibility of the compressors to allow easy maintenance is of considerable importance. Several designs have been considered. In Fig. 3-6a two opposing pistons are connected to a conventional gearbox, and Fig. 3-6b shows two opposing pistons connected by a well-guided drive-frame arrangement. In both cases, the drive rod to the piston is shown, but the piston-cylinder arrangement is not. The arrangement in Fig. 3-6b, the latest design, provides a more compact compressor and is available from several manufacturers.

Figure 3-7 shows an Ingersoll-Rand 15½-4 HHE-2 Slugger hypercompressor used by Cities Service Company at their Lake Charles, Louisiana, plant. A 6,500-hp 200 r/min brushless synchronous motor is directly connected to the compressor. The compressor has four first-stage cylinders, followed by intercoolers, and then four second-stage cylinders.

Units as large as 90,000 to 100,000 lb/h of ethylene are in current demand, and units as large as 150,000 lb/h are already under consideration. Larger motors would of course be required for such units. American manufacturers of high-pressure compressors include Ingersoll-Rand Co. and Clark Brothers Co. Two or three European companies also make such compressors, including the Burkhardt Company.

Consideration has been given to use of a centrifugal compressor in the first step of compression cycle. Ethylene flows are probably too low, even in the largest polyethylene plants, for economic operation. The ethylene produced in a pyrolysis plant leaves the ethylene-purification column as a liquid at a relatively cold temperature (−30°C) and at a moderate pressure (about 20 atm). This liquid ethylene is sometimes used as the feedstock to the compressors. In theory, the liquid ethylene should be easy to compress in a single step to the desired pressure, but complications occur. First only a relatively small amount of ethylene is polymerized per pass, and most of the ethylene is recycled. Recycle ethylene is initially a gas that needs to be liquefied if liquid ethylene is to be pumped. In the purification columns shown in Fig. 3-5, liquefication would result and simultaneously impurities would be removed. Companies that do not pump liquid ethylene probably do not use such a purification column in their plant. At least one American company does not recycle unreacted ethylene, using it instead in the manufacture of other petrochemicals.

Liquid ethylene tends to be quite soluble in the lubricating oils used in the compressor or attenuator that pumps the ethylene to reaction pressures. Hence the compressors must be designed to prevent contamination of the ethylene with lubricating oils.

When a peroxide or other relatively unstable initiator is used, an inert solvent is normally employed to dissolve the initiator (43, 58). Such an arrangement facilitates accurate metering of the amount of initiator added and also permits ready dispersion of it in the reaction mixture. Sometimes the solution is introduced at several locations in the reactor.

Fig. 3-7 High-pressure compressor with a 6,500-hp motor. (*Cities Service Company and Ingersoll-Rand Co.*)

Tubular Reactors (Conventional Process)   Tubular reactors used by many companies were first developed in the United States by Union Carbide Corporation (the largest American polyethylene producer) and in Germany by Badische Anilin- und Soda-Fabrik (BASF). Northern Petrochemical Company at Joliet, Illinois, is currently building four BASF reactors, each of which will produce 125 million lb/year (34).

Ethylene and dissolved initiator (often oxygen) enter the reactor tube at high pressures. Each reactor consists of three zones, the first for preheating, the second for reaction, and the third for cooling (52). These zones are of approximately equal length. Heat transfer is provided by means of oil, Dowtherm, or water in the jacket surrounding the tubes. After the ethylene containing the initiator has been sufficiently preheated, conversion starts and the exothermic heat of reaction causes an additional temperature rise. The rate of polymerization, which is temperature-dependent, increases rapidly, and the temperature inside the tubes becomes higher than that of the heat-transfer fluid in the jacket. The total residence time of the reactants in a tube reactor varies from about 45 to 60 s.

Sufficient initiator is usually added to polymerize several percent of ethylene. Too much initiator would cause too high a conversion, leading to high temperatures that could result in decomps. The polymerization reaction occurs within several seconds. In many cases, additional and relatively cool ethylene and initiator are added at one or more intermediate points along the reactor tube. These additional feeds must be mixed rapidly with the fluids in the tube, otherwise high local concentrations of initiator may cause decomps. Rapid mixing can be obtained by jetting the additional feed in through small-diameter feed lines or by using small turbines driven by the entering ethylene stream (53).

The temperature at which the polymerization rate becomes significant varies of course with the initiator used and with the pressure. When azobis(isobutyro)nitrile is used, the reaction starts at about 110°C, but oxygen requires 170°C and 1,500 atm.

Commercial tubular reactors produce from perhaps 8 to 125 million lb/year (3,500 to 57,000 metric tons) of polyethylene. A small reactor is about 1.0 in (25 mm) ID and 800 ft (245 m) long. The largest reactors are about 3 in. (76 mm) ID and probably at least 2,500 ft (760 m) long.

Table 3-2 indicates several details for four tubular reactors with working pressures from 1,500 to 4,000 atm and with capacities from at least 3,500 to 40,000 metric tons/year (52). These reactors are normally built as a series of loops or coils. Reactor 1 consists of straight sections of 10-m tubing joined by U bends. Often high-pressure flanges are

TABLE 3-2   Data for Four Tubular Reactors (52)

Reactor	Capacity, metric tons/yr	Maximum working pressure, atm	ID, mm	OD, mm	Wall thickness, mm	Total length, m (approx.)	Average heat-transfer coefficient for inside film, Btu/(h)(ft²)(°F)	Material of construction
1	3,500	1,500	34	71	18.5	300	25	DiN-Nc-7707 (30 CrMoV9)
2	......	1,600	80	178	49	...	...	DiN-Nc-7707 (30 CrMoV9)
3	40,000	2,500	70	170	50	500	33	DiN-Nc-7707 (30 CrMoV9)
4	......	4,000	34	121	43.5	...	...	SAE 4340

provided to bolt the straight sections of tubing to the U bends and to complete the loops. A relatively mild steel is used as gasketing material, and high-strength steels are used for the tubes and U bends (see Table 3-2). The tubes should always be smooth on the inside to minimize polyethylene deposits on the wall. Figure 3-8 shows portions of the polyethylene plant of Aquitaine-Total Organico; the tubular heat exchanger shown is for the recycle ethylene, and the blockhouse (or safety barricade) back of the exchanger houses a tubular reactor.

Fig. 3-8   High-pressure plant, showing safety barricade for reactor with tubular heat exchangers for recycle ethylene on the outside. [*Aquitaine-Total Organico*, from (37) *with permission of Chemical Engineering.*]

Heat-transfer resistances are high in these tubular reactors because of the very thick walls required to withstand the high pressures. Only approximately one-third of the heat of reaction was transferred for a specific run in the reaction zone of reactor 1 and only one-tenth in reactor 3 (52). Consequently rises in temperature, often as much as 50 to 100°C, occur with polymerization, and they depend of course on the diameter of the reactor, the pressure, and the amount and type of initiator. The temperature is measured at frequent points along the length of most reactors, and temperatures as high as 280°C are not uncommon at certain lengths.

Several features affecting temperature control are important. High flow velocities are maintained inside the tubes to obtain reasonably high heat-transfer coefficients at the tube wall (see Table 3-2). As a result, pressure drops in the tubes often range from 1,500 to 5,000 lb/in². Selection of the optimum flow velocities is a compromise between obtaining high heat-transfer coefficients and low pressure drops. Pressure drops of course necessitate the use of higher inlet pressures.

Schoenemann (52) has suggested an interesting method for obtaining better heat transfer in tubular reactors. Channels or holes are drilled in the wall of the reactor relatively close to the inner wall. A coolant is pumped at fairly high velocities through these channels. As mentioned previously, cold reactor walls promote polyethylene deposits on the wall. Schoenemann recommends that the coolant in the jacket be 210°C or higher. Such temperatures result in relatively small temperature differences for the cooling zones of the tube reactor.

Viscosity information has been published for mixtures of ethylene and polyethylene (38). The viscosity increases as the temperature is lowered, as the pressure is increased, and as the molecular weight of the dissolved polyethylene is raised. A very large viscosity increase occurs, however, as the concentration of dissolved polyethylene is increased. For pure ethylene at reaction conditions, the viscosity ranges from about 0.05 to 0.08 cP, but it varies from perhaps 4 to 30 cP for product mixtures from commercial reactors. If a polyethylene phase forms, the viscosity of the mixture approximates that of the continuous (ethylene) phase. Flow characteristics in a long tube such as the present reactor are discussed by Carter and Bir (33).

Schoenemann (52) has reported several details for adding ethylene and dissolved oxygen at three intermediate positions of a tubular reactor to obtain an overall ethylene conversion of about 13 to 14 percent. In this case there were four zones in which high reaction rates were obtained, and the temperature increased from about 200 to 280°C in each of the reaction zones. The oxygen concentration in the ethylene fed to the inlet of the reactor was 80 ppm (on a weight basis). Lower oxygen concentrations were employed for each intermediate feed stream. The concentration for the last feed stream was 40 ppm. Because of the increased amounts of flow with increased reactor length, larger-diameter tubes were provided after each addition of ethylene feed.

With multiple additions of ethylene and initiator, up to perhaps 18 to 20 percent of the total ethylene is sometimes polymerized in the reactor. Polymerization levels of 14 to 18 percent are more common, however. When ethylene and initiator are introduced only at the reactor inlet, polymerization levels of 8 to 12 percent are generally realized.

A common, if not standard, method for minimizing the deposits during a run is the practice of pressure-pulsing the reactor. About once a minute, the pressure at the end of the reactor is suddenly dropped for several seconds by partially opening a special valve there. Pressure drop of 1,500 to 4,500 lb/in^2 are common. The pressure pulse then transmits itself through the long tubular reactor, causing sudden increases in the flow velocities which tear the polyethylene deposits from the tube walls. In practice, the ethylene is continuously fed to the reactor at a constant rate, but the product mixture leaves the reactor only when the valve at the exit end is opened.

Occasionally (perhaps every several months), a tubular reactor has to be thoroughly cleaned with a solvent, such as toluene. Smith (25) also reports other cleaning procedures which have been proposed.

Since the temperature, pressure, initiator concentration, and level of polyethylene vary with tube length, the type of polyethylene produced at different locations in the tube varies. Polyethylenes produced in a tubular reactor tend to have wider ranges of molecular weights and of branching than those produced in autoclave reactors.

A tubular reactor is often operated at a single ethylene flow rate and inlet pressure. Variable-flow compressors are usually too expensive to be practical. Changing the type or amount of initiator used is the preferred method for obtaining different types of polyethylene products. Schoenemann (52) reports that a low-molecular-weight product (melt index of 15) is obtained when 100 ppm oxygen is used at 1,500 atm. A higher-molecular-weight product (melt index of 0.2) is obtained if 25 ppm oxygen is used. When azo bis (isobutyro)nitrile is used, a higher-density product is obtained than when oxygen is used; the main reason for this density difference is thought to be the difference in polymerization temperatures employed.

Tubular Reactor (Water-Solvent Polymerization)   Imperial Chemical Industries has publicized (41) a water-solvent polymerization method for use at pressures of about 1,000 atm (15,000 lb/in^2 absolute). It is not known if this method is employed commercially. A typical feed to the reactor has a 1.0:1.0:1.5 weight ratio of ethylene to benzene to water. The ethylene and about 20 ppm of oxygen (used as initiator) are dissolved in the benzene. The two-phase mixture is fed to a tubular reactor, and additional water containing 100 ppm oxygen is injected at multiple points along the reactor tube. Approximately 17 percent of the ethylene is polymerized in the reactor, which is maintained at essentially 190°C (375°F). Stainless steel is used in the tubes to minimize corrosion (39, 41). The operating costs per pound of polyethylene would presumably be significantly greater by the modified process.

Autoclave Reactors Autoclave reactors are widely used. Imperial Chemical Industries (ICI) in Great Britain and E. I. du Pont de Nemours Company in the United States pioneered their early development. ICI has licensed their process, which is used in 60 plants (46). The polyethylenes produced by modifications of autoclave reactors are often quite different.

Autoclave reactors are essentially continuous-flow stirred-tank vessels. The length-to-diameter ratios of these reactors range from about 2:1 to 20:1. The degree of backmixing is especially high at the lower ratios. Some autoclave reactors are compartmentalized or arranged as cascades (52), being essentially a series of reactors with an agitator in each stage. In such an arrangement, the plug flow of a tubular reactor is approximated. Multiple feed points are often used in the longer or compartmentalized reactors. The reactants remain in a reactor for an average of about 25 to 40 s.

Heat-transfer surfaces are sometimes provided in the autoclave and sometimes between stages, to remove part of the heat of polymerization, but autoclave reactors are often operated at essentially adiabatic conditions. The heat of reaction is then almost balanced by the sensible heat required to bring the inlet ethylene to reactor temperature. The system is not quite adiabatic because of heat losses from the hot autoclave to the surroundings and because of the energy added to the reactor by the agitator. These latter two energy terms are both relatively small and essentially balance each other.

A graph like that in Fig. 3-9 is used by producers who use adiabatic autoclave reactors. This graph was constructed using available enthalpy data for ethylene (32) and by assuming a heat of polymerization at reaction conditions of 24,000 cal per gram mole of ethylene. In general, conversions of 10 to 18 percent are obtained in the adiabatic system.

Autoclave reactors used commercially produce from about 10 to 130 million lb/year of polyethylene per reactor. A reactor for 44 million lb has the following features: 200 liters (7 ft^3) capacity; 9.9 in (250 mm) ID; 31.5 in (800 mm) OD; 13.1 ft (4 m) long; and a working pressure of up to 2,300 atm (52). Toyo Koatsu uses a 600-liter (21-ft^3) reactor

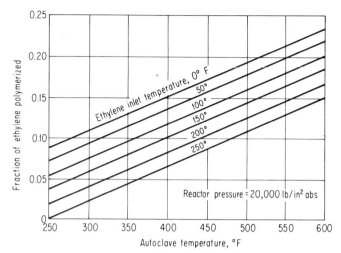

Fig. 3-9   Ethylene conversion as a function of autoclave temperature.   [*From* (*30*) *with permission of Chemical Engineering.*]

for producing 132 million lb/year (35). A 22-ft³ autoclave being sold by an American manufacturer has the following internal dimensions: 20 in diameter and 10 ft long for a 6:1 height-to-diameter ratio; feed inlets (or potential ones) are provided every 6 in in height. High-strength steel is used for fabricating the vessel, and an outer shell is press-fitted over an inner shell to distribute the stresses. The autoclave is designed for 40,000 lb/in², and it was reported several years ago to cost $75,000 without an agitator or any auxiliary equipment. Such a price applied, however, only if several autoclaves were purchased at one time.

The ratios of the outer to inner diameters of these autoclaves generally vary from about 3:1 to 4:1. Often more than one impeller is provided on the agitator shaft positioned at the axis of the autoclave. Three methods are available for transmitting energy to the agitator shaft from an electric motor (52). In two of them the motor is located outside the reactor, either on the top or the bottom. A high-pressure packing gland is provided for the agitator shaft in one method. In the other method, a magnetic drive is provided. In the third method, the motor is located inside the reactor and surrounded with feed ethylene.

Ethylene and dissolved initiator are sometimes partially or completely introduced through the hollow shaft of the impeller (52). Outlets are provided in the impellers. Such an arrangement of course minimizes high local concentrations of initiator in the reactor.

Vaughn and Hagemeyer (58) describe a two-zone reactor that is essentially two continuous-flow stirred-tank reactors positioned one on top of the other. A baffle separates the two zones and essentially eliminates all backmixing between them. The first zone contains a single-paddle or flat-bladed-turbine agitator positioned in the top of the zone near the feed point. Caprylyl peroxide, a low-temperature initiator, is recommended for the first zone. It is introduced in mineral-oil solutions of 5 to 50 percent concentrations or solvent (not specified) solutions of 20 percent concentrations. The first zone is operated at approximately 130 to 190°C (266 to 374°F), and specific examples in the patent are from 160 to 175°C (320 to 347°F). High-temperature initiators require temperatures of 190 to 230°C (374 to 446°F).

In the second zone of the reactor, directional agitation is provided. "Conventional" and higher-temperature initiators, such as lauroyl peroxide and di-*tert*-butyl peroxide, are dissolved in a mineral oil and added to the second stage. The latter initiator is introduced as a 5% solution. An unspecified portion of the ethylene is also introduced directly to the second zone, where temperatures range from 190 to 270°C (374 to 518°F), although examples in the patent show temperatures ranging from 202 to 260°C (396 to 500°F).

Production rate of polyethylene in this two-zone reactor varies from 8 to 15 million lb/year, depending on the operating conditions and the initiator used. Ethylene conversions are 15 to 20 percent per pass. Figure 3-9 is not applicable to this reactor since some cooling is apparently employed. The mineral oil, used as a solvent for the initiator, apparently remains in the polyethylene product and acts as a plasticizer. A comparison (58) has been made of the products obtained when the initiator is dissolved in a mineral oil and in a volatile solvent, which presumably vaporizes and is removed from the polyethylene. A significant difference was noted in the stiffness and softening point of the two products.

The temperatures of the reactor and hence the polymerization rates are controlled by regulating the feed rates of the initiator to the reactor. Controlling an autoclave reactor is not easy, since only relatively small changes of initiator concentrations and/or temperature often lead to unsteady-state operation. An increase in the initiator feed rate results in a higher concentration of free radicals, then a faster rate of polymerization, and finally a higher temperature, which causes a still higher rate of polymerization. Only a small increase in the initiator feed rate sometimes results in a cycle leading to a decomp. A decrease in the initiator feed rate slows down the reaction rate, and even relatively small decreases may rather quickly stop or "kill" the polymerization reaction.

Start-up of a reactor is a critical part of the operation, but relatively little has been published on the subject. The procedure must be carefully followed in order to line out the reactor as soon as possible (to minimize formation of off-grade product) and yet prevent a runaway reaction. Preheating the ethylene, electrically or by steam, is used in some cases if not all. Increased amounts of initiator or low-temperature initiators can also be employed during start-up.

Effective agitation is obviously needed in an autoclave reactor to minimize concentration gradients of reactants and to promote uniform temperatures in the mixture (57). Since the mixture contains appreciable amounts of polyethylene, its viscosity is rather high (37). Other features important in designing an autoclave reactor include the length-to-diameter ratio, the internal design of the reactor, the number and type of agitators, the power input to the agitators, and the design of the feed inlets. The degree of backmixing may be purposely varied to obtain different products. Local mixing intensity and residence-time distribution of the reactants affect the molecular-weight distribution of the polymer. In some cases, the relationship between the character of agitation and the physical properties of the polyethylene has been established by commercial producers.

Buildup of polymer deposits on the surfaces of the autoclave reactor is sometimes a problem, especially if dead spots occur in the reactor. These dead spots are minimized by effective agitation. Polymer deposits contribute to gel formation of fisheyes that decrease the quality of the polyethylene. Such gels occur because polymerization reactions and long-chain branching continue in the surface deposits on the walls of the reactor. When deposits form, they are frequently removed by solvent treatment.

Reactor Operation Mathematical models have been developed for the operation of both tubular and autoclave reactors. One model (52, 55) predicts the amount of polyethylene formed, the characteristics of the polyethylene (molecular weight and branching features), the temperature and pressure profiles in the reactor, and operating expenses. For a tubular reactor, heat-transfer terms must be included in the models. These models are useful for optimizing reactor design for desired types of polyethylene.

Tubular and autoclave reactors have certain advantages and disadvantages. Tubular reactors give a more stable operation, whereas autoclave reactors often tend to be quite unstable. Decomps in autoclave reactors are usually more frequent and have a different character. A well-operated tubular reactor probably has less than two decomps per year. Release of the pressure in a tubular reactor following a decomp is smoother and less of a problem since the materials have to be transmitted considerable distances down the tube. U bends in the tubular reactors are provided with fairly long radii to provide smooth flow of the reactants.

Polyethylene deposition on the walls of the autoclave reactors is considerably less of a problem than for tubular reactors.

Enough flexibility of operation is possible in both types of reactors to permit essentially the same types of polyethylene to be produced in each. In a tubular reactor, the temperature profile along the tube can be changed by modifying the temperatures of the liquid in the jacket and (more important) by varying the type and amount of initiators used. Temperature changes obviously affect both the molecular weight and the branching of the product.

When ethylene and oxygen feeds are added only once in a tubular reactor, production costs are higher than when multiple feeds are used. Lower levels of polymerization are realized in the former case.

As a general rule, polyethylenes with rather wide molecular-weight distributions can be produced cheaper in tubular reactors. Ones with narrow molecular-weight distributions are probably produced cheaper in autoclave reactors.

Programming and planning production schedules for a reactor involve considerable time and effort. Since each producer makes numerous types of polyethylene, a reactor must be shifted from one type to another by changing the operating conditions and perhaps the initiator, modifier, etc. An effort is made to make only those switches in a given reactor which can be done quickly, in order to minimize the production of off-grade material. The changeover can generally be made in 1 h or less.

Separation Steps of High-Pressure Processes Separators following the reactor and the pressure-reducing valves (which are used to control the reactor pressure) are steam-jacketed to prevent excessive heat losses. Pressure reduction of ethylene to intermediate pressures is a constant-enthalpy process, and the temperature increases.

At least two separators are used for the separation of unreacted ethylene from the polyethylene. The first of the two separators operates at a pressure of 900 to 4,000 lb/in² (25), and about 95 percent of the ethylene flashes in this stage. The rate of flashing must be sufficiently low to prevent excessive foaming. Starting with a single phase of ethylene and dissolved polyethylene, a polyethylene "precipitate" forms as flashing starts. As more ethylene flashes, the precipitated granules agglomerate to form a continuous polyethylene liquid. The second separator operates at pressures of 5 to 10 lb/in² gage, and essentially all the remaining dissolved ethylene flashes. The polyethylene "liquid" formed has a viscosity of perhaps 3 million cP.

The flashed ethylene is recovered and generally recycled. Using more than two separators would decrease the amount of energy needed for compressing the recirculated ethylene. In such a case, the ethylene flashed at high pressures could be reintroduced into the compression system at the appropriate location. Two separators are generally considered to be optimum since the capital costs for more cannot be justified by savings in operating costs. Impurities such as low polymers, oxygenated products, and nitrogen are present in the recycle ethylene. Some companies purify part of the recycle ethylene to prevent the buildup of undesired levels of impurities and to liquefy the ethylene. Schoenemann (52) indicates, however, that such purification is not always necessary.

When the ethylene is recycled, it is cooled by air or water as a first step. Partial condensation and separation of the less volatile materials may then result. Often the recycle stream is next condensed by refrigeration, and the condensate is temporarily stored in a hold tank. This liquid is sometimes fractionated to produce a relatively pure ethylene. The required purity for the recycle ethylene depends on the modifier and impurities present. Recirculation of controlled amounts of modifier in the recycle ethylene stream to the reactors can be tolerated.

Processing of Polyethylene   *Extrusion and pelletizing* is the first operation as polyethylene leaves the separator. The semiliquid polyethylene is pumped by a screw inside a barrel or by a gear pump (25), and it is forced through a die with holes of approximately $\frac{1}{8}$ in diameter. The polyethylene is generally extruded directly into a water bath, which freezes it. A rotating knife, or similar device, chops or dices the extruded polymer as it leaves the die, to form pellets. Details on this aspect of the process are presented by Smith (25).

Drying the polyethylene pellets is the next operation. Free water is first drained from the pellets and recycled to the extrusion operation. A shaking separator then removes additional water mechanically. The pellets are finally dried with hot air passed over them in countercurrent direction.

Storing the dried polyethylene pellets is in silos or *make hoppers*. In modern plants, the pellets are pneumatically conveyed to the make hoppers, which hold 30,000 to 100,000 lb of pellets and contain several hours' output from a reactor. Special precautions are taken to prevent polymer contamination. The silos are usually made from aluminum, although stainless-steel-clad steel or glass-lined steel is sometimes used.

Samples of polyethylene are normally obtained every hour from the make hoppers. These samples are analyzed by various quality-control tests, including melt index. If the product is found to be off specification, the operating conditions for the polymerization are adjusted; generally the concentration of the initiator or the modifier to the reactor is changed. Since a make hopper is large, small amounts of off-grade product can be blended with the remaining product, so that the final blended material will meet specifications.

Blending is next used to combine the pellets with certain additives, as follows:

Antioxidants, of which a wide variety are available, including butylated hydroxytoluene and anisole, are used (0 to 300 ppm).

Slip agents are used for extrusion-grade polyethylene. Sometimes fatty amides (0 to 4,000 ppm) are added.

Antiblock agents, such as finely divided calcium carbonate or silica, are added in amounts up to 8,000 ppm to prevent the final film product from sticking.

Antistatic agents are used to prevent dust collection on bottles or other molded objects.

Pigments are sometimes added. When relatively small quantities of a colored polymer are needed, pigments are usually not added in the polyethylene plant because of problems in cleaning the apparatus for another color.

In many cases, if two or more polyethylenes are to be blended, they are combined at this blending stage of the operation.

Tumble blending in a large rotating cylinder is used extensively for batch sizes of 20,000 to 100,000 lb. A more recent development has been an air-blending technique, in which the pellets are fluidized and mixed in a large cylindrical vessel. The blending apparatus normally does not need to be cleaned between batches because only small amounts of materials are left in it from the previous batch. The additives are often prepared in the form of a concentrate made by melt-blending them with small amounts of polyethylene (sometimes using the off-grade material). This mixture is then extruded and pelletized to form the concentrate.

Reextrusion and repelletizing of the polyethylene result in the following:

1. Intimate mixing of the polyethylene and additives
2. Working of the polyethylene, which improves the optical properties of the plastic, particularly the gloss and clarity of the end products
3. Pelletizing of the mixture before packing in forms that are suitable for subsequent molding operations

An effort is often made to provide considerable mechanical working to the polymer to increase its clarity. Additional reextrusions would improve clarity even more, but only one reextrusion is economically justified. Mechanical working of the polymer can be increased during a given extrusion by cooling the barrel of the extruder. This working is thought to distribute grainy particles or gels in the polymer.

In a relatively new type of extruder used widely the plastic is worked before extrusion by two lobes that operate on the same principle used in Banbury mixers.

Packaging the pellets is the last operation in polyethylene production. The pellets from the final extrusion operation are pneumatically conveyed to silos for temporary storage or directly to the packaging equipment. Numerous packaging arrangements are currently used: 50 lb in lined-paper bags; 1,000 lb in cardboard boxes; 10,000 lb in transportable rubber bins; and 100,000 lb in railway cars.

Economics of High-Pressure Processes

Capital costs for high-pressure processes are high (40), in the approximate range of 15 to 20 cents per annual pound of capacity. The lower value is more probable with a very large plant, while the higher value is for a smaller plant. A 500-million-lb plant would then cost about $75 million.

Little information has been published concerning the costs for operating a plant. De Lesquien (37) however, has reported some process economics for a high-pressure plant which uses a tubular reactor and is designed to produce almost 100 million lb/year of polyethylene. Utilities required per ton of polymer include 1,000 to 1,180 kWh of electricity, 1,200 to 2,000 lb of steam, and 35,000 to 48,000 gal of cooling water. Some operating costs per ton are $4 for maintenance, $1.27 for additives including antioxidants, $1.16 for compressor oils, $0.127 for chain-transfer agent, $0.045 for nitrogen, and $0.018 for oxygen used as the initiator.

Total operating costs, including polymerization and finishing operations up to polyethylene storage, are likely to range from 3 to 4 cents/lb. These costs tend to be lower for large plants as compared to small ones. The costs also increase as the pressure is increased above 20,000 or 25,000 lb/in^2 since compression costs are higher and more expensive equipment is needed. At lower pressures, which tend to produce polymers of relatively low molecular weight, the rates of polymerization are relatively low (39), and hence the rate of production for a given reactor is also fairly low. Production costs per pound of product increase as the pressure is decreased in the low-pressure range (probably 18,000 lb/in^2 or less).

Since a bright future is projected for high-pressure polyethylenes (5, 38, 50), developmental research continues on the current processes. Somewhat reduced operating expenses for production of polyethylenes by high-pressure processes are likely to be realized in the future.

III. LOW-PRESSURE PROCESSES

Several processes that operate at relatively low pressures, compared with the high-pressure processes discussed in the previous section, have been developed since 1955. Annual production capacities of these low-pressure processes grew in the United States to 2 billion lb by 1970 (7, 76). Process improvements since 1965 have resulted in much lower operating costs, and some processes now produce polyethylene as cheaply as the high-pressure processes or cheaper. Furthermore, methods have also been developed for production of some low-density polyethylenes that may be competitive with those of the high-pressure processes.

The low-pressure processes are widely used for the production of high-density polyethylenes (HDPE) with densities from about 0.96 to 0.97 g/cm^3. These polyethylenes are highly linear, being essentially chains of methylene groups, and they have crystallinities up to about 90 percent. Lower-density polyethylenes are also produced in large quantities by copolymerizing small amounts of comonomers with the ethylene. Various α-olefins (and particularly propylene, 1-butene, and 1-hexene) are popular comonomers for reducing the density to as low as 0.94 g/cm^3 and in select cases as low as 0.924 (76). Some of these low-density polyethylenes have very high molecular weights (or low melt indexes), and in such cases they crystallize to a much more limited extent. Failure to crystallize is a major cause of the lower densities of some of these polymers.

As the density of the polyethylene is reduced, significant changes in the physical properties occur. Considerable improvement in the stress-cracking resistance and the load-bearing ability of the polyethylenes can be realized. Bottles, cans, and cartons must often be able to bear considerable loads. The stiffness of the polymer is also reduced. In the last few years, several manufacturers have used 1-hexene in preference to 1-butene, claiming improved physical properties, but agreement is not unanimous that the properties are improved.

Organic solvents are used in commercial processes, except for gas-phase processes recently publicized (76, 97, 108). The solvent dissolves the ethylene and suspends the solid catalyst. In some processes, known as solution processes, all the polyethylene dissolves in the solvent too, but in slurry processes, most of the polyethylene is not dissolved (13, 84).

Several types of solid catalyst (9, 13, 14, 20, 23, 81, 87, 104) are used. Metal oxide catalysts derived from transition elements are most important. Over 60 percent of the high-density polyethylene produced in the United States employs chromic oxide catalysts developed by Phillips Petroleum Co. (76). Molybdenum oxide supported on α-alumina is employed in the Standard Oil Company (Indiana) process. Nickel oxide also polymerizes ethylene, but there is no evidence that it is used commercially. Ziegler catalysts are also most important, and they are often prepared by mixing aluminum alkyls and titanium chlorides under controlled conditions (13). Catalysts prepared by depositing certain chromium compounds on silica are also used (89, 97).

Chemistry and Mechanism of Polymerization with Solid Catalysts

The chemistry of polymerizations with solid catalysts has been widely investigated (9, 23, 63, 72, 74, 80, 93, 101), but controversy over the mechanism still exists. Several factors are useful in designing a reactor:

1. Olefin reactants must first transfer to the catalyst, which contains many pores. Whether or not the internal surface areas of the pores are effective often depends on the rates of transfer of reactants into the pores and of polymer out. Solid catalysts are sometimes fragmented by polymer production in the pores. Hence ethylene is transferred into the pores, but polyethylene is not always transferred out (86).

2. The growing polymeric chain sometimes remains attached to the solid catalyst. During chain growth, transfer of the olefin reactants to the catalyst surface probably is impeded by the bulky alkyl groups. For copolymers, transfer of the ethylene may become relatively easier than the transfer of the heavier comonomers when the polymeric chain grows in size. If so, the composition will vary along the chain. For at least one copolymer of ethylene and propylene (96), no such composition changes occurred, indicating that mass-transfer steps are not controlling in that case; but when the relative sizes of the comonomer molecules vary significantly, mass-transfer steps for the larger comonomer molecule may be controlling.

3. Displacement reactions sometimes occur in which the large alkyl group on the surface reacts with ethylene to form a polymer molecule and an ethyl group attached to the catalyst. The catalyst site is then available for the production of another polymer molecule. In slurry processes, polymer molecules are not transferred to any significant extent away from the catalyst surface. In these processes, the polymer has only very limited solubility in the solvent, primarily because of the relatively low temperature employed. Since the polymers are solids, they diffuse to only a minor extent from the catalyst.

4. In solution processes, polymer molecules from the catalyst surface are dissolved in the surrounding solution. Both higher temperatures and "better" solvents are used. Often the reactor is operated at temperatures above the melting point or first-order transition temperature to increase polymer solubility.

Solid catalysts employed for ethylene polymerization are poisoned by trace quantities of oxygen, water, acetylenic compounds, and polar compounds. High-purity reactants and solvents are required. Molecular sieves are sometimes used to remove trace impurities that often have a significant effect on the molecular weight of the polymer. Techniques to obtain high-purity solvent have been reviewed by Cinadr and Schooley (69).

Wisseroth (108) has presented information for ethylene polymerizations using chromic oxide catalysts. Alumina is very effective for removing catalyst poisons from both ethylene and propylene; however, cyclohexane, which is often used as the solvent in such a polymerization, reacts rather slowly with the catalyst to reduce the valence of the chromium from 6 to at least 4, with a subsequent loss in catalyst activity. Such undesired reactions undoubtedly depend on the temperature, but no specific information is available. Information on similar side reactions between various solid catalysts and solvents has generally not been published.

The importance of mass and heat transfer and of fluid mechanics cannot be overlooked in designing polymerization vessels that use solid catalysts. The first step of the overall polymerization reaction is to dissolve the ethylene (plus possible comonomers) in a solvent. Since the polymerizations are generally at temperatures above the critical temperature for ethylene, ethylene dissolves as a gas in the solvent.

If equilibrium solubilities were to occur, the dissolved concentrations of ethylene would be essentially proportional to the ethylene partial pressure. Actual ethylene concentrations may often (if not always) be significantly less, however. The degree of agitation and the equipment used for feeding the ethylene may be such that significant resistances to ethylene transfer occur. Increased ethylene pressures always tend to increase ethylene concentrations in the liquid phase, but increased temperatures reduce them. (Temperature also has an important effect on the kinetic rate constants for the surface reactions.)

The exothermic heat of polymerization is released at the catalyst surface, and heat transfer is from the catalyst surface to the liquid phase. Commercial rates of polymerization are sometimes limited by heat transfer (96).

The solvent has several functions in addition to that of dissolving reactants and polymer. It reduces the viscosity of the solution, controls the rate of olefin consumption, and promotes heat transfer (87). Based on solvent recovery and reuse, high concentrations of polyethylene in the exit product stream are preferred since solvent recycle is decreased.

In most, if not all, types of polymerization, some polymer sticks to the walls of the reactor and other equipment. Wall temperature affects the amount of sticking, and so does surface roughness. Since many reactors are jacketed, the temperature of the heat-transfer fluid is an important design factor.

Polymerization with Catalysts Containing Chromium and Molybdenum

There are at least four relatively different commercial processes using catalysts containing chromium or molybdenum. Two processes use chromic oxide catalysts developed by Phillips Petroleum Company. The gas-phase process of Union Carbide Corp. uses a catalyst prepared by the deposition of various chromium compounds on silica (97). Presumably Cr—O bonds are at least sometimes formed on the silica surface, so in a sense the catalyst may be related to chromic oxide catalysts. Standard Oil Company (Indiana) employs a molybdenum oxide catalyst for their process.

Processes Using Chromic Oxide Catalysts

Preparation of Catalyst: Preparation of the chromic oxide catalyst requires careful attention to details in order to obtain the high catalyst activities required. The catalysts used contain about 1 to 3 percent by weight of chromic oxide supported on silica or silica-alumina (71–73); silica now seems to be the preferred support. The type of support has an important effect on the productivity of the catalyst and on the type of polymer produced. For example, Polypor, a silica gel developed by U.S. Industrial Chemicals Co. and their affiliate National Petro Chemical Corp., results in polymers of significantly lower molecular weight than conventional silica gels. This gel has a surface area of 200 to 500 m^2/g and a narrow-pore-diameter distribution of 300 to 600 Å. Using HF-treated alumina, instead of a silica-alumina support, produces a considerably different catalyst. For comparable runs, the catalyst with the treated-alumina support produced a polyethylene with a molecular weight of 14,000 (73). Catalyst activity increases as the fraction of chromic oxide in the solid catalyst increases, up to about 2 to 3 percent, but further increases do not change the activity.

Chromic oxide catalysts are activated, before being used for polymerization, by contacting the finely divided material with air for several hours at temperatures from about 400 to 975°C; 525°C is specified in an example (23). After activation, the catalyst is cooled in dry air and stored until used. The chromium on the surface of the catalyst should be in a hexavalent state.

When the temperature of activation is increased, the resulting catalysts produce polyethylene with lower molecular weight (73), but higher yields of polyethylene are obtained for a given amount of catalyst (23). Earlier the amount of catalyst added to the reaction mixture was probably about 0.2 to 1 percent by weight, based on the total streams to the reactor (61, 73). With the much more active catalysts now available, much less catalyst is now used; about 5,000 to 50,000 lb of polyethylene is produced per pound of supported catalyst (76, 87).

Chromic oxide catalysts are deactivated with oxygen and moisture under polymerization conditions. Highly purified solvents are needed for the polymerization. Catalysts treated with carbon monoxide, ethylene, light paraffins, other light olefins, or oxygenated compounds are active at temperatures as low as −50°C (72). Without this treatment, the catalyst is active only above 15°C. It is postulated that chemisorption of ethylene occurs on the catalyst surface, and a solid polymer is frequently deposited there. Even the presence of solid polymer on the catalyst does not prevent rapid polymerization.

The effects of the operating variables on the polymerization reaction with chromic oxide catalysts have been reported (75, 77). High pressures decrease the ratio of the weight-average to number-average molecular weight and increase the rate of polymerization. The average molecular weight is controlled primarily by varying the polymerization temperature. Increased temperatures yield lower molecular weights and smaller ratios of the two molecular weights. The rate of polymerization reaches a maximum at about 130°C. The rate depends on the amount of chemisorbed reactants at the surface and on surface polymerization. The concentration of dissolved ethylene in the solvent is most important. Increased temperatures cause lower solubilities and hence decreased amounts of chemisorbed ethylene on the catalyst surface.

A technique employed to reduce the molecular weight of the polyethylene is to add hydrogen to the system. The hydrogen is obviously first adsorbed in the solution, but it is only slightly soluble at moderate partial pressures of hydrogen.

Oxime compounds in the reaction mixture are reported to have a significant effect on the molecular weight of the polymer produced. Molecular sieves are sometimes employed to remove trace amounts of such compounds from the reactants. Controlled amounts of oximes are then sometimes deliberately added to the ethylene to obtain the desired polyethylenes.

Kinetic and molecular-weight results have been used to elucidate the surface mechanisms of chromic oxide catalysts (70, 71), and the Langmuir-Hinshelwood mechanism appears applicable. Polymerization occurs by reactions of adsorbed monomer molecules with growing polymer chains. The resulting polymer has a vinyl group on one end of the chain and a methyl group on the other end. The intermediate groups are primarily methylene.

The chromic oxide catalysts are effective for copolymerizing 1-butene and 1-hexene with ethylene.

Solution Processes of Phillips Petroleum Company   Phillips Petroleum Company has developed an improved, or second-generation, solution process that is cheaper to operate than their first-generation process (76). Both modifications of their solution process have been widely used for production of high-density polyethylenes and especially those with rather low molecular weights (or high melt indexes). In the first-generation process, the catalyst is separated from the polyethylene product, as shown in Fig. 3-10. In the newer process, a much more active catalyst is used. As a result, such small amounts of catalyst are used that the catalyst does not need to be separated from the product. In both modifications, the reactors are stirred-tank autoclaves of at least several thousand gallons' capacity. In one commercial installation, 4,000-gal carbon-steel autoclaves are used (106), and they are jacketed for temperature-control purposes.

Feed streams to a reactor are purified ethylene plus comonomer, solvent (frequently cyclohexane) and a slurry of catalyst and solvent. Pumping a slurry that contains solid (and abrasive) catalyst has caused problems in maintaining mechanical-pump seals (106). A high degree of turbulence and internal recirculation is provided in each reactor. Reaction conditions are reported to be temperatures from 125 to 175°C and pressures from 300 to 450 lb/in² absolute.

However, there is some question whether true "solution" polymerizations occur at lower temperatures of this range. In example XXV of the patent (87), two polymerizations at 127°C and 420 lb/in² were

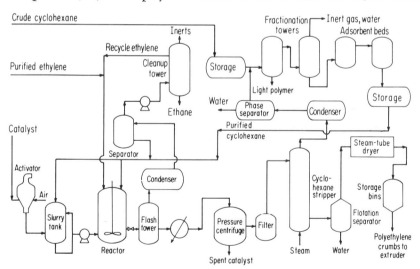

Fig. 3-10   Solution polyethylene process of Phillips Petroleum Company. (Spent catalyst is separated from polyethylene in this version.) [*From* (59) *with permission of Chemical Engineering.*]

described, in which 2,2,4-trimethylpentane was the solvent. In these runs, the effluent stream from the reactor was heated to "dissolve" the polymer. This example seems to indicate a polymerization that was at least partially a slurry type.

The method of introducing ethylene is important to minimize mass-transfer resistances when dissolving ethylene in the solvent phase. Dissolved concentrations of ethylene in the solution vary from about 2 to 5 percent (66, 88), whereas somewhat higher concentrations are possible at equilibrium (59). The polyethylene concentration in the exit solution from the reactor can be much higher. In commercial units, values up to 15 percent are common, and 1.5 h is a likely average residence time. The production capacity of a 4,000-gal reactor is estimated to be between 15 to 20 million lb of polyethylene per year.

The pressure of the exit stream from the reactor is reduced as it enters the ethylene flash tower. Almost all the unreacted ethylene, impurities (such as ethane and dimers and trimers of ethylene), and a portion of the solvent flash. A recovery system consisting of a fractionation column, compressor, condenser, etc., recovers and recycles ethylene, comonomer, and solvent and also separates unwanted inerts.

In the older Phillips process, in which the catalyst is separated from the polyethylene, the cyclohexane solution containing dissolved polyethylene is pumped from the ethylene flash tower to a heater. Heating reduces the viscosity of the solution and ensures that all polyethylene is dissolved. Additional solvent may be added to the solution to reduce the viscosity even further; this step is not shown in Fig. 3-10. Centrifuges operated at speeds greater than 2,000 r/min and developing forces of about 2,000 times that of gravity (61) reduce the catalyst content of the solution to almost zero (less than 8 ppm). A filter, following the centrifuge, is used primarily as a safety device in case the centrifuge develops operational problems. The heater and centrifuges operate from about 75 to 220 lb/in^2 gage (106), which is sufficiently high to prevent flashing.

In the newer Phillips process, in which the catalyst is not separated, the centrifuging step is omitted. In both processes, the cyclohexane solution with dissolved polyethylene enters the cyclohexane stripper. Steam is used to strip the cyclohexane and cool the solution. Polyethylene precipitates, forming a two-phase system of solid polyethylene and water. In this stripper, the polyethylene passes through a sticky intermediate stage. A reliable continuous system has been developed for the stripper (61).

The solid polymer is then separated from the water by a flotation step. The polymer passes over the weir in a vessel with a conical top. The polymer particles contain about 25 percent of volatile materials

(106), mainly water and cyclohexane, which are removed in a rotary steam dryer. Dimensions for a commercial dryer have been indicated as 50 to 60 ft long and 6 ft diameter. The moisture content of the product is reduced to less than 1 percent in the dryer (106). The dried crumbs are irregular granular particles of polyethylene with a diameter of $\frac{1}{32}$ to $\frac{1}{16}$ in, have a bulk density of 16 to 20 lb/ft^3, and contain less than 0.1 percent volatile solvents (23). Crumbs are then conveyed (usually pneumatically) to the finishing operation. The polyethylene can be blended immediately with additives such as antioxidants, and it is then extruded and packaged.

Cyclohexane from the cyclohexane stripper is recovered, purified, and recycled to the reactor system. Drying of cyclohexane is important since water poisons the catalyst.

Plants using Phillips processes are highly instrumented (106). An infrared analyzer determines the purity of the ethylene feed. Moisture and oxygen contents of several streams are measured to protect against catalyst poisoning. The reaction temperature is highly important for controlling the molecular weight of the polyethylene produced. Polymers with a wide range of molecular weights or melt indexes (from 0.2 to greater than 30) can be produced by this process (14).

Particle-Form Processes   The particle-form (PF) or slurry process of Phillips Petroleum Company has achieved considerable importance, and by 1970 its annual production capacity in the United States was over 800 million lb, making it the most important low-pressure process (76). It is most effective for production of high-molecular-weight (or low-melt-index) polyethylenes.

Polymerization of Ethylene: Reaction temperatures for the slurry process are lower than those for the solution processes (84, 14). Temperatures of 20 to 100°C have been reported (94), but the polymerization rates at 20°C are probably very low. Pressures are from about 100 to 500 lb/in^2. Higher pressures are normally required at the higher temperatures to dissolve sufficient ethylene in the liquid phase.

C_3 to C_8 hydrocarbons, sometimes designated as solvents, are the suspending liquids for the solid polyethylene particles in the slurry. *n*-Hexane and *n*-pentane are often preferred. "Very pure" reactants and a "highly active" catalyst are required in the modern slurry process. Any catalyst poisons seriously deactivate the catalyst, which is used in very small quantities.

As polymerization occurs, a solid or at least semisolid polyethylene phase is formed. The original catalyst particles are divided by spalling and splitting, and they form many very fine granules scattered throughout the resulting polyethylene particles (86). Presumably most polymerization occurs at or near the surface of the particles. The low solu-

(a)

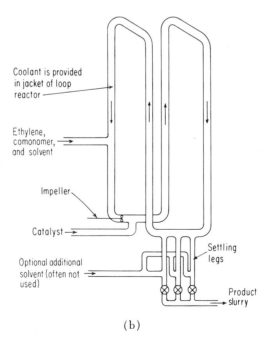

(b)

Fig. 3-11   (a) Plant using Phillips particle-form process; loop reactors shown at right center (*The Badger Co.*); (b) loop reactor for Phillips particle-form process.

bility of the polyethylene in the solvent is caused mainly by the relatively low temperatures used. The portion of the polyethylene dissolved in the solvent has a lower molecular weight than the precipitated polymer.

Stirred autoclaves were initially used as reactors in the Phillips slurry process. Because of polymer deposition on the walls of these reactors, they frequently had to be shut down and cleaned (59, 88). Loop reactors, which are radically different, were then developed and have been successfully used for about 10 years. Polymer deposition is minimal unless a utility failure occurs or an appreciable amount of polyethylene is dissolved in the suspending hydrocarbon liquid. A loop reactor may require cleaning once or twice a year.

Loop reactors and their operation have been described in several patents (91, 92, 94, 98–100, 109). Figure 3-11a shows the two loop reactors in a plant in Belgium. The frontispiece of this book shows the ethylene-polyethylene plant of the Chemplex Company, which also employs such reactors. Figure 3-11b is a sketch of a typical double-loop reactor. The olefins, solvent, and solid catalyst are continuously fed to the reactor, and the resulting slurry of solution and suspended polyethylene solids is circulated around the loop many times. Product slurry is continuously or semicontinuously withdrawn from the loop.

The diameter of the tubing used in these Phillips loop reactors varies from about 12 to 20 in (77, 94). A double loop, as shown in Fig. 3-11b, is a preferred modification. From published pictures (62) and information in the patents, a large loop reactor has tubing 20 in in diameter and four vertical sections in the loop. The vertical sections are each approximately 50 ft long,* and the connecting horizontal sections in the loop are rather short (62). The equivalent length of tubing in the loop probably totals about 280 to 300 ft. The volume of this reactor is about 600 ft^3.

The preferred flow velocities in loop reactors are outlined by Norwood (94). For a loop reactor with a tube diameter of 20 in, velocities of 10 to 30 ft/s are recommended. Sufficient velocity is needed to prevent the polyethylene solids from depositing on the walls. Higher velocities are permissible, but power requirements for the turbine impeller then become excessive. The optimum velocities to be employed depend on the solvent, operating conditions used, and the type of polyethylene. With the above velocities, the slurry recirculates every 10 to 30 s. The highly turbulent flow causes the feed ethylene and comonomer to be rapidly dissolved in the continuous phase, which is the solution.

The slurry in a loop reactor contains 18 to 25 percent solid polyethylene, and the remainder is the solution (94, 107). With an average

* The patent by Norwood (94) describes a loop reactor using sections 50 ft long and 20 in ID.

residence time of 1.5 h in the reactor, the loop reactor described would produce about 40 to 45 million lb of polyethylene per year. These calculations agree well with the announced capacity of the plant with two loop reactors (62).

Temperature control is a critical design consideration for a loop reactor, and loop reactors for the Phillips particle-form process have apparently been limited to tubes 20 in in diameter or less (77). Larger tubes would have poorer temperature control or colder walls, resulting in polyethylene deposition there.

Polymers of extremely high molecular weight can be produced commercially by the slurry process (65). Presumably low temperatures and higher pressures are used to produce such polymers, which have average molecular weights up to 1 million or even higher. These unique materials (64) probably are relatively expensive to produce.

Recovery and Processing of Polyethylene: The method for withdrawing the product stream from the loop reactor is a key step in the operation. Several patents (92, 99, 100) describe methods. Settling legs (as shown in Fig. 3-11b) and screw conveyors are employed to withdraw concentrated slurries containing 50 to 80 percent solids. These techniques minimize polyethylene deposition or plugging in the exit lines, and they also reduce the amount of solution withdrawn, which then has to be processed.

Additional fresh solvent is sometimes added to the concentrated slurry as it is withdrawn to expedite its flow to the flash tanks. A sufficiently high temperature and low pressure (about atmospheric) are used to flash off the solvent. Recycling and superheating a portion of the gaseous solvent provide the energy for vaporization of the solvent.

The polyethylene solid particles from the flash tank may in some cases be processed in auger dryers to remove any remaining solvent (87). The bulk density of the resulting granular polymer is about 25 to 28 lb/ft^3. It may be used directly in this form, or it may be pelletized to form a product with a higher bulk density. The ash or metal content of polymer is low, in the 0.005 to 0.01 percent range.

Over 90 percent of the entering ethylene is converted to polyethylene (76). Low polymers formed in small amounts are separated from the solvent in the solvent-recovery portion of the process.

In earlier versions of the slurry processes, now obsolete, the catalyst was separated from the polyethylene. Methods for separating and recovering the polyethylene have been described earlier (59, 88).

Gas-Phase Polymerization Processes High-density polyethylene is produced by gas-phase processes (75, 83, 90), and two specific units have recently been announced (76, 97, 108). No solvent is employed in these processes, and hence there are no solvent recovery costs. Since

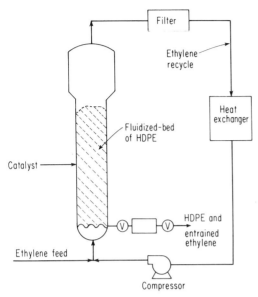

Fig. 3-12 Simplified flowsheet of Union Carbide Corp. process for production of high-density polyethylene (HDPE).

the catalyst is so active, only small amounts are needed, and it is not separated from the final polymer.

Union Carbide Corp. has built or is building gas-phase units with capacities from 30 to 170 million lb of polymer per year (97). Figure 3-12 is a simplified flowsheet for their process, and Fig. 3-13 shows a Swedish plant that produced about 15,000 metric tons (33 million lb)/year. The catalyst used is prepared by the deposition of specific chromium compounds on dehydrated silica (or other supports); supported bis-triphenylsilyl chromate or chromocene catalysts are highly effective for gas-phase polymerizations.* The Union Carbide catalyst produces about 600,000 weights of polymer per weight of metallic chromium. In their process, polyethylenes with densities of 0.94 to 0.966 g/cm³ are produced. Either propylene or 1-butene is the comonomer employed to produce the polymers with the lower level of densities. Polymers can be produced with either a narrow or broad molecular-weight distribution and with essentially any desired average molecular weight, using hydrogen as a molecular-weight modifier as in other low-pressure processes.

The Union Carbide reactors operate adiabatically at a pressure of about 20 atm. The entering rather cool gas fluidizes the solid polyethyl-

---

* See *J. Poly. Sci.*, (A-1) **10**: 2609–2637 (1972) and U.S. Pats. 3,642,749 and 3,687,920.

Fig. 3-13 Polyethylene plant using gas-phase polymerization of ethylene. Two fluidized-bed reactors are in the top center. [*From (97) with permission of Chemical Engineering.*]

ene particles (89) in the bottom portions of the reactors (not shown in Fig. 3-13). On the basis of a picture in a brochure of Constructors John Brown Limited, the bottom portion of each reactor of the Swedish plant is about 8 ft in diameter and 40 ft in height. Pressure drops of the ethylene as it passes upward through the fluidized bed are low, obviously depending on the height of the bed, but likely vary from about 2.5 to 5.0 lb/in². Relatively large quantities of gas are needed to fluidize the particles, and only about 2 percent of the ethylene reacts per pass. Most of the unreacted ethylene leaves at the top of the reactor. The top portions of the reactors are clearly shown in Fig. 3-13; these portions, which act to separate the unreacted ethylene from the solid polyethylene, have a larger diameter than the bottom portion of the reactor. The unreacted ethylene is then recycled using a compressor that probably develops heads of about 10 to 15 lb/in² to overcome the pressure drops in the fluidized bed, gas coolers, and other equipment in the recycle gas line. The compressors must be able to provide extremely uniform flows and heads.

The polyethylene product is withdrawn from a reactor through a pair of valves operated sequentially (89). About 5 percent of the entering ethylene is removed with the polymer product and is separated from the polyethylene solids at low pressures and fairly high temperatures. Approximately 80 percent of separated ethylene is recovered, recompressed, and then recycled. Overall, about 99 percent of the entering ethylene reacts to form polyethylene. The resulting polyethylene powder is ready for compounding and pelletizing or for packaging as a powder. Scale-up of the Union Carbide reactors causes essentially no problems, but provisions are necessary to minimize the static charges built up on the solids in the fluidized bed.

Badische Anilin- und Soda Fabrik (BASF) has built a pilot plant to produce about 4.5 million lb/year of product (108) using chromic oxide or preferably Ziegler catalysts. Ethylene gas (and comonomers when used) are compressed to about 100 atm (1,500 lb/in²). This gas is cooled in a heat exchanger to about 30°C and is then expanded isoenthalpically to 35 atm using a throttling valve at the base of the reactor. The resulting ethylene is a mixture containing about 50 weight percent liquid; the remainder is gas. The reactor contains a mixture of solid polyethylene particles and upflowing ethylene. The solid-gas-liquid mixture is agitated mechanically. Power requirements of the agitator vary from about 20 to 40 W per kilogram of polymer in the reactor; these requirements depend on the type of agitator and on the bulk density of the polymer. The reactors operate essentially adiabatically. The heat of polymerization provides the energy for vaporizing and heating the feed ethylene to reaction conditions. Energy calculations for the BASF reactor indicate that about 10 to 12 percent of the ethylene is polymerized per pass. Recycle of the unreacted ethylene and removal of the polyethylene product from the reactor are presumably accomplished by techniques similar to those employed in the Union Carbide process.

Several features of the two gas-phase processes are similar to those of the particle-form process.

1. Reaction temperatures of about 85 to 100°C are well below the first-order transition temperature of polyethylene. At these temperatures, the powder does not significantly agglomerate or stick to the reactor walls.

2. The ethylene is transferred and adsorbed on the solid particles.

3. The ethylene is transferred in the solid polymer powder to the catalyst sites, where it polymerizes.

4. Spalling and fragmentation of the catalyst occur during polymerization to produce finely divided particles.

As in the particle-form process, the catalyst does not have to be removed from the product, since such small amounts of it are required.

Compression cost for recirculating the unreacted ethylene from a gas-phase reactor is always a major operating expense. Although considerably less ethylene needs to be recycled, on a relative basis, in the BASF process than in the Union Carbide process, pressure drops in the reactors of the latter process are much less. The utility costs, even with large amounts of recycle, are claimed to be relatively very low in the Union Carbide process.

Economics of Polymerization with Chromium-containing Catalysts

The manufacturing costs for producing polyethylene for the particle-form process are certainly much lower than those for the Phillips solution process. Recovery of the polyethylene from a solution is relatively expensive. Development of highly active catalysts was also most important, since expensive removal of the catalyst from the polyethylene is no longer necessary.

The operating costs for the particle-form process of course depend to some extent on operating conditions. The minimum costs generally occur at intermediate operating temperatures. At low temperatures, the rates of polymerization are low, resulting in low production rates. At high temperatures, more polyethylene dissolves in the solvent, complicating the polyethylene-recovery technique.

The operating costs for gas-phase processes are low (97), probably being as low as those of the particle-form process if not lower. Since the gas-phase processes are relatively new, even better cost advantages for these processes are likely within the near future. Elimination of the cost of recovery and recycling of the solvent and the cost of catalyst removal results in substantial savings over other low-pressure processes. Compression costs for ethylene, however, are quite high. Information has not yet been reported on how long the reactors can be operated before cleaning or maintenance is required. If deposition or plugging of the reactor system is a problem, the operating costs are higher and the product quality is reduced. Ethylene-type copolymers, which have lower densities, may result in greater deposition problems since they have lower softening temperatures.

Regardless of the polymerization process used, the production costs for copolymers tend to be higher than those of ethylene homopolymers. The comonomers are generally more expensive than ethylene. The polymerization costs increase as more comonomer is used and the density of the copolymer is decreased. The Phillips slurry process has proved very successful for producing a high-molecular-weight product. Attempts are now being made to modify that process to obtain lower-molecular-weight products. Catalysts developed by U.S. Industrial Chemicals Co. and its affiliate National Petro Chemical Corp. are now available for production of homopolymers with melt indexes up to 10 and of copolymers up to 5. Modified operating conditions are another promising technique (109). Mechanically blending polyethylene obtained by the slurry process with low-molecular-weight polyethylenes of the Phillips solution process particularly is now practiced commercially. If a modified particle-form process could be developed to produce low-molecular-weight polymers, the expensive blending operation would be unnecessary.

Process of the Standard Oil Company (Indiana)   The Standard Oil, or Indiana, process has been described in the literature (95, 110), and its flowsheet has been published (78) (Fig. 3-14). Commercial units have been built in Japan and Italy (105) and more recently in Texas, by Amoco Chemical Company. The polyethylenes of this process have unannealed densities as high as 0.958 to 0.960 g/cm³. They contain small amounts of branching and significant proportions of internal (rather than terminal) vinyl groups (14).

Molybdena Catalysts: Catalysts used in the Standard Oil process are described as commercial hydroforming catalysts containing 8 percent molybdena on alumina (79, 81, 95). Catalyst particles range in size from 6 to 100 mesh, and they are activated with hydrogen at 430 to 480°C within 30 min. Increasing the hydrogen pressure during activation from 0 to 75 lb/in² gage increases the activation rate by 40 percent, but further increases have no effect. Sodium and various hydrides also activate the catalyst and scavenge catalyst poisons in the system. Activation reduces molybdenum from a hexavalent to an intermediate state.

At a temperature of about 230°C and pressure of 900 lb/in², one part of active catalyst produces up to 180 parts of polyethylene. Catalyst

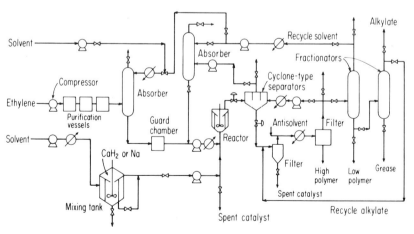

Fig. 3-14   Standard Oil Company (Indiana) process for polyethylene. [*From* (59) *with permission of Chemical Engineering.*]

activity decreases as polymerization continues, but the catalyst can be fully reactivated. Deposits on the catalyst are burned off at 490°C, and the catalyst is reactivated with hydrogen.

Both homopolymers and copolymers of ethylene are produced with these catalysts (78). Propylene, apparently the preferred comonomer, is added in small quantities. It appears that higher α-olefins are not polymerized satisfactorily with these catalysts. A common set of operating conditions for the polymerization is temperature 230°C, pressure 900 to 1,000 lb/in², and amount of catalyst 0.1 g per 50 ml of benzene. The preferred solvent is benzene. At reaction conditions, ethylene reacts to some extent with benzene (or other suitable aromatic such as toluene or xylenes) to produce an alkylation product. Decalin, which will not react with ethylene, can be used as solvent, but for economic reasons probably is not.

Reactant Processing: Ethylene is first compressed to the desired polymerization pressure. Probably two or more compressors are used in series for this step. The compressed ethylene then passes through three vessels, as shown in Fig. 3-14. In the first vessel, it is purified with a granular deoxygenating agent, e.g., metallic copper, maintained at 150°C. In the second one, it is dried with a solid dehydrating agent such as alumina, silica gel, or anhydrous calcium sulfate. In the third vessel, solid sodium hydroxide deposited on asbestos packing removes carbon dioxide.

Purified ethylene enters the bottom of an absorber, where it is contacted with a counterflow of solvent. Typical plate or packed columns are satisfactory for the absorber operated essentially at ambient temperatures of 15 to 35°C. Up to 10 percent by weight of ethylene is dissolved in the solvent (79, 95). Based on equilibrium considerations, a higher concentration of ethylene is possible but would result in high concentrations of polyethylene in the polymerization vessel and would complicate the catalyst-separation step.

The solvent, containing dissolved ethylene, flows through a guard chamber packed with a granular active metal, e.g., sodium or a metal hydride such as calcium hydride. The guard chamber, operating at 100 to 280°C, is used to ensure the purity of the entering feed stream. After passing through the guard chamber, the solution is heated in a heat exchanger to the desired temperature. In the patent (79) it is specified that the mixture is heated to the "polymerization temperature, for example, between about 200° and about 275°C." Probably the exit temperature from the heat exchanger does not have to be as high as that in the reactor. The exothermic heat of reaction should be sufficient to heat the mixture by perhaps 40 to 80°C.

No flashing of ethylene occurs during heating, since ethylene concentrations are about 5 to 10 percent by weight in the solvent. The heated solution containing ethylene is combined with another solvent solution, containing a metal hydride or an alkali metal, which is prepared in a separate mixing tank.

Reactor Operation: The combined mixture of solutions enters the bottom of the reactor, which is provided with an agitator operated at 650 r/min and with baffles. The patent (110) specified that a high degree of intermixing for the catalyst, metal hydride, olefin, and solvent be achieved in the lower section of the reactor.

The method of introducing the hydrocarbon feeds (sparger or jet device), location of the agitator relative to the feed location, agitator design, and type of baffles are important, since each affects local concentrations of reactants and catalyst in the reactor. The preferred method of adding the feed is not specified.

The top portion of the reactor provides a relatively quiescent zone, so that most of the solid-catalyst particles separate and drop toward the bottom of the reactor, where the catalyst is recirculated and reused. The average residence time of the catalyst in the reactor is hence much longer than that for the hydrocarbons. The reaction period, presumably the average residence time of the hydrocarbons in the reactor, varies from about 10 to 100 min, depending on the operating conditions. Longer times would be required at lower temperatures and/or pressures.

The resulting solution in the reactor contains 2 to 5 percent by weight of polyethylene. Apparently about 50 percent of the entering ethylene is polymerized. At higher concentrations, the solution becomes so viscous that considerable amounts of catalyst, particularly the smaller particles, are carried out of the reactor. Furthermore, mechanical handling of the solution becomes more difficult since severe cracking of the solid oxide catalyst also occurs.

The reactor is initially charged before the start of a run with active catalyst and a promoter, through lock-hopper devices. Makeup catalyst is added intermittently during the course of a run, as needed.

A stainless-steel reactor is probably used commercially; such a reactor was used experimentally (95). The surface of the reactor wall is smooth, to minimize any deposition of polymer on it. The reactor is undoubtedly well insulated to minimize heat losses and to prevent reactor walls from becoming cool and encouraging polymer deposit. It is not known how often the reactor has to be cleaned—presumably at least every few months.

Separator Section of Process: Solution from the reactor contains a small amount of solid catalyst and flows through a pressure-reducing valve to a cyclone separator. Pressure in the separator is maintained at 15 to 250 lb/in^2 gage, and a gas stream flashes off. This pressure affects the quantity and composition of the gas stream, which is a mixture of unreacted ethylene and solvent. More gas is evolved if the pressure is lowered, with a resulting increase in viscosity of the remaining liquid solution. A temperature of at least 150°C (higher than the first-order transition temperature of linear polyethylene) is held in the separator to keep all polyethylene in solution, to maintain a relatively low-viscosity liquid, and to promote flashing of ethylene.

Additional hot solvent is sometimes added to prevent deposition of polyethylene on the separator walls and to maintain a low viscosity. Alkylated benzenes, such as ethyl benzene or polyethylated benzenes, are frequently used as this solvent. These alkylated benzenes are formed in the process when the initial solvent is benzene.

The solution and entrained solid catalyst from the bottom of the separator are filtered. (A centrifuge is also practical in some cases.) Sometimes an ultrasonic vibrator is also used to help coagulate the very fine catalyst particles. Part of the recovered catalyst is reactivated and recycled, and the fines in the recovered catalyst are probably separated and discarded during catalyst activation.

The clear solution from the first filter is cooled (20 to 80°C) in a heat exchanger to precipitate polyethylene. Nonsolvents such as low-boiling alcohols or ketones are sometimes added to the clear solution before cooling. These compounds promote precipitation of the polyethylene and destroy the calcium hydride or sodium in the mixture. The temperature to which the mixture is cooled affects the amount and quality of the precipitated polyethylene. If the original mixture was first cooled to 80°C and the resulting precipitate separated, a product with a relatively high molecular weight would be obtained. Further cooling of the remaining solution would yield additional polyethylene with a lower molecular weight. The solid polymer is removed from the solution by filters or centrifuges.

Polyethylene solids are washed—in some cases, first with water and then with acetone. The product is then dried and pelletized in a manner similar to that of other polyethylene processes.

The solution from the filter contains low polymers and greases, alkylate (if aromatic solvents are used), generally nonsolvent, and solvent. Fractionation columns are used to separate the solution into relatively pure streams. Only two columns are shown in Fig. 3-14, but an additional column is needed if any nonsolvent is present. The recovered solvent is recycled to the second absorber, where it is contacted with recycle ethylene from the separator. In the patent (110) the solution from the second absorber is specified as varying from 2 to 10 percent by weight of ethylene.

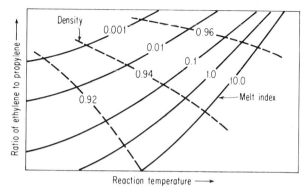

Fig. 3-15   Effect of operating conditions on physical properties of polyethylene produced by Standard Oil process. [*From* (59) *with permission of Chemical Engineering.*]

The following operations have been used in some commercial or pilot-plant units:

1. Removing sodium by reaction with wet alcohol solution
2. Drying polyethylene particles under vacuum at 110°C

Low polymers (grease and waxes) are apparently produced in rather large amounts, especially at higher polymerization temperatures (95). About 10 percent of the ethylene forms such undesired products at 230°C, whereas 5 percent is so involved at 200°C.

Both higher polymerization temperatures and higher feed ratios of ethylene to propylene increase the polymer density. The melt index of the polymer increases at higher temperatures and at lower ratios, as shown in Fig. 3-15.

Because of the higher pressures used in the Indiana process, compared with the Phillips process, stirred autoclaves having smaller volumes are probable. It is estimated that a 1,000-gal autoclave would produce up to about 5 million lb/year of polyethylene.

Production costs for the Indiana process would appear to be relatively high, compared with several other processes, especially the slurry, or particle-form, process of Phillips Petroleum. Rather low yields of ethylene, higher pressures, and a complicated recovery system all contribute to the costs.

Polymerizations with Ziegler Catalysts

Ziegler catalysts are employed in both slurry and solution processes for the commercial production of high-density polyethylenes. The original slurry process using continuous-flow stirred-tank reactors is often referred to as the *Ziegler process.* Major process improvements in the last few years employing new reactors and/or much more active catalysts have resulted in significantly lower operating costs.

Ziegler polyethylenes contain some unsaturated groups such as terminal vinyl, vinylidene, and trans olefinic groups. The resulting unsaturation may result in some long-chain branches or cross-linking during the processing of the polymer, and minor benefits of impact strength and stress-crack resistance can sometimes be realized (13). Homopolymers that are essentially linear or that contain long and short branches can be produced. Copolymers prepared using 1-butene or 1-hexene have short branches and hence lower densities. 1-Butene is sometimes produced by in situ dimerization of ethylene. A Canadian company is reported to be making their low-density polyethylenes with Ziegler catalysts. In general, homopolymers or copolymers can be produced with properties similar to those produced using metal oxide catalysts.

Ziegler Catalysts   Preparation of Ziegler catalysts is still largely an art, but variations in their preparation have a significant effect on the polymers produced (including average molecular weight, molecular-weight distribution, and chain branching) and on catalyst activity. With the extremely active catalysts developed by at least two companies, so little catalyst is used that the catalyst residue is left in the polyethylene.

Organoaluminum compounds used commercially include triethylaluminum, tri(isobutyl)aluminum, tri(n-butyl)aluminum, and alkylaluminum halides. Transition-metal compounds also used include titanium chlorides ($TiCl_3$ and $TiCl_4$) and alkoxy compounds such as $Ti(OR)_4$, where R is an alkyl group. When these two types of compounds are combined, solid complexes form and the substitution groups rearrange on the aluminum and titanium atoms. For example, when aluminum alkyls are combined with titanium chlorides, the complex is a mixture of alkylaluminum chlorides and alkyltitanium chlorides. Catalyst supports including a magnesium compound are also sometimes employed.

The substitution groups or atoms on the titanium or the aluminum atoms affect the linearity of the polyethylene formed. A catalyst prepared from $Ti(OR)_4$ and triethylaluminum produces a polyethylene of significantly higher density than one produced from $TiCl_4$ and triethylaluminum.

The aluminum alkyls act to reduce the valence of the titanium, often to $+3$ or even $+2$. Increased reduction yields higher-molecular-weight polymers. The ratio of Al to Ti affects both the reduction and the rate of polymerization (63, 74). In general, lower ratios result in higher rates.

Zirconium chlorides and vanadium chlorides or oxychlorides are also effective as transition-metal compounds (20). When alkylaluminum halides and vanadium oxychlorides are mixed, the resulting catalyst is soluble in the commonly used solvents. This latter catalyst is used for copolymerizing ethylene and propylene, but it is doubtful if soluble catalysts are used for the commercial production of polyethylenes. A soluble catalyst is of interest in that transfer steps of the reactants to catalyst sites are quite different from those with a solid catalyst.

Polymerization with Ziegler catalysts occurs as result of the growth of the alkyl groups attached to the catalyst surface. The olefin (in this case ethylene) attacks the carbon-to-metal bond (probably C—Ti) and inserts itself between the carbon and metal atoms. Hence the length of the alkyl group increases by two carbon atoms. Additional ethylene molecules react at this carbon-to-metal bond until a long attached alkyl group is present. Some of these alkyl groups are eventually displaced by ethylene molecules. The new ethyl group attached to the surface of the metal is then the site for growth of another long alkyl group.

Slurry Processes Using Ziegler Catalysts Continuous-flow slurry processes using Ziegler catalysts have been described in the literature (77, 82, 51, 25, 102, 103), and at least two types of reactors are widely used. The catalyst is normally introduced to the reactor as a slurry with the solvent used. Comonomers and terminators are also sometimes added to the ethylene feedstock. These terminators, which may be hydrogen, acetylene, or polar compounds, reduce the molecular weight of the product to desired levels of perhaps 10,000 to 70,000. Much higher molecular weights are generally obtained if the terminators are not used (64).

Stirred Autoclaves Used as Reactors: Stirred autoclaves were apparently the first type of reactor used in slurry processes, and glass-lined reactors with capacities of 2,000 to 10,000 gal are still widely employed. These vessels are often provided with a baffle and an impeller having three retreating blades. Agitation is moderate. Reaction conditions normally vary with temperatures of 70 to 90°C and pressures of 15 to 100 lb/in^2 gage; pressures less than 75 lb/in^2 gage are probably most common (25). Polymerization rates increase almost linearly with pressure because of increased ethylene solubility in the solvent.

Up to 1.7 percent by weight ethylene can be dissolved in the solvent, usually a light paraffin, if mass-transfer resistances are not controlling. Actual concentrations are usually somewhat less, however. The solvent most frequently used is *n*-hexane, which has a reasonably low volatility and is cheap; *n*-pentane is often too volatile; and *n*-heptane is sometimes used especially at higher reaction temperatures. All solvents must be highly purified to remove water and other catalyst poisons.

Ethylene is generally introduced to the reactor by means of a sparger located close to the impeller. Such a technique promotes rapid transfer of ethylene to the liquid phase. In some reactors, the ethylene is introduced into the gas phase of the reactor, and in others the entering solvent is saturated with ethylene.

Operating variables of importance in the reactor system include the degree of agitation, the specific catalyst used, temperature, pressure (mainly the sum of the partial pressures of the reactants and the solvent), and polar impurities in the system (83). In addition to affecting the chemical structure of the polymer molecules, these variables also affect the porosity and size of the particles formed.

Polyethylene solids form around the solid-catalyst particles, and the concentrations of solids in the solid-liquid slurries are as high as 30 percent (25). Lower concentrations are probable in most reactors, and these concentrations depend on the average residence time (1 to 2 h is common) in the reactor. Production capacities of autoclave reactors are estimated at about 5 to 25 million lb/year of polyethylene as the reactor size increases from 2,000 to 10,000 gal.

There is a slow buildup of polymer on the walls of the reactor. Minimizing this buildup is an important factor in choosing the desired operating conditions. In general, a reactor must be cleaned every 1 to 10 weeks. The concentration of polyethylene in the slurry is important because it affects the viscosity, agglomeration, and sticking of the polyethylene particles to each other and to the wall.

Loop Reactors Used in Solvay Process: The relatively new process developed by Solvay et Cie (68, 77, 103) uses a loop reactor that is similar to the one used in the Phillips particle-form process. Solvay has developed a highly active Ziegler-type catalyst (68, 102) supported on a magnesium compound; so little catalyst is employed for polymerization that it does not have to be removed from the product.

Claims have been made that polymer deposition is not a problem in this reactor (77), perhaps because of the relatively low reaction temperatures used. As a result larger-diameter tubes can be used in the loop reactors of this process (compared with Phillips particle-form process units). A loop that has a tube diameter of 24 in (61 cm) has been employed for several years to produce 50 million lb/year of polyethylene. At least one reactor that has a significantly larger production capacity is positioned in a horizontal, rather than a vertical, position.

Polymerization occurs at temperatures from 60 to 90°C and at pressures of 25 to 35 atm (350 to 500 lb/in² absolute). Hexane is used as the solvent in at least some reactors. Comonomers employed with ethylene to produce lower-density polymers are 1-hexene, 1-butene, and propylene. Hydrogen is often added as a molecular-weight modifier (77). Partial pressures of hydrogen of at least several pounds per square inch gage are often required in the reactor to dissolve enough hydrogen (it is only sparsely soluble in hexane).

High-density homopolymers or lower-density copolymers can be produced with melt indexes from very low values up to 30 (77, 103). Thus this process apparently has an important advantage over the Phillips particle-form process, which cannot readily produce polymers with such high melt indexes. Solvay claims that three catalysts effective for their process allow production of polymers with both narrow or broad molecular-weight distributions.

Separation Section of Slurry Processes: The slurry of solvent and polyethylene solids from a reactor normally enters a flash separator. For the Solvay process, the pressure is reduced to 1.5 atm (103). Two or more separators are used in some cases (88). Most of the unreacted ethylene and other light ends (such as hydrogen, if present) flash here. The gas stream from the separators is compressed, condensed, fractionated, and dried. Recovered ethylene and solvent are recycled to the appropriate process units.

The slurry leaving the flash separator of the Solvay process is centrifuged and then steam-stripped to kill the catalyst and to remove residual hexane. The resulting polymer, referred to as fluff (77), is primarily 250 to 1,000 μm in size. The fluff has a bulk density of 0.35 to 0.4 g/cm^3 (22 to 25 lb/ft^3).

For the older, more conventional Ziegler processes, the solid particles obtained by centrifuging are treated with an alcohol and then water to kill the catalyst. The alcohol reacts with the catalyst; the alkyl group is displaced, forming a polymer molecule; and an alkoxy group (from the alcohol) is attached to the catalyst. The resulting organometallic compounds are dissolved in the alcohol, and a polymer-solvent slurry is formed. Methanol, *n*-propanol, and isopropanol have been used in this step, but isopropanol is generally preferred.

Sometimes the alcohol is contacted with the slurry in the flash separator. In other cases, it is added to the slurry in the first stage of the stage washer. Such a washer is apparently used in all commercial processes for the subsequent treatment of the slurry with water and also caustic. The stage washer is provided with vigorous agitation and is probably always a series of stirred tanks operated with continuous-flow feed streams. This type of washer performs several functions.

1. It provides a means of contacting the slurry first with an alcohol, next with a neutralizing agent (for HCl), and finally with water.

2. It provides sufficient time and agitation for digestion of the organometallic compounds. Diffusion is often the rate-controlling step of this digestion process.

3. It controls the temperatures to the desired levels. Increased temperatures would be expected to increase the rates of digestion.

Following the stage washer, the solid polyethylene particles are separated from the slurry by filters or centrifuges. The polyethylene solids contain about 20 percent solvent, and they are then flash-dried in some commercial units by using steam. Air is not used until enough solvent has been removed to ensure that explosive mixtures will not form. Dried polyethylene is finally blended and extruded in a manner similar to that in other polyethylene processes. The solvent is recovered from the steam by condensation and then distillation.

The alcohol solution contains some dissolved polymer, which primarily has a rather low molecular weight. The alcohol is recovered by suitable distillation, and perhaps 2 to 5 percent of the original ethylene is recovered as a low polymer (88).

Catalyst removal from the polymer, when required, is one of the most important and also most expensive steps of the entire polymerization process. Companies that still use separation steps often develop their own specific variations, which are considered highly proprietary in nature. With the very active catalysts now available, catalyst-removal steps will be used to a lesser extent in the future, with the subsequent significant reduction in the overall operating costs for the polymerization processes.

Solution Processes   The Du Pont solution process uses considerably higher temperatures and pressures for polymerization than the slurry process. Exact temperatures and pressures for the commercial unit are not known, but the patent (60) indicates that the temperature must be above the melting point of the polymer, in the range of 180 to 270°C. Relatively high pressures are needed to dissolve up to 12 to 20 percent by weight of ethylene in the solvent. Pressures reported in the patent range from about 800 to 3,000 lb/in².

Conversions per pass of ethylene to polyethylene were indicated to be above 10 percent and preferably about 50 percent, and concentrations of polyethylene in the product solution should be about 6 to 10 percent. Ethylene is apparently dissolved in the solvent before introducing the mixture into the reactor.

Catalysts mentioned in the patent include those prepared from the following combinations: titanium tetrachloride and lithium aluminum tetradecyl; titanium tetrachloride and triisobutylaluminum; tetraisopropyl titanate and diethylaluminum bromide; and diisopropoxytitanium dichloride and triisobutylaluminum. The importance of the Ti-to-Al ratio in the catalyst and the effect of operating conditions on the type of polyethylene produced are reported. Suitable solvents are benzene, n-paraffins, and cyclohexane. The last was used in the examples of the patent.

Details of the commercial plant are not available, but it seems to be similar in several respects to the Standard Oil process. The polyethylene produced is said to have the following advantages over the Ziegler slurry process. It has a narrower molecular weight, is tougher, and has slightly higher densities. Removal of the catalyst is claimed to be simple. Since the polyethylene is already dissolved in the solvent, the washing steps are fast and easy to control. (Methanol was used in example 1 of the patent for the separation of polyethylene and catalyst residue.) High yields (based on ethylene feed) of polyethylene are also obtained.

The autoclave-type polymerization reactor of the Stamicarbon solution process (67) operates in the range of 130 to 225°C. The solvent employed is either n-hexane or a C_6 isoparaffin. The catalyst produces over 100,000 weight parts of polymer per part of transition metal, such as titanium.

Economic Comparisons of Processes Using Ziegler Catalysts   Large differences in both the capital and operating costs occur between plants using the conventional slurry process (requiring separation of the catalyst) and the relatively new Solvay slurry process. Cost information for plants of the Solvay type located in the Gulf Coast area of the United States was estimated in 1970 (77). Capital costs varied from $5.2 to 16 million as the plant size increased from 40 to 270 million lb/year. Total operating costs for the polymerization and the finishing operations, i.e., all operations to go from the starting ethylene to polyethylene storage, in more modern plants vary from about 2.5 to 3.0 cents/lb, depending on the plant.

Operating costs like those mentioned above are lower by 2 to 3 cents/lb than older slurry processes requiring extensive separation steps. Separation of polyethylene from a solution process as in the Du Pont process is a relatively expensive operation. Operating costs for the Du Pont process are undoubtedly higher than those of the Solvay process.

Montecatini (85) also claims low operating costs for their slurry process (which uses an autoclave reactor), since separation of the catalyst from the polymer is unnecessary. Operating costs of a Montecatini plant may be similar to those of the Solvay and Phillips particle-form processes. Table 3-3 outlines the amounts of raw materials, chemicals, and utilities required for the Montecatini process. Chemicals and utilities apparently cost about 1 cent per pound of polyethylene.

Economics of Polyethylene Production

A polyethylene manufacturer obviously wishes to produce the desired polyethylene at the lowest cost. Technology is now available so that low-, medium-, or high-density polyethylenes can be produced with either a low-pressure catalytic process or a high-pressure free-radical

TABLE 3-3   Raw Materials, Chemicals, and Utilities for Montecatini Edison Process (85)*

Basis: 1,000 kg (1 metric ton) polyethylene produced in this particle-form process

Ethylene, kg...........................	1,060
Diluent, kg............................	30
Catalyst, $...........................	4
Chemicals, $..........................	3
Long-pressure steam, kg................	2,350
Cooling water, metric tons.............	200
Electricity, kWh......................	600
Labor for 60,000-metric-ton unit........	8 men and 1 foreman

* See also E. Susa, *Hydrocarbon Process.*, July 1972, pp. 115–116.

process. Furthermore the molecular weights and the type of chain branching can be modified, at least to some extent, in each process. The following discussion considers factors of importance in choosing the preferred polymerization route.

Low-density polyethylenes are produced in the so-called low-pressure processes, as described earlier, by copolymerization techniques. Short-chain branching, as obtained in high-pressure processes, can be closely duplicated by using mixtures of 1-butene and 1-hexene as comonomers. More difficulties are experienced, however, in duplicating long-chain branching, which affects the swelling characteristics, melt viscosity, and general processibility when the semimolten polymer is extruded or molded. Some long-chain branching occurs when a polymer molecule with a terminal vinyl group copolymerizes with the ethylene. Since 1-butene and 1-hexene are more expensive than ethylene, raw-material costs increase as more comonomers are used to produce lower-density polyethylenes.

The operating costs for production of copolymers by low-pressure processes are generally more expensive than for homopolymers. The copolymers are more soluble in the solvents since they have lower degrees of crystallinity. Separation and recovery of these copolymers tend to be more difficult than for the homopolymers. Recovery and recycling of the unreacted ethylene or comonomer are also a more complicated operation. Differences in the operating costs for copolymers and homopolymers depend of course on the specific process being used.

An almost linear polyethylene can be produced in a high-pressure process if extremely high pressures, above 100,000 lb/in^2, and low temperatures are used for polymerization. The manufacturing costs for such polymers are of course higher than those of the conventional polymers produced at about 15,000 to 45,000 lb/in^2. Capital costs also increase as the operating pressures increase.

In general, lower manufacturing costs are possible with higher-density polyethylenes if a low-pressure process is used. A high-pressure process, however, is generally more economical for low-density polymers. The preferred switchover for high-pressure and low-pressure processes depends on many economical and technical factors, but it is apparently at about 0.94 g/cm^3. Phillips Petroleum has announced that they can modify their low-pressure processes for production of low-density products. They claim to be currently producing polyethylenes with 0.924 and 0.939 g/cm^3 densities (76), but these materials have properties quite different from those of polymers produced in the high-pressure processes.

An important consideration for all polyethylene plants is the size of a line, i.e., the complete system for producing and processing all the way from the ethylene feed to the final product. Numerous plants contain up to several lines. The rated annual capacities of a line perhaps vary from as low as 5 to 10 million lb/year to at least as high as 125 million lb.

The optimum size of a line and the number of lines to be installed are related to the complicated economics of the operations and are affected by several variables, including the following:

1. Expected sales. Sales and actual plant capacity frequently do not agree, and a plant is often run at partial capacity. Many plants are designed on a basis of expected sales after several years. Adding more lines is an obvious way of increasing plant capacity.

2. Operating costs per pound of polyethylene. These costs are generally lower as the line capacity increases.

3. Dependable operation. This is especially important for larger lines since a failure of any portion of the line might shut the entire line down with a subsequent large loss of production.

4. Converting from one type of polyethylene to another. Careful planning and programming of the plant are required to minimize or eliminate off-grade polyethylene while a change is in progress. Factors to consider involve the amount of reactants in the system at any given time (or the average residence time of the reactants in the polymerizations vessel), the time required to change from one steady-state operation to another, and the characteristics of the polyethylene produced in the interim. Interim polyethylenes can often be blended with other polyethylenes to produce salable products. Blending operations require careful planning and are expensive, however.

5. Types and numbers of polyethylenes being manufactured by the company involved. In general, the number of polyethylenes produced in a given line should be minimized so that only a few changeovers are necessary. Similar polyethylenes should be produced, if possible, in a given line.

6. Storage capacity for polyethylene products. A company producing many types of polyethylenes will need more storage capacity.

Significant differences exist in the total production costs for polyethylene produced by the different commercial processes. Costs as low as 6 to 6.5 cents/lb (including the cost of ethylene) are probable in large plants using either the Phillips particle-form process or the latest Solvay process. Costs higher by 2 to 3 cents/lb are probable in other processes, which involve various complicated steps for catalyst removal, polymer recovery, and solvent recovery.

Future process improvements are believed possible for slurry processes in the recovery and recycling techniques employed for the solvent and possibly the ethylene. A modified Solvay process has been claimed with a significant improvement in this area. In addition, gas-phase processes such as those used by Union Carbide Corp. and by BASF seem to offer considerable possibilities for reducing total operating costs to perhaps as little as 5.5 to 6.0 cents/lb. Such a conclusion is based on the assumption that gas-phase reactors can be operated for extended periods without shutdown. If the reactors must be cleaned at frequent intervals, or if they require expensive maintenance, the operating expenses will of course be higher. In any case, gas-phase processes are potentially most attractive, and considerable effort will be devoted to their development in the future.

Literature Cited

1. Albright, L. F.: Polymerization of Ethylene, *Chem. Eng.*, Nov. 21, 1966, pp. 127–131.
2. Billmeyer, F. W.: "Textbook of Polymer Science," Interscience, New York, 1962.
3. Campbell, P. E., R. V. Jones, and E. D. Caldwell: High-Molecular-Weight Ethylene Polymers, in N. M. Bikales (ed.), "Encyclopedia of Polymer Science and Technology," vol. 6, pp. 332–336, Interscience, New York, 1967.
4. Canterino, P. J.: Derivatives of Ethylene Polymers, in N. M. Bikales (ed.), "Encyclopedia of Polymer Science and Technology," vol. 6, pp. 431–454, Interscience, New York, 1967.
5. Outlook for Polyethylene Is Healthy, *Chem. Eng. News*, July 21, 1969, pp. 14–15.
6. Clegg, P. L.: High-Pressure (Low and Intermediate) Polyethylene, in "Kirk-Othmer Encyclopedia of Technology," 2d ed., A. Standen (ed)., vol. 14, pp. 217–241, Interscience, New York, 1967.
7. Fedor, W. S.: Outlook Seventies, *Chem. Eng. News*, Dec. 15, 1969, pp. 98–99.

8. Fork, R. W.: Physical Properties of Ethylene Polymers, *J. Appl. Polym. Sci.,* 9: 2879 (1965).
9. Gaylord, N. G., and H. F. Mark: "Linear and Stereospecific Addition Polymers," Interscience, New York, 1959.
10. Ceil, P. H.: Polymer Morphology, *Chem. Eng. News,* Aug. 16, 1965, pp. 72–84.
11. Geil, P. H.: "Polymer Single Crystals," Interscience, New York, 1963.
12. Ceil, P. H., Polymer Crystallization, *J. Polym. Sci.,* C20: 109 (1967).
13. Gloor, W. E.: Polyethylenes Made by the Ziegler Process, in "Kirk-Othmer Encyclopedia of Technology," 2d ed., A. Standen (ed.), vol. 14, pp. 259–282, Interscience, New York, 1967.
14. Hogan, J. P., and R. W. Myerholtz: High-Density (Linear) Polyethylene, in "Kirk-Othmer Encyclopedia of Technology," 2d ed., A. Standen (ed.), vol. 14, pp. 242–259, Interscience, New York, 1967.
15. Plastics properties chart, in "Modern Plastics Encyclopedia," vol. 45, no. 1A, pp. 29–46, McGraw-Hill, Inc., New York, 1967.
16. Parker, G. R.: Strong Demands Bring Bright Earnings Outlook, *Chem. Eng. News,* Sept. 1, 1969, pp. 64A–68A.
17. Potter, W.: Cross-linked Polyethylene, in "Modern Plastics Encyclopedia," vol. 42, no, 1A, pp. 253–254, McGraw-Hill, Inc., New York, 1965.
18. Pritchard, J. E.: Polyethylene Resins, in "Modern Plastics Encyclopedia," vol. 45, no. 1A, pp. 205–210, McGraw-Hill, Inc., New York, 1967.
19. Southern, J. H., and R. S. Porter: The Properties of Polyethylene Crystallized under the Orientation of Pressure Effects of a Pressure Capillary Viscometer, *J. Appl. Polym. Sci.,* 14: 2305 (1970).
20. Raff, R. A. V.: Ethylene Polymers, in N. M. Bikales (ed.), "Encyclopedia of Polymer Science and Technology," 2d ed., vol. 6, pp. 275–332, Interscience, New York, 1967.
21. Raff, R. A. V., and K. W. Doak: "Crystalline Olefin Polymers," pts. I and II, Interscience, New York, 1965.
22. Rees, R. W. (to E. I. du Pont de Nemours & Company): Ionic Hydrocarbons Polymers, U.S. Pat. 3,264,272 (Aug. 2, 1966).
23. Renfrew, A., and P. Morgan: "Polythene," 2d ed., Interscience, New York, 1960.
24. Sittig, M.: "Polyolefin Resin Processes," Gulf Publishing, Houston, Tex., 1961.
25. Smith, W. M.: "Manufacture of Plastics," vol. 1, Reinhold, New York, 1964.
26. Speradi, C. A., W. A. Franta, and H. W. Starkweather: The Molecular Structure of Polyethylene, V: The Effect of Chain Branching and Molecular Weight on Physical Properties, *J. Am. Chem. Soc.,* 75: 6127 (1953).
27. Traflet, R. F.: Low-Molecular-Weight Polyethylene, in "Modern Plastics Encyclopedia," vol. 42, no. 1A, pp. 251–252, McGraw-Hill, Inc., New York, 1965.
28. Waters Associates: *Tech. Inf. Bull.,* Framingham, Mass., 1970.
29. Williams, F. R., M. E. Jordon, and E. M. Dannenberg: The Effect of the Chemical and Physical Properties of Carbon Black on Its Performance in Polyethylene, *Cabot Corp. Tech. Serv. Rep.,* Boston, Mass.
30. Albright, L. F.: High-Pressure Processes for Polymerizing Ethylene, *Chem. Eng.,* Dec. 19, 1966, p. 113.
31. Beasely, J. K.: The Molecular Structure of Polyethylene: Kinetic Calculations of the Effect of Branching on Molecular Weight Distribution, *J. Am. Chem. Soc.,* 75: 6123 (1953).

32. Benzler, A., and A. V. Koch: Ein Zustandsdiagramm fur bis zu 10,000 ata Druck, Chem.-Ing.-Tech., 27: 71 (1955).
33. Carter, D., and W. G. Bir: Axial Mixing in Tubular High Pressure Reactor, Chem. Eng. Prog., 58(3): 40 (1962).
34. Chementator, Chem. Eng., Sept. 8, 1969, p. 57.
35. Japanese Polyethylene Producers Think Big, Chem. Eng. News, May 15, 1967, p. 38.
36. Clegg, P. L.: High Pressure (Low and Intermediate Density) Polyethylene, in "Kirk-Othmer Encyclopedia of Chemical Technology," A. Standen (ed.), vol. 14, pp. 217–241, Interscience, New York, 1967.
37. De Lesquien, P.: Low-Density Polyethylene made in Tubular Reactors, Chem. Eng., May 29, 1972, pp. 42–43.
38. Ehrlich, P., and J. C. Woodbrey: Viscosities of Moderately Concentrated Solutions of Polyethylene in Ethane, Propane, and Ethylene, J. Appl. Polym. Sci., 13: 117 (1969).
39. Grimsley, F. N., and E. R. Gilliland: Continuous Oxygen-initiated Ethylene Polymerization, Ind. Eng. Chem., 50: 1049 (1958).
40. Haselbarth, J. E.: Updating Investment Costs for 60 Types of Chemical Plants, Chem. Eng., Dec. 4, 1967, pp. 214–215.
41. Polyethylene, Hydrocarbon Process., November 1961, p. 286; November 1967, pp. 221–223; November 1969, p. 225.
42. Ingersoll-Rand: Scope in Recipocating Process Compressors, Bull., New York, 1968.
43. Kresser, T. O.: "Polyolefin Plastics," Van Nostrand Reinhold, New York, 1969.
44. Lenoir, J. M.: Compressibility Factors for Super Pressures and Temperatures, Pet. Refiner, August 1960, pp. 135–138.
45. Lyderson, A. L., R. A. Greenkorn, and O. A. Hougen: Univ. Wis. Eng. Exp. Stn. Rep. 4, October 1955.
46. Miller, R. I.: Process Technology for License or Sale, Chem. Eng., Apr. 20, 1970, p. 113.
47. Mortimer, G. A.: Chain Transfer in Ethylene Polymerization, J. Poly. Sci., (A-1)4: 881 (1966).
48. Mullikin, R. V., P. E. Parisot, and N. L. Hardwicke, Analog Computer Studies for Polyethylene Reaction Kinetics, 58th Annu. Meet. AIChE, Philadelphia, Dec. 5–9, 1965.
49. Nichols, L.: The Molecular Structure of High-Pressure Polyethylene, J. Chim. Phys., 3: 177, 185 (1958).
50. Parker, G. R.: Five Polymers and Fibers, Chem. Eng. News, Sept. 1, 1969, pp. 64A–68A.
51. Platzer, N.: Design of Continuous and Batch Polymerization Processes, Ind. Eng. Chem., 62(1): 6 (1970).
52. Schoenemann, K.: Zur technischen Ausführung der Hochdruck-Polyaethylen-Synthese, personal communications, 1968.
53. Schoenemann, K.: personal communications, 1970.
54. Takehisa, M., S. Machi, and S. Sawayannazi: Polyethylene via Radiation, Hydrocarbon Process., 47(11): 169 (1968).
55. Thies, J., and K. Schoenemann: The Calculation of High Pressure Polyethylene Reactors by Means of a Model Comprehending Molecular Weight Distribution and Branching of the Polymer, 1st Int. Symp. Chem. React. Eng., Carnegie Inst., Wash. D.C., 1970.

56. "Ullmans Encyclopedia of Technical Chemistry," 3d ed., vol. 14, pp. 137–164, Urban and Schwarzenburg, Munich and Berlin, 1963.

57. Molen, T. J. van der, and C. van Heerden: The Effect of Imperfect Mixing on Initiator Productivity in the High Pressure Radical Polymerization of Ethylene, *1st Int. Symp. Chem. React. Eng., Carnegie Inst., Wash., D.C., 1970.*

58. Vaughn, S. B., and J. J. Hagemeyer (to Tennessee Eastman Co.): Preparation of Polyethylene in Two-Zone Reactor Employing Capryl Peroxide in the First Zone as Catalyst, U.S. Pat. 3,178,404 (Apr. 13, 1965).

59. Albright, L. F.: Commercial Low-Pressure Processes for Polymerization of Ethylene, Chem. Eng., Jan. 16, 1967, pp. 169–174; Feb. 13, 1967, pp. 159–164.

60. Anderson, A. W., J. M. Bruce, and E. L. Fallwell (to E. I. du Pont de Nemours & Company): Polymerization of Ethylene, U.S. Pat. 2,862,917 (Dec. 2, 1958).

61. Bacon, J. B.: Linear Polyethylene via the Phillips Process, *Chem. Eng.,* Apr. 4, 1960, pp. 110–113.

62. Badger Company, Cambridge, Mass.: *Badger Briefs,* 3(1), 1969.

63. Boor, J.: The Nature of the Active Site in the Ziegler-Type Catalyst, *Macromol. Rev.,* **2**: 115–268 (1967).

64. Caldwell, E. D.: High Molecular Weight Ethylene Polymers, in A. Standen (ed.), "Kirk-Othmer Encyclopedia of Chemical Technology," 2d ed., vol. 14, pp. 336–338, Interscience, New York, 1967.

65. Campbell, P. E., and R. V. Jones: High Molecular Weight Ethylene Polymers, in A. Standen (ed.), "Kirk-Othmer Encyclopedia of Chemical Technology," 2d ed., vol. 14, pp. 332–336, Interscience, New York, 1967.

66. *Chem. Eng. News,* Apr. 23, 1956, p. 197.

67. Stamicarbon Develops New High Density Polyethylene Process, *Chem. Eng. News,* Apr. 20, 1970, p. 36.

68. New Superactive Catalysts for High-Density Polyethylene, *Chem. Week,* June 3, 1967, p. 61.

69. Cinadr, B. F., and A. T. Schooley: How to Treat Solvents for Solution Polymerization, *Chem. Eng.,* Jan. 26, 1970, pp. 125–129.

70. Clark, A.: Some Reactions of Olefins on Solid Catalysts, *Ind. Eng. Chem.,* **57**(7): 29 (1967).

71. Clark, A., and G. C. Bailey: Formation of High Polymer on Solid Surfaces, *J. Catal.,* **2**: 230–247 (1963).

72. Clark, A., J. N. Finck, and B. H. Ashe: Mechanism of Polymerization of Ethylene on Supported Chromium Oxide Catalysts, *Proc. 3d Int. Cong. Catal. Amsterdam, 1965.*

73. Clark, A., J. P. Hogan, R. L. Banks, and W. C. Lenning: Marlex Catalyst Systems, *Ind. Eng. Chem.,* **48**: 1152 (1956).

74. Coover, W. H., R. L. McConnell, and F. B. Joyner: Relationship of Catalyst Composition to Catalytic Activity for Polymerization of α-Olefins, *Macromol. Rev.,* **1**: 91–118 (1967).

75. Czenkusch, E. L., and Fawcett, W. L. (to Phillips Petroleum Co.): Gas Phase Polymerization Utilizing a Free-setting, Fluidized Catalyst and Reactor System Therefor, U.S. Pat. 3,002,963 (Oct. 3, 1961).

76. Davison, J. W., J. E. Pritchard, and S. Renaudo: Phillips' Polyolefin Process, *Pap. Natl. Meet. Am. Inst. Chem. Eng., Houston, Tex., 1971.*

77. Dingle, J., and J. F. Franklin: Solvay H.D. Polyethylene Offers Flexibility, *Eur. Chem. News, Large Plant Surv.,* Sept. 25, 1970.

78. Field, E., and M. Feller (to Standard Oil Co., Indiana): Group VI-A Oxide-

Alkaline Earth Metal Hydride Catalyzed Polymerization of Olefins, U.S. Pat. 2,731,452 (Jan. 17, 1956).

79. Field, E., and M. Feller: Promoted Molybdena-Alumina Catalysts in Ethylene Polymerization, *Ind. Eng. Chem.*, 49: 1883 (1957).

80. Friedlander, H. N., and K. Oita: Organometallics in Ethylene Polymerization, *Ind. Eng. Chem.*, 49: 1879 (1957).

81. Friedlander, H. N., and W. Resnick: Solid Polymers from Surface Catalysts, *Adv. Pet. Chem. Ref.*, 1: 526–570 (1958).

82. Gilbert, W. I., G. H. Gwynn, and J. G. McNulty (to Goodrich-Gulf Co.): Process for Polymerizing Olefins Utilizing Ziegler Catalyst, U.S. Pat. 3,160,622 (Dec. 8, 1964).

83. Goins, R. L. (to Phillips Petroleum Co.): Olefin Polymerization, U.S. Pat. 2,936,303 (May 10, 1960).

84. Harban, A. A., E. Field, and H. N. Friedlander: Polyolefin Production with Preferred Solid Catalysts, *J. Polym. Sci.*, 41: 157 (1959).

85. Heath, A.: Polymer Purification Made Easier in HDPE Route, *Chem. Eng.*, Apr. 3, 1972, pp. 66–67.

86. Hogan, J. P.: Olefin Copolymerization with Supported Metal Oxide Catalysts, chap. 3, pp. 89–113, in G. E. Ham (ed.), "Copolymerization," Interscience, New York, 1964.

87. Hogan, J. P., and R. L. Banks (to Phillips Petroleum Co.): Polymers and the Production Thereof, U.S. Pat. 2,284,721 (Mar. 4, 1958).

88. Polyethylene Processes, *Hydrocarbon Process.*, November 1961, p. 288; November 1967, p. 224; November 1969, pp. 226–227.

89. Johnson, R. M. (to Union Carbide Corp.): Polymerization of Olefins with Silane Modified Catalyst System, U.S. Pat. 3,687,920 (Aug. 29, 1972).

90. Lanning, W. C., J. P. Hogan, R. L. Banks, and C. V. Detter (to Phillips Petroleum Co.): Polymerization Process Using a Solid Catalyst Dispersion Medium, U.S. Pat. 2,970,135 (Jan. 31, 1961).

91. Marwil, S. J., and R. G. Wallace (to Phillips Petroleum Co.): Polymer Solid Recovery from Flowing Stream, U.S. Pat. 3,374,211 (Mar. 21, 1968).

92. Mudd, J. F., and E. H. Casey (to Phillips Petroleum Co.): Apparatus for the Recovery of Solids from Pressure Vessels, U.S. Pat. 3,203,766 (Aug. 31, 1965).

93. Natta, G., and J. Danusso: "Stereoregular Polymers and Stereospecific Polymerizations," Permagon, Oxford, 1967.

94. Norwood, D. D. (to Phillips Petroleum Co.): Method and Apparatus for the Production of Solid Polymers of Olefins, U.S. Pat. 3,248,179 (Apr. 26, 1966).

95. Peters, E. F., A. Zeitz, and B. L. Evering: Solid Catalysts in Ethylene Polymerization, *Ind. Eng. Chem.*, 49: 1879 (1957).

96. Phillips, G. W., and W. L. Carrick: Transition Metal Catalysts: Random Ethylene-Propylene Copolymers with a Low Pressure Polymerization, *J. Am. Chem. Soc.*, 84: 290 (1962).

97. Rasmussen, D. M.: High Density Polyethylene Polymerized in the Gas Phase, *Chem. Eng.*, Sept. 18, 1972, pp. 104–105.

98. Rohlfing, R. G. (to Phillips Petroleum Co.): Reactor Inlet with Feed Inlet Along Shaft, U.S. Pat. 3,226,205 (Dec. 28, 1965).

99. Scoggin, J. S. (to Phillips Petroleum Co.): Method and Apparatus for the Recovery of Solid Olefin Polymers from a Continuous Path Reaction Zone, U.S. Pat. 3,242,150 (Mar. 22, 1966).

100. Scoggin, J. S., and H. S. Kimble (to Phillips Petroleum Co.): Separation of Solid Polymers and Liquid Diluent, U.S. Pat. 3,152,872 (Oct. 13, 1964).

101. Smith, D. A.: "Addition Polymers: Formation and Characterization," Plenun Press, New York, 1968.

102. Solvay et Cie: Catalysts for Low-Pressure Polymerization of Ethylene, French Pat. 1,427,204 (Apr. 4, 1966); British Pat. 1,059,430 (Feb. 22, 1967).

103. Stevens, J.: Here's How Solvay Makes HDPE, *Hydrocarbon Process.*, November 1970, pp. 179–182.

104. "Ullmans Encyclopedia of Technical Chemistry," 3d ed., vol. 14, Urban and Schwarzenberg, Munich and Berlin, 1963.

105. Weisemann, G. H.: Making High-Density Polyethylene Using the Indiana Process, *Hydrocarbon Process.*, November 1965, p. 146.

106. Weyermuller, G., B. E. Cask, and S. T. Ross: Easy Catalyst Removal, *Chem. Process.*, April 1958, p. 87.

107. Whittington, W. H. (to Phillips Petroleum Co.): Apparatus for the Removal of Solids from Pressure Vessels, U.S. Pat. 3,172,737 (Mar. 19, 1965).

108. Wisseroth, K.: Grunlagen and Verfahrenstechnick der Gasphasen-Polymerisation, *Angew. Makromol. Chem.*, 8: 41–60 (1969).

109. Witt, D. R. (to Phillips Petroleum Co.): Catalytic Polymerization of 1-Olefins, U.S. Pat. 3,378,540 (Apr. 16, 1968).

110. Zletz, A.: Ethylene Polymerization with Conditioned Alumina-Molybdena Catalysts, U.S. Pat. 2,692,257 (Oct. 19, 1954).

CHAPTER FOUR

Production of Polypropylene

Polypropylene was first commercialized in the late 1950s, and the annual production growth rate since then has been large on the average. By 1972, about 1.6 billion lb/year was produced in the United States, and large production capacities were also available in several other countries.

Polypropylene plastics can be divided into three broad categories (27, 58). First, the unmodified homopolymer, which is the oldest variety, has the largest production rates. Second, modified polypropylenes are widely used; materials such as asbestos, talc, fiber glass, and graphite fibers (5) are dispersed in amounts up to 20 or even 40 percent throughout the plastic matrix. Third, propylene-type copolymers probably account for 30 percent or more of the polypropylene market; these copolymers are frequently produced using at least several percent of ethylene as a comonomer. Both random and block types of copolymers are available. They are frequently referred to as medium- or high-impact polypropylenes, depending on the structure and amount of ethylene used to prepare the copolymer.

Polypropylene homopolymers have substantially higher tensile strength, surface hardness, flexural strength, and heat resistance than high-density polyethylenes. Melting points of about 167°C for many commercial grades permit their use at fairly high temperatures. In addition, they have somewhat better chemical resistance, lower thermal expansion, improved clarity and gloss, and lower gas permeability. Unmodified polypropylene and the copolymers especially have the lowest densities of all the major polyolefins, with values of about 0.9 g/cm^3. Homopolymers, however, have relatively poor impact strength, especially at temperatures lower than room temperature. Improved impact properties can be realized with propylene copolymers, but a somewhat lower flexural modulus and hardness results. Modified (or filled) polypropylenes have improved stiffness, less heat distortion, and lower thermal expansion. Fire-resistant propylenes are also available.

The two biggest applications for polypropylene and related copolymers are for injection-molded items and for numerous extruded materials, especially fibers (28). Some common injection-molded items include toys, laboratory and medical supplies, appliances, closures, and housewares. Modified (or filled) polypropylenes are molded to produce various appliances and automotive fan shrouds. Automotive kick panels, trim, protective shielding, appliances, tote boxes, luggage, seating parts, and battery cases are produced by molding high-impact polypropylenes.

Propylene homopolymers are extruded in large quantities to produce fibers and filaments. A major step was to develop techniques for dyeing the fibers now widely used in indoor and outdoor carpeting. Although the problem has not been completely solved, at least fair success has been realized by the following two approaches. First, dye receptors such as organic nickel, aluminum, or other metal compounds can be mechanically blended into the polypropylene. Second, dye-receptive groups can be incorporated or chemically bonded to the polypropylene molecule by copolymerizing or grafting techniques. Fibers that are acid-dyeable have recently been announced. The first entry of a polyolefin fiber into the large knit-apparel market is a polypropylene yarn manufactured by Hercules, Inc. (8). This bicomponent yarn is produced by combining two fibers with different molecular weights. The yarn develops a three-dimensional, helical, irreversible crimp when immersed in boiling water or steam or when heated.

Polypropylene monofilaments are employed in the production of rope, twine, cordage, webbing, and bagging. Polypropylene is also extruded to produce film (39), sheeting, coatings for wire, and piping or ducts. Most polypropylene film, however, is produced by casting. Extruded film has high clarity and is suitable for food packaging and for textile wrapping. A polymer blend containing 75 to 95 percent crystalline polypropylene and the remainder low-density polyethylene and an amorphous ethylene-propylene copolymer has been found to make a very clear film (13). Film with a microporous structure is available from the Celanese Plastics Company; passage of bacteria and colloidal particles is prevented (11), but gas and even certain liquids can permeate it.

Blow-molded polypropylene bottles appear to have a promising future for pharmaceuticals, cosmetics, mouthwashes, detergents, and similar uses. Phillips Petroleum Company has pioneered manufacture of bottles in which the high-clarity polypropylene is biaxially oriented as a result of a cold-parison blowing technique (40). Polypropylene bottles have an advantage over the polyvinyl chloride bottles frequently used for similar products in that no pollution problems occur when the bottles are disposed of by burning.

Polypropylene printing plates have recently been developed for printing newspapers. It is claimed that these plates speed up printing by 30 percent (9).

Structure and Molecular Weight

Several structural features of polypropylene can be varied to at least some extent in the polymerization process, and these features have a significant effect on the properties of the final polypropylene product. Specifically they are the stereospecificity of the repeating units ($-C_3H_6-$) in the polymer chain, the average molecular weights of the polymer, and, for copolymers, the arrangement of the two or more repeating units in the polymer molecule. It was not until the 1950s that suitable polymerization procedures were devised to produce polypropylenes useful as plastics.

Stereospecificity of Polypropylenes Isotactic, syndiotactic, and atactic configurations are shown in Fig. 4-1 for polypropylene. Similar stereospecific configurations occur for all linear polymer molecules that have an asymmetric carbon atom in the repeating unit of the chain. As will be discussed in later chapters of this book, polyvinyl chloride and polystyrene also exhibit similar types of stereospecificity. Polyethylenes, however, do not contain asymmetric carbon atoms except at the occasional atoms where branches are attached to the chain.

Isotactic polypropylene molecules are those in which the methyl groups are all arranged on the same side of the principal chain of carbon atoms (see Figure 4-1). The main chain is assumed to be the idealized fully extended zigzag of carbon atoms, i.e., without twists between the repeating units. Actually rather uniform twists normally do occur. In any case, two types of isotactic stereoisomers form; one has the methyl groups on one side of the chain, and the other has them on the alternate side. For syndiotactic polypropylenes, the methyl groups are arranged with the adjacent groups on alternate sides. For atactic polymers, the groups are arranged randomly.

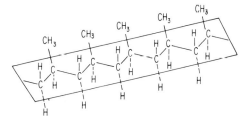

(a) Isotatic

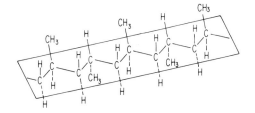

Fig. 4-1 Structures of polypropylene with isotactic and syndiotactic configurations. The atactic configuration has a random arrangement of —CH₃ groups.

(b) Syndiotactic

The stereospecificity of polypropylene (or other linear polymers containing asymmetric carbon atoms) has a significant effect on the packing characteristics of the molecules and hence on the intermolecular attractive forces. The packing abilities are as follows: isotactic > syndiotactic > atactic. The best way to measure the amounts of isotactic and syndiotactic stereoisomers in polypropylene is probably by high-resolution nuclear magnetic resonance (NMR) (25). Commercial polypropylenes usually contain about 75 to 98 percent isotactic stereoisomers. Most but not all isotactic stereoisomers crystallize, and commercial polypropylenes often contain 50 to 60 percent crystalline isotactic, 20 to 30 percent amorphous isotactic, and 20 to 30 percent amorphous atactic polymers. With annealing at 145 to 150°C, an additional 10 percent crystallize, and up to perhaps 70 percent can often be attained. Syndiotactic polypropylenes are currently prepared only in the laboratory.

Crystallinity in polypropylenes is generally measured by both the classical techniques of x-ray diffraction and infrared methods. Values reported by the latter methods are often about 10 percent higher, presumably because infrared measurements register crystalline regions of less perfect order than x-ray diffraction techniques.

Portions of isotactic polypropylene that are noncrystalline are not truly amorphous or completely disorganized in packing arrangements. Instead they exist in a variety of paracrystalline forms. One such form, designated as the smetic form, is obtained by quenching a melt to low temperatures or by cold-drawing the α (or crystalline) form (42). It is not surprising then that the densities of amorphous polypropylenes vary according to the orderliness of packing (or the approach toward crystallinity) and on the stereospecificity. Densities of amorphous, crystalline syndiotactic, and crystalline (α) isotactic polypropylenes are about 0.85 to 0.858, 0.898, and 0.936 g/cm^3, respectively (31).

Partial separation of the various stereoisomers is often accomplished by selective solubility techniques, using one or more solvents at various temperatures. Acetone, diethyl ether, n-pentane, n-hexane, n-heptane, 2-ethylhexane, and n-octane have all been employed. Diethyl ether, liquid propylene, and n-pentane are sometimes used to remove most atactic and considerable amorphous fractions and to leave essentially pure isotactic fractions.

Molecular Weight of Polypropylenes Commercial polypropylenes have number-average molecular weights ranging from about 75,000 to 200,000 and weight-average molecular weights from 300,000 to 700,000. The ratios of these two molecular weights for specific polymers range from 2 to 10 but generally from 3 to 7 (25). The molecular weight has a significant effect on melt flow rate during extrusion or molding operations and also on the physical properties of the finished product. A standard test is to measure the melt flow rate at 230°C. This measurement is made in the same equipment used to measure the melt index of polyethylene; the melt-index measurement, however, is at a significantly lower temperature (namely 190°C) because of the lower softening temperatures of polyethylenes compared with polypropylenes. Sometimes the melt flow rate of polypropylene is referred to as the melt index.

Repka (42) reported for several polypropylenes the relationship between the melt flow rate and the weight-average molecular weight. A polypropylene with a melt flow rate of 60 had a molecular weight of 156,000, whereas with a melt flow rate of 0.4 the molecular weight was 550,000. Molecular-weight measurements are normally by either solubility fractionation or gel-permeation chromatography. The latter method is reported by Ogawa and associates (37) as being the more accurate, and the resulting chromatograms can be used to calculate both number-average and weight-average molecular weights.

Repka (42) has also summarized how important physical properties of polypropylenes change because of variations of the average molecular weight, the molecular-weight distribution, i.e., measured as a ratio of the weight-average to the number-average molecular weights, and the density (or level of crystallinity). Higher-molecular-weight polypropylenes tend to have lower levels of crystallinity, with the resulting change of properties expected of lower-density polymers. Apparently short chains orient themselves more rapidly to crystallize.

Propylene-Type Copolymers Copolymers in which propylene is the main comonomer are widely used since they have significantly improved impact properties, especially at lower temperatures, compared with homopolymers of propylene. Often 5 to 30 percent ethylene is copolymerized with the propylene, and both block and random copolymers are produced commercially in large quantities.

Block copolymers contain segments of largely isotactic polypropylene that are chemically bonded to segments of polyethylene or propylene-ethylene random copolymers. These copolymers have fairly high crystallinity, the high-temperature properties of polypropylene, and the low-temperature performance of polyethylene. In manufacturing the block copolymers (discussed in detail later), some random copolymerization also occurs, especially in the portion of the linear chain where the two blocks are joined. The level of this random copolymerization affects the physical properties of the final plastic material. There is also some question whether the polyethylene and polypropylene blocks are always chemically bonded. Mechanical blends of the two homopolymers also give some improved low-temperature properties. Tusch (55) reports that the impact strength of propylene-ethylene block copolymers increases rapidly as the melt-flow index decreases in range from 6 to about 1.5; that is, the average molecular weight of the polymer increases.

Random copolymers also have increased impact strengths compared with propylene homopolymers. They have lower strength properties, lower crystallinities, and higher clarity than block copolymers at similar ethylene levels since the repeating units ($-C_3H_6-$ and $-C_2H_4-$) are rather uniformly distributed in the plastic.

Graft copolymers in which specific acrylate monomers are graft-copolymerized with polypropylene have much improved low-temperature properties (24). An oxidation technique is used to produce polypropylene hydroperoxide. Then a redox catalyst and an acrylate such as n-butyl acrylate are added to the reaction vessel. Free-radical polymerization results in the grafts.

Repka (42) has reported values for many mechanical properties of typical molding grades of polypropylenes. Tensile yield strengths at 2 in/min are 5,300; 4,300 to 4,500; and 3,000 to 3,200 lb/in^2, respectively, for a propylene homopolymer, medium-impact copolymers, and high-impact copolymers. The impact strengths measured by a drop weight at $-29°C$ on a smooth plaque for the three plastics are <5, 28 to 55, and 80 to 95 in drop of a 4-lb weight respectively.

Polymerization Considerations

The preparation of the catalyst for propylene polymerization is a most important step in the overall process. With the proper catalyst, the following features can be obtained:

1. At least fairly rapid rates of polymerization
2. High isotactic content in the polypropylene product
3. High yields of polymer expressed as parts of polymer produced per part of catalyst used
4. Desired morphology of final polymer product from the polymerization process

Heterogeneous catalysts are required to obtain polypropylenes with the desired high levels of isotacticity, and they are normally prepared using various titanium chlorides, aluminum alkyls, and perhaps other compounds. Several specific combinations of reactants used to prepare successful catalysts are reported below.

1. Titanium trichloride $TiCl_3$, and various dialkylaluminum chlorides are popular combinations. The so-called violet crystalline modifications of $TiCl_3$, however, are required. Natta et al. (34) have compared the results when several dialkylaluminum chlorides were employed. Diethylaluminum chloride, $Al(C_2H_5)_2Cl$, which is readily available and is one of the cheaper aluminum alkyls, is quite widely used. $Al(C_2H_5)_2Cl$ is a relatively weak alkylating agent, and the resulting $Ti—C_2H_5$ groups formed are generally considered to be the active sites for polymerization. Some $Al—Cl$ bonds are also formed during this alkylation (or activation) of the titanium atom. Catalysts produced with these types of reactant result in polypropylenes with greater than a 90 percent isotactic content. The activity of the catalyst prepared from a mixture containing highly subdivided δ-$TiCl_3$ and cocrystallized $AlCl_3$ is increased as the size of the $TiCl_3$ crystals is decreased.

2. Combinations of $TiCl_3$ and triethylaluminum, $Al(C_2H_5)_3$, have been found to give a highly active catalyst, i.e., one that results in rapid polymerizations. Unfortunately the polypropylenes produced often have relatively low levels of isotacticity. $Al(C_2H_5)_3$ is a stronger alkylating agent than $Al(C_2H_5)_2Cl$.

3. Combinations of $TiCl_4$ and $Al(C_2H_5)_3$ have been used to prepare catalysts that produce isotactic polypropylenes. Vandenberg (56) reports the production of polypropylene that is 97 percent insoluble in n-heptane. He prepared the catalyst using n-heptane solutions of $TiCl_4$ and $Al(C_2H_5)_3$.

4. Highly satisfactory catalysts can be produced using alkylaluminum dihalides, transition-metal halides, and a third compound containing strongly complexing ligand elements such as phosphorus or nitrogen (14, 30). A specific combination is ethylaluminum dichloride, $Al(C_2H_5)Cl_2$; either α- or δ-$TiCl_3$; and hexamethylphosphorictriamide. The polypropylene produced with this catalyst has a 97 percent isotactic content, but the rate of polymerization was somewhat less than half that of the catalyst prepared from $TiCl_3$ and $Al(C_2H_5)_2Cl$. As a general rule, catalysts that are very active give rather low levels of isotacticity. The more exposed titanium sites that cause high rates of polymerization apparently do not give high isotactic contents. The compounds added for complexing purposes appear to deactivate these more active titanium sites rather selectively. Trace amounts of water deliberately added to propylene may also deactivate these sites (12).

The mechanism and factors affecting polymerization, including stereo-specificity and the rates of reaction, have been discussed and reviewed by numerous investigators (1, 2, 3, 4, 6, 15, 21, 57). The shape and physical characteristics of the catalyst particles have several important effects on the overall polymerization process. The position of the active sites on or near the bulk surface of the particles affects both the rate of polymerization and the type of stereospecificity. Quantum-chemical principles and crystal chemistry have been used to explain the type of stereospecificity obtained. Although considerable differences in the postulated mechanisms still exist, heterogeneous catalysts are apparently required to obtain mainly isotactic stereoisomers. When a homogeneous catalyst is used, such as that prepared from vanadium tetrachloride and dialkylaluminum halides, syndiotactic polypropylenes are obtained with low-temperature polymerizations (62). For heterogeneous catalysts, it has been suggested that there are essentially two types of active sites. One type of site produces atactic polypropylene; isotactic polypropylene is produced at the other type (6).

The structure of the starting catalyst particle is also important for the condition and fate of the catalyst upon completion of polymerization. The initial catalyst particles are usually relatively large, being about 5 to 30 μm in diameter (6, 21). These particles are actually composed of many thousand cohering primary particles of much smaller dimensions, perhaps 100 to 400 Å. For most if not all commercial polymerizations, the starting particles fragment as propylene diffuses to the active sites and as polypropylene forms, causing the primary particles to be forced apart. Such action exposes and makes available additional active sites. With continued polymerization, the catalyst particles decrease in size and become well dispersed throughout the resulting polymer particles. If sufficient fragmentation occurs (and with the simultaneous high yields of polymer per given amount of catalyst), a separation step to remove the residual catalyst particles from the polypropylene product is unnecessary in commercial units. Such a result that offers large savings in operating costs apparently has been realized to date only in the Badische Anilin- und Soda-Fabrik (BASF) vapor-phase polymerization process, described later.

The morphology of the polypropylene particles produced by the normal slurry process also depends on the type of catalyst particle used (2, 21). Some catalyst particles result in porous polypropylene flakes, whereas other types result in dense, relatively nonporous materials. It has been suggested that during vapor-phase polymerization the initial rate of polymerization on a catalyst surface is more rapid than the rate of crystallization. Consequently the initial hemispherical mass of polypropylene is amorphous. As the particle grows in size, the rate of polymerization decreases and at the same time folded-chain lamellae (one type of crystallite) are formed.

Although preparation of the catalyst is still partly an art, it is obvious that several important characteristics of the catalyst include the size and shape of the primary particles, cohesive forces between these particles, the chemical structure of the particles, and the overall size of catalyst particles at the start of polymerization. It has been suggested, for example, that $AlCl_3$ is not really an inherent part of the catalyst, its role being instead to open up the crystal structure (6).

The molecular weight of the polypropylene is controlled by the type of catalyst used or by the operating conditions (4). Adding hydrogen as the chain-transfer agent, as was done in production of high-density polyethylenes, is a common control method. Catalysts prepared using diethylzinc have also been found effective for lowering the molecular weight. In addition, polymerizations at higher temperatures result in somewhat lower-molecular-weight polymers.

Commercial Processes for Polypropylene

Modern polypropylene plants are always large, the plant of Amoco Chemical Co. shown in Fig. 4-2 being typical. Three relatively different types of process are employed for polypropylene production. The first, and by far the most important, type involves a slurry polymerization. Some of the slurry processes for polypropylene are very closely related to slurry processes used for the production of high-density polyethylenes, discussed in Chap. 3. In some cases, the similarity is so great that a given plant can be used interchangeably (with only minor modifications) to produce either polypropylene or polyethylene. Other polypropylene processes include a solution polymerization with a solid catalyst and also a gas-phase polymerization process. The solution pro-

Fig. 4-2   Polypropylene plant.   (*Amoco Chemical Corporation.*)

cess operates at higher temperatures and pressures than the slurry process. These last two processes are quite similar in many respects to comparable polyethylene processes.

Storage of reactants must be carefully planned since all reactants are potentially very dangerous. Propylene and all hydrocarbon solvents are highly flammable, and explosive mixtures can be formed with air. Propylene is often stored in pressure vessels at essentially ambient temperatures. Pressures as high as 10 atm may occur in these vessels, especially during the summer. Solvents, if used, are stored in conventional tanks; generally an inert atmosphere such as nitrogen is used in order to minimize air contamination.

Aluminum alkyls and other catalysts react explosively with water or air. Aluminum alkyls are frequently diluted with hydrocarbons, and the resulting mixtures are much less dangerous to handle. Storage tanks, however, must be carefully designed in any case. Mixtures of titanium trichloride and aluminum chloride are generally slurried with a hydrocarbon liquid such as a C_5 to C_7 paraffin, cycloparaffin, or a mineral oil. The slurry is stored in agitated tanks and is then pumped, as needed, to the reactor.

The propylene and solvent, if one is used, must both be highly purified to reduce catalyst poisons such as water, oxygen (from air), carbon oxides, sulfides, various dienes, and acetylenic compounds to less than 20 ppm. A typical analysis (26) of the propylene feed indicated the following composition: 99.7 percent propylene, 10 ppm water, 2 ppm total sulfur, 4 ppm oxygen, and 1 ppm carbon oxides. Equipment for purifying the entering and recycle propylene and solvent will be discussed later.

Slurry Processes for Polypropylene   Several modifications of slurry processes have developed within the last 15 years, and current processes are in general much simpler and cheaper to operate than earlier versions. These processes are either operated batchwise or with continuous-flow operation. In addition, some slurry processes use a hydrocarbon solvent, but others do not. In all cases, however, a slurry is present in the polymerization reactors, and solid particles that are primarily polypropylene are suspended in the liquid phase. The solid particles also contain finely dispersed catalyst particles and adsorbed or entrapped propylene, comonomers, and solvent (if one is used). The liquid phase contains primarily solvent and propylene. If a solvent is not used, the reaction is controlled so that only part of the propylene is polymerized, and the polymerization can be characterized as bulk (or mass) polymerization. The unreacted propylene then acts as the liquid to suspend the polypropylene particles. In all slurry processes, some low-molecular-weight and amorphous polypropylenes are dissolved in the liquid phase.

Process steps that have been or still are employed in polypropylene manufacture include the following:

1. Catalyst preparation that may in some cases be completed in situ in a polymerization vessel.

2. Polymerization of propylene and possible comonomers to produce the product slurry. This step is obviously always the most important one.

3. Recovery and recycling of the unreacted propylene.

4. Recovery and recycling of the solvent if one is used.

5. Destruction and separation of the catalyst or its residue from the final polymer.

6. Removal of amorphous (or atactic) and low-molecular-weight polypropylene from the solid-polypropylene particles formed in the reaction vessel. The objective frequently is to produce polypropylene that is almost completely crystalline (and isotactic).

7. Finishing operations for the polypropylene product, such as drying, blending additives, and extrusion and pelletizing.

If a catalyst could be developed which resulted in essentially complete isotacticity in the product and which was extremely active, steps 5 and 6 could be simplified, if not eliminated, with a significant savings in operating expenses. Although no such catalyst has apparently been developed, chances of doing so in the next few years seem good. By

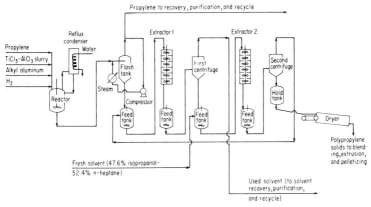

Fig. 4-3 Flowsheet of Rexall polypropylene process, in which reactor contains a slurry of solid polypropylene suspended in liquid polypropylene.

proper choice of posttreatment techniques, steps 5 and 6 are often accomplished simultaneously. Furthermore, step 4 is also eliminated if no solvent is used and liquid propylene is used instead as the suspending liquid.

Several patents and other literature (7, 12, 16, 18, 22, 25, 26, 29, 35, 42, 43, 46, 56, 59) report enough information for many key features of current processes to be known. Figure 4-3 shows the basic flowsheet of a Rexall process (43) that is probably fairly typical for a process using propylene for the liquid phase.

Catalyst Preparation: Preparation of the catalyst is probably the least publicized of all phases of the process, and the exact techniques used are often closely guarded secrets. Mixtures of the $TiCl_3$ and $AlCl_3$ often used in the preparation of the final catalyst particles are frequently prepared from aluminum and $TiCl_4$. Aluminum granules are ground in a ball mill or similar device to produce fresh (and highly reactive) aluminum surfaces. These activated surfaces react readily with $TiCl_4$ to produce essentially 3:1 mixtures of $TiCl_3$ and $AlCl_3$. When combined (often in the reactor) with aluminum alkyls, e.g., diethylaluminum chloride, $Al(C_2H_5)_2Cl$, such mixtures result in highly active polymerization catalysts.

Reactor Design and Operation: Several types of reactor systems have been discussed in the literature (36, 38, 41, 43, 60):

1. A single stirred autoclave is often employed in either a batch or a flow process.
2. Several autoclaves are often connected in series so that the reaction mixture (or slurry) flows continuously from one autoclave to the next.

3. A loop reactor like that described in Chap. 3 for the production of high-density polyethylene is also suitable for polypropylene production.

Regardless of the type of reactor used, temperatures in slurry processes are usually in the range of 45 to 85°C, which is significantly below the first-order transition temperature of polypropylene. Temperature has a significant effect on the polymerization. Higher temperatures promote faster rates of polymerization, result in lower-molecular-weight products, and necessitate higher operating pressures (to maintain adequate concentrations of propylene in the liquid phase). Pressures of 20 to 30 atm are common in units using propylene liquid as the suspending liquid for the solid polypropylene, but they can be as high as 40 atm when high operating temperatures are used.

At higher temperatures, especially at 80°C or higher, considerable amounts of polypropylene dissolve in the liquid phase. In such a case, strands of polypropylene may form in the reaction mixture, causing mechanical problems in maintaining a slurry (25). In extreme cases, runaway reactions result in a veritable ball of polypropylene thread that has to be mechanically removed from the reactor.

AUTOCLAVES USED AS REACTORS: Autoclaves used as reactors are always provided with efficient agitators, and they can be used either for batch or continuous-flow operation. For polypropylene production they are glass-lined or have polished stainless-steel inner surfaces. The former have been widely used, and the latter have been employed by several companies in recent installations.

Temperature control is always one of the key design features of a reactor. Since propylene polymerization is highly exothermic (42, 45), provisions must be made to remove most of this heat. Several techniques are used, but jacketing the reactors is common. Higher coefficients of heat transfer are obtained through the walls of the reactor with polished stainless-steel inner walls than with glass-lined ones. In all reactors, polymer deposition on the jacketed (and relatively cold) inner walls is of concern, and the skin temperatures affect this deposition. Little tendency to foul occurs, however, with glass-lined or well-polished stainless-steel reactors. The Brighton Corp. electropolishes the inner stainless-steel walls of their reactors to produce an almost mirror finish.

Complicated baffles that are internally cooled have been installed in many recently built autoclaves. External reflux condensers are also provided on some reactors, especially for slurry processes that employ essentially pure liquid propylene as the suspending liquid for the solid polypropylene. Propylene, which is readily vaporized from the reaction mixture, is condensed and returned to the reaction mixture. Such a condenser is employed in the Rexall process (Fig. 4-3).

The level of agitation in an autoclave is important because agitation promotes good heat transfer with cooled surfaces and because the character and size of solid polypropylene particles suspended in the slurry are affected. Turbulence at the wall also affects the rate of polymer buildup on the wall (45).

Batch polymerization is particularly applicable when speciality and generally low-volume polypropylenes are to be produced. A single autoclave is used in such a case. For flow processes, a single autoclave, i.e., a continuous-flow stirred-tank reactor (CSTR), may be used. Sometimes several (often three to six) autoclaves are connected in series. The type of product obtained with multiple reactors will often differ from that produced by a single reactor. The rate of polymerization varies to a considerable extent with the level of polymerization and the solids content of the slurry. As polymerization progresses, the degree of fragmentation of the catalyst particles increases and the size of polypropylene solid particles increases. When several reactors are arranged in series, the design and operating conditions can be varied (and usually are) as a function of the level of polymerization in the reactor train. For example, when five stainless-steel reactors were used in series, three of the five contained more cooling surfaces than the other two. With stainless-steel or stainless-steel-clad reactors used in series, the reactors are sometimes jacketed and sometimes unjacketed; jacketed reactors appear to be preferred. As discussed later, arranging the reactors in series is common when block copolymers are to be produced.

Reactors are designed to maintain careful control of flows, temperatures, and pressures. Use of automatic controls for autoclave reactors has been described by Schlegel (45). In the case of emergency or runaway polymerizations, water can be added in small amounts to destroy the catalyst and stop the polymerization. Pressure-relief valves are provided for emergencies involving excessive pressure buildups.

Product streams to the reactors vary of course with the process. In the Rexall process (which does not use a solvent but instead uses unreacted propylene as the suspending liquid for the solid polypropylene particles), four streams are introduced continuously to a single autoclave reactor (Fig. 4-3) to produce almost 2,500 lb/h of polypropylene product. The amounts of these streams are shown in Table 4-1. The reactor used in this process is glass-lined, has a capacity of 2,500 gal, and is operated at 66°C. The average residence time of the reaction mixture

TABLE 4-1   Reactants and Product Stream in Rexall Process to Produce 2,497 lb/h of Polypropylene (43)

	Pounds per hour
Reactants to reactor:	
(1) Propylene (as liquid)...	4,162
(2) Hydrogen (as gas)...	0.35
(3) Catalyst mixture of $TiCl_3$ and $AlCl_3$ (suspended in 1.5 lb of mineral oil)...	2.5
(4) $Al(C_2H_5)_2Cl$ (dissolved in 96 lb of liquid propylene).............	4
Product stream from reactor:	
(1) Solid polypropylene (melt index $\sim$ 3.0)........................	2,497
(2) Propylene liquid...	1,611.5
(3) Dissolved and amorphous polypropylene.......................	50
(4) Catalyst residues...	6.5

in the autoclave is 1.5 h.   A reflux condenser is employed for removal of most of the heat of polymerization, and the pressure in the reactor is about 420 lb/in² gage (26 atm).

The slurry product in this example contained 60 percent solids, but slurries with solid contents varying from 30 to 60 percent are common. The production capacity of a specific reactor (or autoclave) tends to be higher with higher solid contents, but mechanical problems in handling the slurries increase rapidly, especially with solid contents of about 60 percent or higher.   If the slurry product is removed by intermittent opening (and then closing) of the exit valve on the bottom of the autoclave, plugging of the exit line with solids tends to be minimized.   In one example using the Rexall unit, the slurry was discharged for 1 s every 10 s.   The exit flow, although intermittent, can be considered for the rest of the process as if it were at steady-state conditions.   The exit product in the example reported for the Rexall process contains the materials shown in Table 4-1.   In particular, 2,497 lb/h of solid polypropylene (eventually recovered as product) is obtained.

Hydrogen is commonly used to regulate (and to decrease) the molecular weight of the polypropylene, i.e., to increase the melt flow rate (or melt index).   The Rexall Company (43) has reported the following relationship:

$$\text{Log melt index} = 2.4 \log [H_2] + 2.3$$

where $[H_2]$ is the hydrogen concentration in mole percent in liquid propylene.   Presumably this relationship applies only to the specific catalyst employed in their system.

Slurry processes that employ hydrocarbon liquids (often called solvents) for slurrying the polypropylene particles are very similar to the processes used for high-density polyethylenes. In these polypropylene processes, the propylene must be dissolved in the solvent before polymerization occurs. Solvents that are frequently used include C_5 to C_8 paraffins of cycloparaffins. Cyclohexane and n-heptane are two popular solvents. Aromatics are also suitable from a polymerization standpoint, but because of their toxicity they are not widely used. For the polymerization of polypropylene to be used for synthetic fibers, C_9 to C_{11} paraffins (and especially n-decane) or xylenes have been recommended (7).* Pressures normally employed in the reactor are in the range of 50 to 100 lb/in² gage, which is significantly less than the pressures used when liquid propylene is the suspending fluid for the solids.

When a solvent is employed, temperature control is usually maintained primarily by heat transfer through the reactor walls to a jacket or to cooled baffles installed internally. Propylene is generally sparged into the reaction mixture of each reactor in order to dissolve it in concentrations up to several percent in the solvent phase. Dissolved concentrations depend of course on the type of sparger used and on the conditions in the reactor, including pressure, temperature, and agitation. In addition, the concentrations depend on the properties of the liquid phase since resistances to mass transfer of the propylene from the gas to the liquid phase may be significant (and controlling). The latter properties, including viscosity, vary to a considerable extent with the amount of polypropylene dissolved. As mentioned earlier, low polymers (or waxes) and amorphous polypropylene are relatively soluble; hence the amount produced of these undesired polymers may have a significant effect on the properties of the liquid phase.

LOOP REACTORS: Loop reactors can be used instead of autoclave reactors in processes like that in Fig. 4-3; they are similar if not identical to the loop reactors used for the production of high-density polyethylene by the particle-form (PF) process developed by Phillips Petroleum Co. (36, 41, 48). These loop reactors were discussed in considerable detail in Chap. 3. The reactor, produced from large tubing, is commonly constructed to form two loops, through which the slurry is recirculated numerous times. The product is exhausted from the reactor by intermittent opening and closing of the exit lines. With flow velocities inside the reactor loop of 10 to 30 ft/s for a reactor constructed of tubing 20 in ID, polymer deposition on the walls of the reactor is essentially nonexistent.

A loop reactor used for production of about 3,800 lb/h of polypropylene has been described (36). This reactor was constructed from tubing that was 20 in ID. Sections of tubing 50 ft long were connected

* In this process, fibers are formed by extrusion of a hydrocarbon solution of polypropylene.

TABLE 4-2   Reactants and Reactor Effluent for
20-in ID Loop Reactor*

	Pounds per hour
Reactants:	
Propylene	10,700
Propane (impurity in propylene)	740
Hydrogen	1.7
TiCl₃ (slurried with equal weight n-pentane)	4.15
Diethylaluminum chloride dissolved in 10 lb of n-pentane	3.3
Reactor effluent stream:	
Propylene	6,700
Propane	740
Hydrogen	1.7
Solid polypropylene	3,800
Soluble polymers	200

* Information from (41); flow rates are very similar to those in (36) that indicate volumetric capacity of loop reactor.

by 20-in elbows to form a two-loop reactor with an internal volume of 3,620 gal. In an example reported, the reactor was operated at 49°C, at 25 atm (355 lb/in² gage), and with a flow velocity inside the reactor of 20 ft/s. Table 4-2 shows the flow rates of the reactants and of the product stream. In this run, no solvent was used, and liquid propylene was used to slurry the polypropylene product. The solids content in the slurry was 35 percent, and the average residence time of the reactants in the reactor was calculated to be about 1.5 h.

Straightening vanes (or baffles) were installed both before and after the impeller used to recirculate the slurry through the loop reactor. These vanes dampen out the vortex or cyclonic flow characteristics of the slurry as it is pumped. The cooling water fed to the jacket of the reactor was heated because of heat transfer through the tube walls from about 30 to 36°C.

Based on the example reported (36), production of polypropylene is slightly less in loop reactors of comparable size than the production assumed to occur for high-density polyethylene in the Phillips PF process. Since polypropylene has a lower heat of polymerization on a weight basis (and hence easier temperature control), and since it has a higher first-order transition temperature (and hence perhaps fewer deposition problems), somewhat greater polypropylene production capacities than reported in the example seem possible.

OPERATING CONSIDERATIONS: The operation of the reactors is important to the subsequent separation and recovery steps for obtaining crystalline polypropylene, noncrystalline polypropylene, unreacted propylene, and solvent. In particular, the choice of solvent (or use of liquid propylene instead) and of the temperatures has an appreciable effect on the amounts of polypropylene dissolved in the liquid phase of the slurry. In general, higher-molecular-weight solvents and higher temperatures promote higher polypropylene solubilities. Ideally, all or at least most of the amorphous and low-molecular-weight polypropylenes would be dissolved, and most crystalline polypropylenes would not. Then simple centrifuging (or filtering) of the slurry would separate the amorphous and crystalline polypropylenes. Step 6 (mentioned earlier) for separation of these two types of polypropylenes would be greatly simplified and in a sense eliminated since it would have occurred in the reactor itself. Unfortunately a sufficiently clean separation usually does not occur in the reactor although one approaching it can be obtained, so that subsequent separation steps are greatly simplified.

In choosing a solvent (or suspending fluid) for the slurry process, the role of the solvent in the reaction system and the ease of recovery in the separation phase of the process must both be considered. In general, solvents that dissolve large amounts of polypropylene (especially the noncrystalline variety) are desired. On solubility considerations alone, higher-molecular-weight solvents are as a rule preferred, but lower-molecular-weight (and hence more volatile) solvents are easier to recover in the separation phases. Furthermore, there is evidence that some solvents have chemical effects, particularly on the activity of the catalyst (60). In any case, several factors that have a large effect on the overall process have to be considered in selecting the solvent or type of fluid to be used for slurrying the polypropylene solids. Although relatively large amounts of solvent are used in some processes, only small amounts are used in others.

Separation and Recovery Steps of Slurry Process: Procedures employed in separating and recovering various materials in the slurry from the reactor vary widely in different plants. Variations in the nature and composition of the product slurry (as mentioned earlier) are in part responsible for differences in the separation and recovery techniques used. In addition, these latter techniques can result in small but sometimes significant differences in the properties of the polypropylene product. Some recovery techniques are preferred for certain types of products. At least two relatively different separation techniques have been reported.

1. The solid polypropylene particles are first separated from the liquid phase. Subsequently the catalyst residue and amorphous polypropylene are removed from the solid polypropylene. The liquid phase is then processed to recover propylene and solvent, if one is used.

2. The catalyst residue and atactic polypropylene are first dissolved in the liquid phase before the solid particles and the liquid phase of the slurry are separated.   After separation of the two phases, each phase is processed to complete the recovery.

The first technique, as shown in Fig. 4-3, is probably more common, and details, as reported in the Rexall patent (43), are thought to be rather typical.   This technique will now be discussed for a process using propylene as the suspending liquid for the solid polypropylene.

RECOVERY OF CRUDE POLYPROPYLENE: Rexall has reported that they use a cyclonic separator (or flash tank) (see Fig. 4-3).   The slurry and superheated propylene gas enter this separator operated at about 90°F (30°C).   The superheated gas provides the heat required to vaporize most of the propylene, leaving essentially dry polypropylene solid.   Typically 1 part by weight of propylene vapor at 300°F (150°C) is adequate to cause essentially complete vaporization of liquid propylene from 7.1 parts by weight of slurry at 130°F (54°C) containing 40 weight percent solids.   The resulting gaseous propylene stream of about 5.3 parts is divided into two portions.   Then 1 part by weight is compressed, reheated to 300°F, and recycled to the cyclonic separator. The remaining propylene (4.3 parts) is sent to the propylene-repurification section (discussed later) before it is recycled to the reactor.

EXTRACTION STEPS FOR PURIFICATION: The polypropylene solids from the cyclonic separator contain the atactic polymer and as a rule slightly more than 1 percent propylene.   These solids are transferred to the polypropylene-purification portion of the process, where the catalyst residue and atactic polymers are removed by extraction.   Acidified mixtures of an alcohol and n-heptane are extraction agents.   The alcohol reacts with the catalysts to produce alkoxides that are soluble in the solvent. Atactic polypropylene is dissolved primarily because of n-heptane.   An azeotropic solvent mixture containing 47.6 weight percent isopropanol and the remainder n-heptane is recommended in the Rexall patent (43). About 0.17 percent HCl was added to this particular mixture.   If it is unnecessary to extract atactic polymer from the polypropylene product, the hydrocarbon, e.g., n-heptane, is not used.   (In some cases, the amount of atactic polymer is sufficiently low; in other cases, some atactic polymer is desired in the final product.)

Total contact times required to obtain the desired extraction generally vary from about 0.5 to 1.0 h, depending to some extent on the temperature used, the size of the polypropylene particles, and the degree of extraction desired.   Higher temperatures tend to increase the rate of extraction, and temperatures of about 180°F (80°C) are probably common with total contact times of about 1 h.   Frequently the weight ratios of solvent to solid polypropylenes used in the extraction units are about 4:1 to 6:1. Examples reported in the Rexall process indicated ratios of about 4:1. With higher ratios, somewhat higher extraction efficiencies are obtained, but operating costs increase since more solvent must then be processed. Vigorous agitation is required in order to promote rapid rates of extraction.

Agitated, cocurrent contactors containing several compartments are highly suitable as extraction vessels, and two such vessels are frequently used, with average residence times of about 0.5 h in each. Several compartments in each vessel minimize the short circuiting (or low residence times) of some solid particles as they pass through the units. Because of the presence of hydrogen chloride, special alloys, such as Hastelloy B, are required in these contactors and are often used to clad the vessels.

As shown in Fig. 4-3, the almost dry polypropylene solids from the flash tank are first slurried with solvent in the feed tank to extractor 1. The slurry from this extractor is maintained in a feed (or surge) tank before entering the first centrifuge. Pure nickel is often recommended for construction of the internal parts of this centrifuge. The centrifuges are operated cyclically and with automatic controls. After the extract is separated from the solids, the polymer is rinsed with clean solvent and the wash liquid is added to the extract. The combined extract from the centrifuge is then processed in equipment not shown in Fig. 4-3, as will be discussed later.

The solid cake (containing perhaps 20 percent by weight of residual extract) from the first centrifuge is combined with a fresh (or new) azeotropic mixture of isopropanol and n-heptane in the feed tank to extractor 2. In this tank, the polypropylene solids are reslurried. In the second extractor, most of the remaining catalyst residues and atactic polypropylene go into solution, and most of the residual extract from the first extractor is washed from the pores of the polypropylene solid. The second slurry is separated in the second centrifuge. The extract from this centrifuge is heated somewhat before being used as the feed solvent to produce the slurry feed for extractor 1.

DRYING AND FINAL PROCESSING OF POLYPROPYLENE: The wet solids cake from the second centrifuge contains about 20 percent extract, mainly solvent. In some plants, equipment such as squeeze rolls or porous belting is provided to allow some of this extract to be separated (and to be collected) from the solids. The wet solids are then dried. In the Rexall process, a steam-jacketed unit provided with a screw conveyor is used. When the materials are maintained at about 180°F (80°C) for about 20 min, polypropylene solids containing less than 0.1 percent of solvents are obtained. In this drying operation, care must be taken to prevent the formation of combustible gas mixtures and the degradation (or oxidation) of the rather hot polypropylene. As a result, a noncombustible gas, e.g., carbon dioxide or nitrogen, is heated and then used in the dryer to expedite removal of the solvent and propylene from the solid polypropylene.

The dry polypropylene powder is then processed by techniques essentially similar to those used for polyethylene and already described (Chap. 3). In summary, the polypropylene powder is generally blended with various additives (including stabilizers), extruded, pelletized, and then prepared for shipment. A stabilizer is often needed since polypropylenes rapidly degrade when exposed to ultraviolet light (or sunshine). Carbon black is an effective stabilizer since it converts the ultraviolet rays to relatively harmless wavelengths; the resulting polymer, however, is black. Zinc oxide, also an effective stabilizer, renders polypropylene opaque and white. Stabilizers have been compared (10, 19), and a combination of zinc oxide and diethylthiocarbamate was found especially effective. This combination in an accelerated stability test increased the time before failure from 168 to over 7,000 h; stabilizers increase the cost of the polymer by about 3.5 percent.

PURIFICATION OF SOLVENT: The solvent extract from the first centrifuge (see Fig. 4-3) is processed to recover the solvent, soluble polymer and waxes, and waste solids. Equipment for this purification is not shown in Fig. 4-3. The extract is first neutralized using a relatively concentrated caustic solution in a stirred tank. A pump is frequently adequate for contacting the two liquids. The two liquid phases can then be separated by decanting or centrifuging. The caustic is recycled until it is depleted, when it is replaced with a fresh solution. The wet solvent, containing perhaps 0.5 weight percent propylene and some water, is then fed to a distillation column. Here all the propylene and most of the isopropanol, n-heptane, and water are vaporized, and the top product of the column is partially condensed. The gaseous product, containing most of the propylene and a small amount of the solvent, is often flared. Part of the top condensate is used as liquid reflux to the column, and the remainder is dried in conventional equipment using desiccants such as molecular sieves or alumina. These desiccants also remove catalyst poisons from the solvent before it is sent to storage and later to the reactor.

The bottoms from the column contain amorphous polypropylenes, catalyst residues, salts, and enough solvents for the mixture to be pumped. The bottoms are sent to waste.

EXTRACTION TECHNIQUE USING LIQUID PROPYLENE: Phillips Petroleum Company (41) has described a technique for purifying crude polypropylene that uses liquid propylene as an extraction agent for amorphous polymers. The slurry from a reactor that contains polypropylene solids suspended in liquid propylene is first heated to a temperature somewhat higher than that in the reactor. A reactant (an extractant) for the catalyst is then added to the heated slurry. Acetylacetone or propylene oxide is recommended, and it is added so that 1 to 5 times the stoichiometric amount based on the catalyst is present. Contacting the solid polypropylene for about 20 to 30 min at about 150 to 170°F (65 to 77°C) usually provides adequate time for effective extraction (or removal) of both the catalyst residues and the atactic polypropylene. After this treatment, the slurry is separated, and the polypropylene solids are washed with clean propylene liquid. The washed polypropylene solids are recovered by centrifuging, dried, and processed into pellets.

RECOVERY OF UNREACTED PROPYLENE: Some unreacted propylene is always present in the exit product stream from the polymerization reactor of a polypropylene plant. In processes like that in Fig. 4-3, where liquid propylene is the suspending agent, large amounts of propylene are unreacted. Less propylene is present, however, in the exit product from processes in which an inert solvent is employed as the suspending fluid for the polypropylene product.

Gaseous propylene in the recovery portion of the process, as from the flash tank of Fig. 4-3, is first compressed and then condensed by cooling. Crude liquid propylene is then purified by distillation to remove impurities, including light paraffins and other olefins. Recycle and feed propylene are combined in a storage tank before being sent to the reactor. Propylene from this tank must generally be purified to remove catalyst poisons. Alumina (60) or molecular sieves are effective for the adsorption (and removal) of polar impurities such as water, carbon oxides, and sulfur compounds. Light diolefins (such as propadiene) or acetylenic compounds are also catalyst poisons. If present, selective hydrogenation techniques, as described in Sec. III of Chap. 2, are often used for their removal.

MODIFICATIONS USED IN RECOVERY OPERATION: Recovery, separation, and recycling of solvent and unreacted propylene account for a significant fraction of the total capital and operating costs of a polypropylene plant. Many modifications have been tried in an effort to reduce the costs and improve the separations. The solvent chosen, for example, has an important effect on the costs, and there has been a trend toward lower-molecular-weight (and more volatile) hydrocarbons as solvent or suspending fluid. In such a case, the solvent flashes (or vaporizes) at temperatures close to its critical temperature. As a result, the heats of vaporization are low, and relatively small pressure drops are needed to obtain flashing. Staged pressure reductions may be used in some cases to minimize compression costs for recovery and recirculating the solvent. If a "solvent" is employed above its critical temperature, there is no latent heat of vaporization. In this respect, propylene (which is in a sense a solvent) has a critical temperature of 91.4°C (197°F); this is often in the general range of temperatures used commercially.

With volatile solvents, a slurry of polypropylene solids and solvent can be separated rather easily (51). An elongated tubular heat exchanger and a gas-solid separator have been used, for example. The solvent vaporizes in the heat exchanger, and the solid, relatively dry powder is entrained in the gas stream flowing at high velocities. The gas and solids are then separated, e.g., in a centrifugal separator. Spray-drying techniques have also been employed for separating polypropylene and solvent.

For contacting the polypropylene solids with solvents used for extracting amorphous polymers and catalyst residues, more than one contacting vessel has sometimes been provided. Countercurrent flow of the solids and extracting solvent between vessels has been tried. Rather complex mixtures of solvents have also been tested. A mixture containing a polyhydric alcohol in which at least one of the hydrogen atoms is replaced with a metal atom (of the alkali-metal or alkaline-earth family) has been found very effective for neutralizing acid (generally HCl) formed by destruction of the catalyst.

Polymer plugging and deposition throughout the equipment are always potential problems, and they can be minimized with proper equipment design and operating procedures. Pulsed flow of slurries from various extraction vessels and pieces of equipment is often used to minimize the plugging of lines with polypropylene solids. Keeping a polymerization reactor full at all times is said to minimize deposition in the reactor (16).

Solution Processes for Polypropylene Although of lesser importance than the slurry processes just described, solution processes are used for the production of both homopolymers and copolymers of propylene. In these processes, the polymer produced is dissolved in the liquid phase of the reactor. Details have been reported (20) of two different versions of a process and catalyst system developed by Eastman Kodak Co. In the first version, a paraffinic hydrocarbon boiling in the range of about 180 to 200°C is employed. The second version really employs a bulk- (or mass-) polymerization technique, and propylene is the "solvent" for polypropylene. This process, however, differs from the bulk process (described earlier in this chapter) in which polypropylene was not dissolved.

In order to dissolve polypropylene, fairly high temperatures (160 to 250°C) are required, and hence higher pressures are needed in order to dissolve appreciable amounts of propylene in the liquid phase and to obtain high rates of polymerization. When a solvent is employed in the Eastman Kodak process, pressures of 400 to 1,180 lb/in² gage (27 to 80 atm) are specified. When no solvent is employed, much higher pressures are needed to obtain a suspending fluid (mainly propylene) of reasonably high density. Propylene has a critical temperature of 91.4°C and a critical pressure of 45.4 atm; hence propylene is not a liquid in the second version of the process but a compressed gas or fluid. At pressures of 1,200 to 1,500 atm (17,700 to 22,000 lb/in²), the propylene has densities of about 0.5 to 0.65 g/cm³ (based on calculations using a generalized compressibility graph for high pressures).* The propylene fluid therefore has densities comparable to those of many liquid hydrocarbons and is capable of dissolving polypropylene; conversely, it is completely adsorbed in the polypropylene to form a liquid solution.

Polymerization catalysts used in the slurry processes are not generally suitable for the solution processes since they are rather quickly destroyed at the higher temperatures. A stable catalyst is prepared using lithium metal, a lithiumaluminum hydride, and titanium tetrachloride; sometimes catalyst promoters are used. The solid catalysts formed produce polypropylene with crystallinities of 70 to 90 percent.

Several examples of the large-scale production of polypropylene are reported by Hagemeyer and Edwards (20). When a paraffinic hydrocarbon was used as the solvent, two 500-gal autoclaves connected in series were used as reactors. Propylene, solvent, and catalyst were continuously added to the first reactor, and additional propylene was introduced (probably through a sparging device) to the second reactor. The reaction mixture was maintained at about 160 to 173°C and 1,000 lb/in² gage (about 70 atm). The exit solution from the second reactor contained (in examples reported) 30 to 38 percent dissolved polypropylene, 18 to 23 percent absorbed propylene, and the remainder solvent. It is estimated that residence times of about 1 h are employed in the reactors, and hence approximately 2,000 lb/h of polypropylene are produced in these two autoclaves.

The exit solution from the second reactor was sent to a flash tower at 50 lb/in² gage, where unreacted propylene was flashed and sufficient recycle paraffinic solvent was added to produce a 10 percent polymer solution. The additional solvent reduced the viscosity of the solution, which was then filtered to remove catalyst residue. The filtered solution was concentrated using hot (200°C) propylene gas as a stripping agent. The exit gas from the stripping column was cooled, and the solvent

---

* Similar calculations were made in Chap. 3 for ethylene used in high-pressure polyethylene processes.

was condensed by means of suitable cooling. The concentrated solution from the stripping column contained some residual solvent and propylene. Devolatilizers (see Chap. 8) are useful for removal of residual amounts of volatiles, including light hydrocarbons, from the polypropylene. After mixing with suitable additives, including stabilizers, the devolatilized polypropylene is generally pelletized to complete the overall polymerization process.

Polypropylenes with melt flow rates of 1.56 to 2.42 produced by this solution process had much higher impact strengths and lower brittleness temperatures than a conventional polypropylene with a melt flow rate of 4.62 produced by a slurry process. The solution-type polypropylene was 88 percent crystalline based on extraction with hexane.

Details of the reactor and the operating conditions for the process modification in which polypropylene is dissolved in the unreacted propylene are presented later in this chapter. Polypropylene with crystallinities as high as 90 percent can be produced by this modification.

Because of the higher temperatures employed in the solution processes or the bulk process developed by Eastman Kodak, residence times of the reactants in the reactors are apparently low (since the kinetics of polymerization is high). Reactors with relatively small internal volumes seem probable, compared with those used in slurry processes. The solution processes, however, have significantly higher polymerization costs than slurry processes, for several reasons. First, higher-pressure reaction vessels are required. Second, and most important, recovery of polypropylene from a solution is rather expensive compared with recovery steps for a slurry process. Further, in a slurry process, there is the opportunity, during the catalyst-removal steps, to remove at least most of the atactic polypropylene. In a solution process, atactic polypropylene can be removed only if additional (and expensive) extraction steps are incorporated in the overall process scheme. Polypropylenes produced by solution processes are likely to have overall production costs at least 1 to 2 cents/lb higher than those of slurry processes.

Gas-Phase Polymerization of Propylene   Propylene can be polymerized by gas-phase processes (32, 33, 52, 53, 54, 61), and at least one such process is now being operated commercially (38). BASF is producing 24,000 metric tons/year of polypropylene at their Rheinische Olefinwerke in Wesseling, West Germany. This polypropylene is quite different from conventional polypropylenes. It has a relatively low level of crystallinity since about 25 to 28 percent of it is heptane-soluble, but its impact strength is much better than that of conventional polypropylene. Some of its properties are related to those of low-density polyethylenes. This polypropylene is a preferred material for coarse fibers.

Gas-phase polymerization of propylene involves considerations similar to those discussed in Chap. 3 for gas-phase polymerization of ethylene. The simplified flowsheet of the BASF process (38) is similar to the one described in the previous chapter for polyethylene. Feed and recycle propylenes are purified by distillation, and the distillate is dried with alumina. Other catalyst poisons are also removed by this technique.

The dry liquid propylene is then continuously added to the bottom of the polymerization reactor, and a modified Ziegler-Natta catalyst is added at the top. A mixture of solid particles and upflowing propylene gas is present inside the reactor. The solid particles are primarily polypropylene containing finely divided catalyst. The solid catalyst fragments as the propylene polymerizes and as the size of the polypropylene granules (or powder) grows. The phenomena occurring during polymerization are essentially the same as those occurring during gas-phase polymerization of ethylene and were discussed in detail in Chap. 3.

Although BASF does not report details of the reactor design or operation of the commercial unit being used, a fluidized-bed reactor or an agitated and expanded powder-bed reactor is probably used. In several patents (53, 54, 61), a reaction temperature of 90°C and pressures of 30 to 36 atm were reported; the product obtained appears to be identical in at least one case to the commercial BASF product. Agglomeration of the polypropylene particles or of polymer deposition on the walls of the reactor would probably become a problem at higher temperatures. Temperature control in the reactor is accomplished by using liquid propylene, which partially evaporates in the reactor. In one example, 18 percent of the propylene is in the reactor while the remaining 82 percent is being recycled. Unreacted propylene and powder polypropylene are removed from the reactor. This propylene is recovered, compressed, repurified, and then recycled to the reactor. The polypropylene product contains some adsorbed propylene that has to be removed. The resulting polymer is then combined with suitable additives, including stabilizers, and it is then generally pelletized before being stored or packaged.

To be economically competitive, gas-phase processes require a catalyst for which essentially no separation steps are required after the polymerization reactor. Separation steps for catalyst residues or atactic (or noncrystalline) polypropylene would increase the operating costs excessively. Although such separation steps are not practiced in the BASF process, it should be emphasized that BASF makes a rather unconventional polypropylene. As yet a suitable catalyst has apparently not been developed for production of conventional polypropylenes by gas-phase processes, i.e., for production of polypropylene with only low concentrations of atactic polypropylene. There is hope, however, that such catalysts can be found in the near future. If such a catalyst is developed, gas-phase processes will probably find large-scale commercial applications. Presumably these processes would have polymerization and capital costs as low as those of the slurry processes discussed earlier and probably lower. Major costs of such gas-phase processes would probably be those related to the recovery, compression, purification, and recycling of unreacted propylene. Efforts could then be expected to find ways to reduce the amount of recycle propylene, i.e., to increase the fraction of propylene polymerizing per pass.

Propylene-Type Copolymers Both random and block copolymers produced primarily from propylene and using relatively small amounts of ethylene (or possibly another comonomer) are polymerized either by slurry or solution processes. The copolymers are somewhat more soluble in the solvent (or light hydrocarbon that is sometimes the unreacted propylene) than the more crystalline homopolymers of propylene. Hence slurry processes are somewhat less applicable for copolymer production.

For completely random copolymers, the ethylene (or other comonomer) is mixed with the propylene in the reactor, and the two comonomers react randomly to form the polymeric chain. Reactors such as the autoclave, loop, or gas-phase types described earlier for the production of propylene homopolymers are used. The processes for these copolymers are basically the same as those described earlier. Minor differences result, however, because both unreacted propylene and ethylene must be recovered and recycled. Solubility differences between the copolymer and the homopolymer often cause small changes in the operating conditions and in the steps to separate the polymers, waxes, solvent, and catalyst residues. A solution process, for example, has been found to be useful for production of a copolymer containing 5 to 15 percent ethylene. This product is especially useful for production of wastebaskets, bottles, tubing, and film using blow-molding, injection-molding, or extrusion techniques. As an example, a mixture of propylene and ethylene was fed to a reactor maintained at 110 to 115°C. A catalyst prepared from $TiCl_3$ and triethylaluminum was employed, and cyclohexane was the solvent for the copolymer.

Block copolymers are produced by means of two (or more) step processes. For continuous-flow processes, two or more reactor compartments or units must be provided. For batch processes, a two-step polymerization procedure is employed. Numerous combinations and modifications have been used for the multistep operation, and several examples are discussed.

Production of four block copolymers by means of an Eastman Kodak

Co. process has been reported (20). The specific unit described produced 2,170 to 4,020 lb/h of copolymer by a high-pressure mass- (or bulk-) polymerization technique. This copolymer contained 2 to 9 percent of reacted ethylene and the remainder of propylene. The vertical reactor was 14 ft long and had an internal diameter of 20 in. A centrally located baffle divided the reactor into two reaction zones, and essentially no backmixing occurred between zones as the reactant mixture passed downward in the reactor. An agitator shaft positioned at the axis of the reactor extended throughout the entire length. A four-bladed paddle-type agitator was positioned in the top zone, and several mixing paddles were provided in the second zone. In general, the reactants increased in temperature as they passed through the reactor, indicating at most only partial backmixing in each reaction zone.

The reactor used in the above runs was designed for extremely high pressures, up to at least 1,500 atm, and it appears to have the internal dimensions and design similar to two-stage autoclave reactors used for high-pressure polymerization of ethylene, described in Chap. 3. Very possibly this reactor could be used interchangeably for either polyethylene or polypropylene production. Temperature control to the first reaction zone was controlled by the feed rate, by the temperature of the inlet propylene, and by external cooling. Four examples reported for this reactor are summarized as follows:

Propylene feed to first zone, lb/h......... 7,300–10,800
Catalyst to first zone, lb/h.............. 0.23–0.41
Ethylene feed to second zone, lb/h....... 230–380
Reactor pressures, atm 1,270–1,500
Temperature range in reactor, °C........ 168–190 to 178–206
Production rate, lb/h................... 2,170–4,020

In the first reaction zone, only homopolymers of propylene are formed, but copolymerization occurs in the second zone. Block copolymers form to at least some extent, and the blocks formed in the second zone are produced as propylene and ethylene react randomly. Residence times in the reactor were calculated to be about 5 to 10 min. Approximately 22 to 33 percent of the entering propylene reacted, whereas 30 to 70 percent of the ethylene did. The resulting copolymers (as indicated by extraction with boiling hexane) had crystallinities in excess of 70 percent, and brittleness temperatures were low, ranging from $-21°C$ to less than $-60°C$.

The copolymers of this bulk process are recovered by a procedure similar to that described earlier for propylene homopolymers in a solution process. Most of the propylene is rather easily vaporized, but a devolatilizer or devolatilizing extruder is needed for removal of the residual amounts of propylene from the polymer product.

A batch process has been described for the production of a different type of block copolymer (17). In this case, the two blocks of the polymer chain are formed either from all propylene or all ethylene. Homopolymers of propylene and a few of ethylene are also produced. A solvent (specified as a mineral oil), $(C_2H_5)_3Al$, and $TiCl_3$ are added first to the reactor. Propylene gas containing a trace of hydrogen is bubbled through the liquid containing entrained catalyst. Mechanical agitation is also provided, and the reactants are maintained at 75°C and a total pressure (caused primarily by propylene) of 31.2 atm for 6 h. Propylene homopolymers are produced during this initial period of polymerization. Then the propylene flow is stopped, and the reactor pressure drops within a few minutes to essentially atmospheric as the propylene reacts. Nitrogen gas is introduced, with agitation, to a pressure of 3.5 atm; the pressure is then released. This flushing procedure is repeated two or three times to remove all unreacted propylene and to complete the first stage of the batch run.

Ethylene is then introduced to the reactor to a pressure of 3.5 atm to start the second stage. As a rule, ethylene pressure is maintained only a relatively short time; the time depends on the amount (generally 0.5 to 5.0 percent) of ethylene to be incorporated in the block copolymer. At the end of the second stage of polymerization, the reaction mixture is heated to 160°C to lower the viscosity of the mixture, and it is then filtered to remove catalyst residues. The copolymer is then recovered by techniques similar to those employed for recovery of homopolymers from solution. A batch process of Imperial Chemicals Industries, Ltd. has also been reported (23). When two (or more) reactors are used in series (49) and a stripping tank or unit is installed between the reactors, a continuous-flow process can be operated to produce copolymers similar to those produced batchwise. Batch processes were relatively more important several years ago than they are today.

Slurry-polymerization processes have been developed for production of block copolymers using at least two reactors in series. In one process (47), liquid propylene at −32°C (−28°F), catalyst, and a small amount of ethylene are introduced into the first stirred reactor, which is operated essentially adiabatically. A small fraction of propylene and presumably most of the ethylene are polymerized in the first reactor, operated at about 16°C (60°F). The reaction slurry and a small amount of hydrogen (to control the molecular weight of the polymeric chains) are introduced into the second reactor, maintained at 16 to 60°C (60 to 140°F). In one example reported, 1,939 lb of copolymer was produced when a feed of 5,450 and 29 lb of propylene and ethylene, respectively, was provided. The resulting slurry from the last reactor then contained approximately 35 percent by weight of copolymer suspended in liquid propylene. In another modification of a slurry process, the ethylene is introduced into the last reactor instead of the first.

Several other processes have been developed for the addition of ethylene-type blocks to polypropylene chains in which the first step is a slurry-type polymerization. In one such process, propylene is first batch-polymerized in first-stage reactors (50). Two such reactors are provided and operated alternately. While one is being used for preparing a homopolymer of propylene, the second is being unloaded into a continuous-flow second-stage reactor employed for adding ethylene blocks. Upon completion of the batch polymerization in a first-stage reactor, the excess propylene is flashed off, and the homopolymer, containing active catalyst particles, is transferred to the second-stage reactor. In this transfer step, an inert diluent such as heptane or isooctane is employed to protect the polypropylene powder from the air and to act as a carrying agent. In the second-stage reactor, a small amount of ethylene is added, and ethylene blocks and some polyethylene molecules are formed.

Another process employing a slurry polymerization for the first stage has been reported (44), but in this case, a gas-phase polymerization technique is used for the second stage. The first-stage polymerization is a continuous-flow method essentially identical to the one described for the production of propylene homopolymers (Fig. 4-3). The polypropylene powder (still containing active catalyst) from the flash tank is transferred by a screw conveyor to the second reactor; propylene gas acts as a seal to prevent contact of the powder with air. Ethylene is introduced into the second reactor, controlled at 66°C (150°F) and 20 lb/in² gage. A ribbon blender is provided in the horizontal second reactor to promote solid-gas contact and hence adsorption and polymerization of the ethylene on the solid powder. Up to 10 percent ethylene can be incorporated into the copolymer, which is then ready for further processing similar to that employed for homopolymer powder. Polymer deposition may be a problem, especially in the second reactor, and the length of time this reactor can be used without cleaning is an important operating factor.

Selection of the preferred process for production of copolymers involves consideration of both the operating costs and of the properties of the polymeric products. Block copolymers, which in all cases require at least two reaction steps, tend to have higher operating costs than random copolymers and (especially) propylene homopolymers, which require only a single reaction step. Handling of a comonomer and recovery and recycling steps that often are somewhat more complicated also tend to increase the operating costs for copolymers compared to homopolymers. In addition, copolymer production is generally significantly smaller than that of homopolymers. As a result of these several factors, operating costs for production of copolymers are generally higher. Furthermore the type of copolymer produced by various processes is generally somewhat different, resulting in small but nevertheless significant differences in the properties.

Although most block copolymers use ethylene as the minor comonomer, 1-butene, 1-hexene, or mixtures of these two olefins have also been employed. Numerous other comonomers have also been used for production of experimental copolymers.

Future of Polypropylene

In the 1960s, polypropylene production had the most rapid relative growth rate of all major addition-type plastics (polyethylenes, polyvinyl chlorides, and polystyrenes are also in this group). Furthermore, predictions for the 1970s like those shown in Table 1-1 indicate that polypropylene production is likely to continue to have the most rapid relative growth rate. By 1980, polypropylene will probably account for 9.3 percent of all plastics produced in this country. Its past performance is all the more outstanding since it is competing in many applications with the other major plastics which have a substantially lower selling price. Clearly polypropylene's unique properties make it the preferred polymer for numerous applications.

The relatively high selling price of polypropylene is currently caused both by higher direct and indirect operating costs for polymerization and by large research and developmental costs. Information has been published (22) about the solvent used, catalyst and chemical costs, and utilities required. Total operating costs—including propylene, chemicals, labor, utilities, maintenance and operating, plant overhead, and depreciation—probably are often in the range of 8 to 9 cents/lb. Variations depend on the plant location, cost of propylene feedstock, process being used, and specific type of polypropylene being produced. As indicated earlier, all major polypropylene processes are very similar or closely related to major processes used for polyethylenes, especially high-density polyethylenes. Polypropylene processes tend to have high operating costs for the following reasons:

1. More separation steps are presently necessary in all polypropylene processes, except for the gas-phase BASF process, which still is of rather low relative importance. Modern high-density polyethylene processes do not require special separation steps for removal of the catalyst or undesired polymer fractions.

2. Polypropylene processes appear in general to be less developed than low-pressure polyethylene processes. Two or three polyethylene

processes have been widely adapted, whereas there appear to be more variations between polypropylene plants.

3. Relatively more developmental work is often being devoted to finding new and improved polypropylene products for specific uses.

Decreased operating expenses can be expected in the future for polypropylene processes. Currently the major objective is to develop an improved catalyst so that catalyst removal and separation of amorphous and low-molecular-weight polypropylenes will be unnecessary. The desired catalyst should produce polypropylenes with high isotactic contents and also have high activity, resulting in large amounts of polypropylene per given amount of catalyst and rapid rates of polymerization. Currently highly active catalysts are not satisfactory since only moderate isotactic levels are produced. Consequently polypropylene plants now require either larger reactors and/or higher levels of catalyst than polyethylene plants.

When (and if) improved catalysts are developed and employed, polypropylene and polyethylene plants will be even more similar in design and operation. With improved catalysts of the future, presumably the operating costs for both slurry and gas-phase polypropylene plants will be similar to those of comparable polyethylene plants. Homopolymers of propylene with reasonable expectancy could be produced for as little as 6.0 to 6.5 cents/lb. Propylene-type copolymers and especially block copolymers will also have reduced operating expenses, but they will have slightly greater costs than the homopolymers.

Literature Cited

1. Arlman, E. J.: Ziegler-Natta Catalysis: Surface Structure of Layer-Lattice Transition Metal Chlorides, *J. Catal.*, **3**: 89 (1964).
2. Arlman, E. J., and P. Cosee: Ziegler-Natta Catalysis: Stereospecific Polymerization of Propene with the Catalyst System $TiCl_3$-$AlEt_3$, *J. Catal.*, **3**: 99 (1964); **5**: 178 (1966).
3. Boor, J.: Nature of Active Site in Ziegler-Type Catalyst, *Macromol. Rev.*, **2**: 115–268 (1967).
4. Boor, J.: Review of Recent Literature on Ziegler-Type Catalysts, *IEC Prod. Res. Dev.*, **9**: 439 (1970).
5. Brody, H., and I. M. Ward: Modulus of Short Carbon and Glass Fiber Reinforced Composites, *Polym. Eng. Sci.*, **11**: 139 (1971).
6. Buls, V. W., and T. L. Higgins: A Uniform Site Theory of Ziegler Catalysis, *J. Polym. Sci.*, (A-1)**8**: 1025–1053 (1970).
7. Buntin, R. R. (to Esso Research and Engineering Co.): Solution Spinning of Polypropylene, U.S. Pat. 3,507,948 (Apr. 29, 1970).
8. New Olefin Fiber, *Chem. Eng. News*, May 3, 1971, p. 11.
9. Polypropylene Printing Plates Replace Metal, *Chem. Eng. News*, Aug. 21, 1972, p. 16.

10. Stabilizing Polyolefins, *Chem. Eng.*, Sept. 7, 1970, p. 94.
11. Polypropylene Film Boasts Microporous Structure, *Chem. Eng.*, Nov. 1, 1971, p. 42.
12. Cheney, H. A. (to Shell Oil Co.): Polymerization Process in Presence of a Ziegler Catalyst, Water, and Amine, U.S. Pat. 3,311,603 (Mar. 28, 1967).
13. Combs, R. R., D. F. Slonaker, and W. C. Wooten (to Eastman Kodak Co.): Polyolefin Blends for Films and Sheeting, U.S. Pat. 3,515,775 (June 2, 1970).
14. Coover, H. W., and F. B. Joyner: Stereospecific Polymerization of α-Olefins with Three-Component Catalyst Systems, *J. Polym. Sci.*, (A-3) **3**: 2407 (1965).
15. Cosee, P.: Ziegler-Natta Catalysts: Mechanism of Polymerization of α-Olefins, *J. Catal.*, **3**: 80 (1964).
16. Eastman Kodak Co. (inventors, W. D. Carter, T. F. Reid, and G. A. Akin), Process for Polymerizing Olefins, British Pat. 1,088,035 (Oct. 18, 1967).
17. Eastman Kodak Co. (by H. J. Hagemeyer and M. B. Edwards): Crystalline Polypropylene and Its Ethylene-Propylene Copolymers, Belg. Pat. 624,523 (Feb. 28, 1963).
18. Giachetti, E., R. Serra, and R. Moretti (to Montecatini Societa Generale): Process for Polymerization Alpha Olefins in the Presence of Activated TiCl₃, Aluminum Dialkyl Monohalides, and a Zinc Alkyl Halide, U.S. Pat. 3,252,958 (May 24, 1966).
19. Guillory, J. F., and C. F. Cook: Mechanism of Stabilization of Polypropylene by Ultraviolet Absorbers, *J. Polym. Sci.*, (A-1)**9**: 1529 (1971).
20. Hagemeyer, H. J., and M. B. Edwards (to Eastman Kodak Co.): Lithium, Lithium Aluminum Hydride, Titanium Trichloride, and Promoter for Olefin Polymerization, U.S. Pat. 3,304,295 (Feb. 14, 1967).
21. Hock, C. W.: How TiCl₃ Catalysts Control the Texture of As-polymerized Polypropylene, *J. Polym. Sci.*, (A-1) **4**: 3055 (1966).
22. Polypropylene, *Hydrocarbon Process.*, November 1969, p. 230; November 1971, p. 201.
23. Imperial Chemical Industries, Ltd.: Ethylene-Propylene Copolymerization, Neth. Pat. Appl. 6,413,070 (May 12, 1965).
24. Jabloner, H., and H. Mumma: Heterogeneous Grafting of Polypropylene and Physical Blends of Graft Blends, *J. Polym. Sci.*, (A-1)**10**: 763 (1972).
25. Jezl, J. L., and E. M. Honeycutt: Propylene Polymers, in N. M. Bikales (ed.), "Encyclopedia of Polymer Science and Technology," vol. 11, pp. 597–1969, Wiley, New York, 1969.
26. Jezl, J. L., T. J. Kelley, H. M. Khelghatian, and E. M. Honeycutt: Polypropylene, in W. M. Smith (ed.), "Manufacture of Plastics," vol. 1, pp. 194–215, Reinhold, New York, 1964.
27. Junghans, P.: Polypropylene, in "1970–1971 *Modern Plastics* Encyclopedia," pp. 213–215, McGraw-Hill, Inc., New York, 1970.
28. Kelley, R. T.: Polypropylene, *Plast. World*, August 1969, pp. 58–60.
29. Labine, R. A.: Polypropylene by the Ziegler Process, *Chem. Eng.*, June 27, 1960, pp. 96–99.
30. McConnell, R. L., M. A. McCall, G. O. Cash, F. B. Joyner, and H. W. Coover: Interaction of Ethylaluminum Dichloride with Organic Nitrogen and Phosphorous Compounds in Three-Component Polyolefin Catalysts, *J. Polym. Sci.*, A3: 2135 (1965).
31. Miller, R. L.: Crystallinity, in N. M. Bikales (ed.), "Encyclopedia of Polymer Science and Technology," vol. 4, pp. 449–528, Wiley, New York, 1966.

32. Montecatini Societa Generale (inventors, A. V. Kreisler, K. Schönwald, T. Meyer): Verfahren zur Herstellung von Katalystoren für die stereospecifische Polymerisation von α-Olefins, German Pat. 1,253,244 (Nov. 2, 1967).
33. Moretti, G., and V. D'Allessandro (to Montecatini Societa Generale): Process for Preparation of Supported Catalyst for the Stereospecific Polymerization of α-Olefins in Vapor Phase, U.S. Pat. 3,252,959 (May 24, 1966).
34. Natta, G., I. Pasquon, A. Zambelli, and G. Gatti: Highly Stereospecific Catalyst Systems for the Polymerization of α-Olefins to Isotactic Polymers, *J. Polym. Sci.*, **51**: 387 (1961).
35. Natta, G., P. Pino, G. Mazzanti (to Montecatini Societa Generale): Isotactic Polypropylene, U.S. Pats. 3,112,300 and 3,112,301 (Nov. 26, 1963).
36. Norwood, D. D. (to Phillips Petroleum Co.): Method and Apparatus for the Production of Solid Polymers of Olefins, U.S. Pat. 3,248,179 (Apr. 26, 1966).
37. Ogawa, T., Y. Suzuki, and T. Inaba: Determination of Molecular Weight Distribution for Polypropylene by Column Fractionation and Gel-Phase Chromatography, *J. Polym. Sci.*, (A-1)**10**: 737 (1972).
38. Novelen Process for Polypropylene, *Oil Gas J.*, Nov. 23, 1970, p. 64.
39. Park, W. R. R.: "Plastic Film Technology," pp. 44–45, Van Nostrand Reinhold, New York, 1969.
40. Oriented Polypropylene: Strong Contender Enters Bottle Race, *Plast. World*, September 1970, pp. 56–58.
41. Phillips Petroleum Company: Recovery of Solid Olefin Polymers from Slurries Thereof, British Pat. 991,089 (May 5, 1965).
42. Repka, B. C.: Polypropylene, in A. Standen (ed.), "Kirk-Othmer Encyclopedia of Chemical Technology," 2d ed., vol. 14, pp. 282–309, Interscience, New York, 1967.
43. Rexall Drug and Chemical Co.: Improvement in and Relating to the Production of Polyolefins, British Pat. 1,040,669 (Sept. 1, 1966).
44. Rexall Drug and Chemical Co.: Polymerizing an α-Olefin Monomer, British Pat. 1,092,665 (Nov. 29, 1969).
45. Schlegel, W. F.: Design and Scaleup of Polymerization Reactors, *Chem. Eng.*, Mar. 20, 1972, pp. 88–96.
46. Schwaar, R. H., and E. G. Foster (to Shell Oil Co.): Polymerization Process, U.S. Pat. 3,502,633 (Mar. 24, 1970).
47. Scoggin, J. S. (to Phillips Petroleum Co.): Olefin Copolymerization, U.S. Pat. 3,437,646 (Apr. 8, 1969).
48. Scoggin, J. S. (to Phillips Petroleum Co.): Polymerization of Mono-1-olefins, U.S. Pat. 3,454,545 (July 8, 1969).
49. Scoggin, J. S. (to Phillips Petroleum Co.): Olefin Block Copolymerization, U.S. Pat. 3,454,675 (July 8, 1965).
50. Scoggin, J. S. (to Phillips Petroleum Co.): Block Copolymers Prepared by Feeding Batch Prepolymers in a Continuous Reactor, U.S. Pat. 3,525,781 (Aug. 25, 1970).
51. Scoggin, J. S., and H. S. Kimble (to Phillips Petroleum Co.): Separation of Polymers from Liquid Diluents, U.S. Pat. 3,152,872 (Oct. 13, 1964).
52. Trieschmann, H. G., W. Rau, T. Jacobsen, and H. Pfannmueller (to Badische Anilin- und Soda-Fabrik): Catalytic Gas-Phase Polymerization of Propylene, German Offen. 1,805,765 (May 21, 1970).
53. Trieschmann, H. G., W. Rau, H. Mueller-Tamm, and H. Pfannmueller (to Badische Anilin- und Soda-Fabrik), Granulated Polypropylene, German Offen. 1,901,264 (Apr. 16, 1970).

54. Trieschmann, H. G., K. Wisseroth, R. Scholl, and R. Herbeck (to Badische Anilin- und Soda-Fabrik): Propylene Polymers, S. African Pat. 68 05,083 (Jan. 23, 1969).
55. Tusch, R. L.: Properties of Propylene-Ethylene Block Copolymers, *Polym. Eng. Sci.*, **6**: 255 (1966).
56. Vandenberg, E. P. (to Hercules, Inc.): Process and Catalyst for Polymerization of 1-Olefins, U.S. Pat. 3,261,821 (July 19, 1966).
57. Van Looy, H. M., L. A. M. Rodriguez, and J. A. Gabant: Study of Ziegler Catalysts, *J. Polym. Sci.*, (A-1)**4**: 1905–1991 (1966).
58. Walton, R. J.: Polypropylene, in "1969–1970 *Modern Plastics* Encyclopedia," pp. 185–194, McGraw-Hill, Inc., New York, 1969.
59. Winkler, D. E., and K. Nazaki (to Shell Oil Co.): Polymerization Catalysts, U.S. Pat. 2,971,925 (Feb. 14, 1961).
60. Wisseroth, K.: Grundlagen und Verfahrenstechnik der Gasphasenpolymerisation, *Angew. Makromol. Chem.*, **8**: 41–60 (1969).
61. Wisseroth, K., K. Herbert, R. Scholl, and H. G. Trieschmann (to Badische Anilin- und Soda-Fabrik): Gas-Phase Polymerization of Propylene, German Offen. 1,905,835 (Sept. 3, 1970).
62. Youngman, E. A., and J. Boor: Syndiotactic Polypropylene, *Macromol. Rev.*, **2**: 33–69, (1967).

CHAPTER FIVE

Production of Vinyl Chloride

New vinyl chloride processes developed since 1962 have essentially made all earlier ones obsolete and uneconomical. As a result, the selling price of vinyl chloride has decreased significantly from about 10 cents/lb in 1960 to 4.5 cents/lb in the early 1970s. Such a price is available to large customers on long-term contracts. The price reduction for vinyl chloride has in turn been responsible for a significant decrease in the selling price of polyvinyl chloride resins (commonly called PVC resins or polymers). Typical analysis of vinyl chloride as sold is shown in Table 5-1.

Production of vinyl chloride monomer (VCM) has increased rapidly within the last few years. The vinyl chloride production capacity in the United States probably totaled over 4 billion lb/year at the beginning of 1970. Actually the sum of announced capacities was 4.8 billion lb (20), but some companies probably exaggerated somewhat. Actual production of vinyl chloride was over 3 billion lb in 1970, and predicted production for 1980 is 8 to 9 billion lb (33).

Production capacities for vinyl chloride in other parts of the world have also grown rapidly. European production capacity is now greater than that of the United States (72). Japanese capacity was over 2

TABLE 5-1   Typical Analysis of Vinyl Chloride

Component	%	ppm Molar	ppm Weight
Vinyl chloride.................	99.9 min.*		
Butadiene.......................	....	<10	
1,2-Dichloroethane.............	....	<50	
1,1-Dichloroethane.............	....	<50	
Hydrogen chloride.............	....	....	<1.0
Water........................	....	....	<100
Iron.........................	....	....	<1.0
Sometimes the following specifications are added:			
Acetaldehyde................	....	....	<5.0
Peroxides....................	....	....	<0.06
Sulfur.......................	....	....	<3.0

*General specification; 99.98% minimum, however, is required by certain producers.

billion lb/year in 1967 (24) and has undoubtedly grown significantly since then.

Fairly recent trends of vinyl chloride production are as follows:

1. Ethylene is replacing the significantly more expensive acetylene as the major hydrocarbon feedstock. Plants using only acetylene as feedstock are now rather uneconomical in most cases in the United States, but some plants using a combination of ethylene and acetylene are still competitive. Ethane, which is significantly cheaper than ethylene, is used in a process announced in 1971 (21, 25, 50) and likely to be commercialized soon.

2. The manufacturing cost (60) and consequently the selling price of ethylene has also dropped significantly. The selling price on a large-scale basis ranges perhaps from 3.0 to 3.2 cents/lb.

3. To an increased extent, only large plants can be competitive. In the United States, the capacities of new plants must be at least 0.5 billion lb/year (72).

Balanced Processes for Vinyl Chloride

All modern vinyl chloride plants employ ethylene or acetylene as the hydrocarbon feedstock. In 1971, the Transcat process, which employs ethane as a feed, was announced (21, 25, 50). Ethane is significantly cheaper than ethylene and (especially) acetylene. Apparently no commercial plants of this process have yet been built, however. Chlorine or hydrogen chloride is also always a feed material for all vinyl chloride

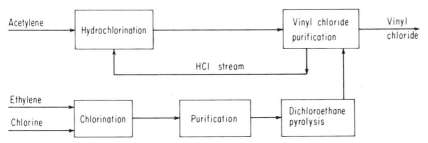

Fig. 5-1 Balanced process using concentrated acetylene, ethylene, and chlorine. [*From (2) with permission of Chemical Engineering.*]

process. Although hydrogen chloride is frequently a waste product from many processes, it is sometimes relatively expensive in an anhydrous state. Yet some vinyl chloride processes cannot use wet hydrogen chloride, or hydrochloric acid. Chlorine is available at about 3 cents/lb.

Numerous process modifications are possible for obtaining vinyl chloride from the above-mentioned reactants. The basic chemical steps are outlined below. In each case, the yields of products and the conversions per pass are high, about 95 percent or greater. Common methods (1, 2) of combining these steps will be outlined first, and a discussion of the commercial units will follow.

Acetylene and anhydrous hydrogen chloride react readily in the presence of mercuric chloride to form vinyl chloride:

$$(1) \quad CH{\equiv}CH + HCl \rightarrow CH_2{=}CHCl$$

Ethylene can first be chlorinated to produce 1,2-dichloroethane, which is then pyrolyzed to obtain vinyl chloride and anhydrous hydrogen chloride:

$$(2) \quad CH_2{=}CH_2 + Cl_2 \rightarrow CH_2Cl{-}CH_2Cl$$
$$(3) \quad CH_2Cl{-}CH_2Cl \rightarrow CH_2{=}CHCl + HCl$$

In this sequence of steps, at most only half the chlorine is incorporated in the vinyl chloride. Several companies who use hydrogen chloride have made vinyl chloride by a combination of reactions 2 and 3. For other companies, the question is how to use the hydrogen chloride, which, at least in the United States, has a lesser value than chlorine.

One of the first solutions was to react the hydrogen chloride with acetylene to form more vinyl chloride (reaction 1). Figure 5-1 indicates the balanced process which has been used for several years. This balanced process is represented essentially by adding reactions 1 to 3 to obtain:

$$(4) \quad CH{\equiv}CH + CH_2{=}CH_2 + Cl_2 \rightarrow 2CH_2{=}CHCl$$

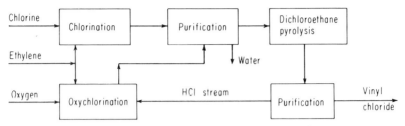

Fig. 5-2   Balanced process using separate oxychlorination step.   [*From* (2) *with permission of Chemical Engineering.*]

The highly successful balanced process shown in Fig. 5-2 was developed in the 1960s, when the following oxychlorination of ethylene to produce 1,2-dichloroethane was perfected:

$$(5)\quad CH\!\!=\!\!CH_2 + 2HCl + \tfrac{1}{2}O_2 \rightarrow CH_2Cl\!\!=\!\!CH_2Cl + H_2O$$

This balanced process (Fig. 5-2) is a combination of reactions 2, 3 (twice), and 5, and is represented essentially as follows:

$$(6)\quad CH_2\!\!=\!\!CH_2 + Cl_2 + \tfrac{1}{2}O_2 \rightarrow 2CH_2\!\!=\!\!CHCl + H_2O$$

A third method for effective utilization of the hydrogen chloride is to oxidize the hydrogen chloride to chlorine by the so-called Deacon process:

$$(7)\quad 2HCl + \tfrac{1}{2}O_2 \rightarrow Cl_2 + H_2O$$

A balanced process involving reactions 2 (twice), 3 (twice), and 7 results in the overall chemistry shown in reaction 6. This balanced process is shown schematically in Fig. 5-3.

Modern balanced processes employing both acetylene and ethylene have also been devised, as shown in Fig. 5-4. In these cases, a hydrocarbon stream is cracked to produce a C_2 fraction containing essentially equal numbers of moles of acetylene and ethylene. The hydrocarbon feed stream to the pyrolysis unit is usually a naphtha, but ethane or

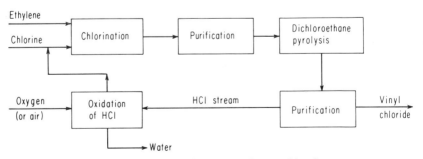

Fig. 5-3   Balanced process using oxidation of hydrogen chloride as a separate step.

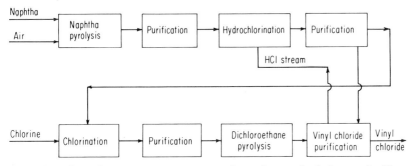

Fig. 5-4 Balanced process using a mixture of acetylene and ethylene with chlorine. [*From* (2) *with permission of Chemical Engineering.*]

propane can be used. In such a process, the pyrolysis products, which are heavier (and perhaps also lighter) than the C_2 fraction, are separated from the C_2 cut. The C_2 fraction is combined first with hydrogen chloride, and vinyl chloride is formed by reaction 1. The vinyl chloride is rather easily separated, and the ethylene is then reacted with chlorine to form 1,2-dichloroethane (reaction 2). Pyrolysis of 1,2-dichloroethane (reaction 3) yields more vinyl chloride.

The overall chemistry for this latter balanced process (Fig. 5-4) is also shown by reaction 4; its major advantage over the one shown in Fig. 5-1 is that the expensive separation of pure acetylene and pure ethylene from a pyrolysis stream is replaced by the rather easy separation of vinyl chloride and 1,2-dichloroethane from the product streams.

Both oxychlorination and chlorination of ethylene can in some cases be practiced in the same reaction vessel. M. W. Kellogg Co. (37, 42) has combined these two steps in a single reactor in their process (see Fig. 5-5). The overall stoichiometry of the process is also represented by reaction 6.

Direct chlorination of ethylene to produce vinyl chloride in high yields was unsuccessful in the past. In 1971, a process using a palladium chloride catalyst at 100 to 130°C (212 to 266°F) was announced (22); the only by-product is vinylidene chloride. Insufficient information is currently available to determine the probable commercial feasibility of the process.

Processes for Hydrochlorination of Acetylene

Plants in which relatively pure acetylene is reacted with hydrogen chloride still produce significant amounts of vinyl chloride. No further plants of this type, once the predominant one, are likely to be built, at least in the United States, since other processes are now more economical.

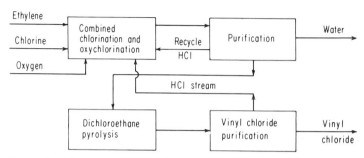

Fig. 5-5 Balanced process with combined oxychlorination and chlorination steps. [*From* (*2*) *with permission of Chemical Engineering.*]

A plant[*] for the production of 30 million lb/year of vinyl chloride has been described in detail (Fig. 5-6). Information on other units is also available (68).

Production of Acetylene   Up to several years ago, production of vinyl chloride accounted for 30 percent of all acetylene manufactured in this country. Since ethylene has been replacing acetylene as a feedstock for vinyl chloride production and also for other processes, it is unlikely that new acetylene plants will be built for several years unless a substantially cheaper method is found.

The more recent acetylene plants have used hydrocarbons as the feedstock instead of the calcium carbide of the older plants (52, 73). Acetylene from these hydrocarbon processes is thought to be considerably cheaper. Claims, based on research information, have been made that acetylene can be produced at from 3 to 5 cents/lb (58). However, several companies using hydrocarbon processes have experienced considerable difficulties and expense in reaching design capacities and yields. Present cost in many if not all units may be as high as 7 to 8 cents/lb. In 1965, about 53 percent of the acetylene produced in this country was by the carbide process, but these plants are gradually being phased out.

Care must be taken in handling acetylene because it can detonate under some conditions. General safety procedures include (52):

1. Avoidance of alloys or solders containing copper, silver, or mercury
2. Using small-diameter piping, up to 6 in ID, at relatively low pressures and with flame arresters
3. Compression by means of centrifugal, liquid-sealed compressors

---

[*] Details on this particular plant were presented by General Tire and Rubber Co. at a seminar sponsored by AIChE at Akron, Ohio, Oct. 16, 1964. R. R. Mattiko and P. R. Sayre were responsible for presentation of the material on vinyl chloride synthesis.

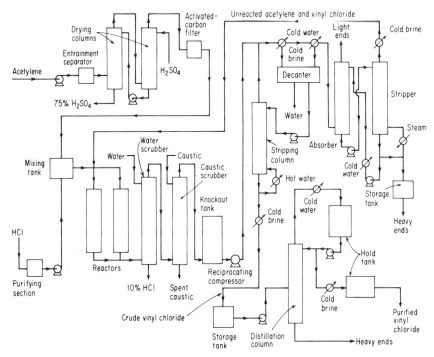

Fig. 5-6 Flowsheet for commercial hydrochlorination of acetylene to produce vinyl chloride. [*From* (*1*) *with permission of Chemical Engineering.*]

When hydrocarbon streams, e.g., the effluent stream from a naphtha cracker, contain only a small fraction of acetylene, less stringent safety precautions are necessary.

Mechanism of Hydrochlorination   Mercuric chloride, the usual catalyst for hydrochlorination, reacts with acetylene to form the intermediate compound, *trans*-2-chlorovinylmercuric chloride (56).

$$(8) \quad CH\equiv CH + HgCl_2 \rightarrow \begin{array}{c} Cl \quad H \\ | \quad\quad | \\ C=C \\ | \quad\quad | \\ H \quad HgCl \end{array}$$

This compound reacts with hydrogen chloride to form vinyl chloride and to regenerate mercuric chloride. As indicated above, acetylene is chemisorbed on the catalyst. Vinyl chloride and possibly hydrogen chloride are also strongly adsorbed on the catalyst (80). The controlling chemical step is probably the surface reaction between adsorbed reactants. At start-up, hydrogen chloride should contact the catalyst bed before the acetylene flow is slowly started (54).

Mercuric chloride is, as a rule, impregnated on activated-carbon pellets. Transfer of the acetylene and hydrogen chloride to the catalyst surface and of vinyl chloride away from the surface are important steps of the reaction sequence. The surface available in the carbon pellets affects the resistance of the transfer steps, and one or more of these steps partly control the kinetics of the overall reactions.

Feed Purification   Purification of the acetylene is of major concern and is often a significant factor in capital and operating expenses. Moisture, higher acetylenes, sulfides, and phosphorous and arsenic hydrides are all highly deleterious (68).

In the plant used by General Tire and Rubber Co. (1), acetylene at about 1.5 atm absolute pressure is first compressed to 2 atm with a Nash compressor using water as the sealing medium. Acetylene is then purified by passing it through an entrainment separator that contains refrigerated water. The temperature of the acetylene is lowered to 59°F (19°C), and a considerable amount of water is condensed and removed from the gas. A rotary positive-displacement meter measures the gas flow to within 0.7 percent.

Next the acetylene is scrubbed in two columns with sulfuric acid to complete the drying. The first scrubber uses relatively dilute acid and is 2 ft in diameter by 24 ft high. The steel tower is lined with acid brick. The second tower uses a more concentrated acid and is 1.6 ft in diameter by 22 ft high. Each column is packed with 2-in Raschig rings. The acid in each column is recycled until its strength has dropped to a low level, when it is replaced with strong acid. An external heat exchanger maintains the temperature of the recycle acid at ambient conditions. (This part of the process is not shown in Fig. 5-6.) Acetylene from the second scrubber is passed through an activated-carbon filter to remove traces of catalyst poisons such as sulfides. The filter is 3 ft in diameter by 7 ft high. A second filter is provided as a standby after the first filter has been depleted.

The purifying section for HCl eliminates water, chlorine, or chlorinated organic materials that may be present from previous chlorinations. Refrigerated condensers are generally used. Purified HCl is then compressed to about 2.25 atm absolute using a reciprocating compressor. The HCl at 158°F (70°C) is then passed through a separator tank to remove any entrained liquid. An automatic ratio controller maintains the desired molar ratio of acetylene to HCl, usually about 1:1 although slight variations are sometimes provided. The HCl and acetylene streams are combined in a mixing tank.

Reactor Design   Five reactors are used in the 30 million lb/year plant, but only two are shown in Fig. 5-6. The five reactors are arranged as follows: two in parallel, followed by two more in parallel, and then the last one.   Each reactor is essentially a shell-and-tube heat exchanger containing 200 tubes, each 2 in ID by 16 ft long.   Bigger reactors used in other plants contain up to 3,000 tubes, which may have diameters as large as 3 in (68).   The steel tubes are packed with activated-carbon pellets impregnated with about 10 weight percent mercuric chloride. The reaction is highly exothermic (about 24,500 cal per gram mole of vinyl chloride).   Heat-transfer fluid on the shell side maintains the desired temperatures.

The temperature used in the reactor varies from about 90 to 140°C (194 to 284°F), depending on the age and condition of the catalyst. Lower temperatures are used with fresh catalyst.   At higher temperatures, mercuric chloride begins to sublime from the hotter inlet of the reactor (where most of the reaction occurs) and then recondenses near the cooler outlet.   As a result, the locations of the highest rates of reaction vary somewhat (79).   The suggestion has been made that the life of the catalyst could be increased if the direction of flow to a reactor were occasionally reversed, but it is not known if such a technique is used commercially.   In such a case, the mercuric chloride would tend to move from one end of the reactor tube to the other.   Maintaining catalyst temperatures of less than 120°C gives little or no sublimation loss (5), and the amount of vinyl chloride produced per pound of catalyst consumed is much greater than it is at higher temperatures.

Temperatures of the catalyst in the tubes also vary radially because of the cooling fluid on the shell side of the tubes.   The temperature and rate at which the cooling fluid is pumped through the shell of the reactor affects the temperature gradients in the reactor.   A batch of catalyst has a life from about 6 months to somewhat over 1 year, depending on the catalyst poisons present and on the effectiveness of temperature control (68).

Pressures in the General Tire reactors are maintained at 1.5 to 1.6 atm absolute, but other reactors sometimes operate at lower pressures (54).   In such cases, explosive dangers of acetylene are minimized, and better temperature control can be maintained since reaction rates are decreased.

Oil is used as the cooling fluid for the process described here, but presumably other heat-transfer liquids would work satisfactorily.   A water-cooled heat exchanger was used several years ago to cool the oil, but now an air-cooled one is used.   At start-up, the oil is heated with steam to raise the catalyst to reaction temperatures.

About 98 to 99 percent conversions of each reactant occur in the reactor.   The product stream is primarily vinyl chloride containing trace amounts of trichloroethylene, dichloroethylene, and aldehydes.   In addition, small amounts of unreacted acetylene and HCl are present.   A continuous infrared on-stream analyzer monitors the HCl content of the exit gases. Kinetic information of the hydrochlorination reaction has been presented by Wesselhoft et al. (80). A mathematical model based on these kinetics and on catalyst decay reactions has been devised for optimizing the operation of the reactor unit (59). In the reactor units used by General Tire and Rubber Co., faster reaction rates are obtained, however, presumably because of a more active catalyst.

Recovery and Purification Sections   Exit gases from the reactor are fed directly into the bottom of the water scrubber, which is packed with 2-in Raschig rings and built of brick-lined steel. Cooling water (absorbent) is recirculated several times countercurrent to the gases, but this feature is not shown in Fig. 5-6. The exit solution from the scrubber contains from 3 to 10 percent HCl. As the concentration increases, more HCl is carried over to the caustic scrubber, also packed with Raschig rings. The gases leave the scrubbers essentially free of HCl and flow through a suction knockout tank to remove entrained liquids. The gases are then compressed to about 7 atm absolute.

These gases are next cooled with normal cooling water and then cold brine to condense most of the vinyl chloride, water, and chlorinated hydrocarbon by-products. A decanter separates the water layer from the organic layer, which is fed to a stripping column. The column contains 20 bubble-cap trays and is 27 ft high. The bottom of the column is maintained at 115°F (46°C), using hot water in the reboiler of the stripper. Hot water is used instead of steam in order to obtain lower surface temperatures and minimize decomposition of vinyl chloride or unstable by-products. The bottom product of the stripper is crude vinyl chloride, which after cooling with brine is fed to a storage tank. The top gaseous product of the stripper is combined with the exit gas stream from the knockout tank. Mild steel is used for construction of the column, but a nonferrous metal is used for the reboiler tubes (68).

The uncondensed gases from the brine cooler, primarily acetylene, vinyl chloride, and inert gases, are fed into the bottom of an absorber, where trichloroethylene, a by-product of the process, is used as the solvent. The absorber operates at about 4 atm absolute and 86°F (30°C). The stripper employed in conjunction with the absorber operates with a reboiler temperature of about 194°F (90°C) and 2.25 atm. Both columns are packed with 1-in saddles. The vent gases from the absorber contain 90 percent inerts, and the remainder is primarily acetylene.

The gas stream from the stripper is cooled with brine to condense most of the trichloroethylene. Uncondensed gas contains 40 percent vinyl chloride plus 20 percent acetylene, representing most of the unreacted acetylene. These gases, recycled to the reactors, also contain small amounts of chlorinated hydrocarbons, primarily trichloroethylene. As additional trichloroethylene (and other chlorinated by-products) are recovered, they are fed to the heavy-end storage tank.

Chlorinated organic materials and aldehydes are removed from the vinyl chloride in a distillation column that contains 30 bubble-cap trays and is 48 ft high. It is operated at 4.0 to 4.5 atm absolute so that a high-purity vinyl chloride overhead stream (Table 5-1) can be condensed at 95°F (35°C) with normal cooling water. Feed to this column enters at the twentieth tray from the bottom. The top trays and parts of the equipment contacting vinyl chloride are built from stainless steel to minimize iron contamination of vinyl chloride. Hot water is used as the heating fluid in the reboiler to minimize decomposition of the bottom product, which is discarded.

For every 2,000 lb (32 lb mol) of vinyl chloride produced, the following are required: 1,183 lb (32.5 lb, mol) of HCl and 877 lb (33.7 lb mol) of acetylene. Heavy ends produced are 23 lb (0.18 lb mol if they are assumed to be pure trichloroethylene).

Modified Hydrochlorination Processes Nair (55) has described a Czechoslovakian plant that produces about 5 million lb/year of vinyl chloride. The hydrochlorination reactor contains about 1,000 tubes, each 5 cm in diameter, 3 m long, and packed with catalyst. Calculations indicate that the rate of production for a given bulk volume of the catalyst is only about one-quarter to one-third that in the General Tire process.

A hydrochlorination reaction is also part of several balanced processes that use a feed mixture containing only relatively dilute concentrations of acetylene. Balanced processes of the type in Fig. 5-4 are in this category. The feed stream for a plant using the process developed by Kureha Chemical Industry Co., Ltd., of Japan, is reported to contain 9.1 percent acetylene (38). In this process, the partial pressure of acetylene was less than 0.6 atm. Hence the total pressure in the reactor was less than 6.6 atm. Temperatures for the reactor were specified as 248 to 356°F (120 to 180°C).

The Kureha process is well instrumented and is computer-controlled (79). Acetylene content of the feed stream to the hydrochlorination unit is continuously measured by a stream analyzer (38). Hydrogen chloride flow is metered to provide about 1 percent less than the stoichiometric amount. Yields of product are 95 to 98 percent based on the acetylene in the feed stream and more than 99 percent based on hydrogen chloride.

The design for the hydrochlorination reactor of the Kureha process (38) is similar to the one just described. Temperature control with a dilute acetylene feed stream is easier than with essentially pure acetylene. As a result, at least one of the following modifications seems possible:

1. Large-diameter tubes for the packed catalyst bed.
2. Higher gas temperatures since diluents in the gas stream minimize local hot spots in the catalyst bed. Gas temperatures probably are more uniform throughout the catalyst bed.

Somewhat longer reactor tubes and lower space velocities may be required to obtain essentially complete reaction with the more dilute acetylene streams.

The catalyst used in the Kureha process (38) is described as a "mercuric chloride catalyst" and is probably similar to those used with pure acetylene. Reactions with ethylene, carbon oxides, hydrogen, and other gases in the feed stream (from the naphtha cracker) are "hardly observed."

Although fewer details of the reactor design are available for other balanced processes of the type shown in Fig. 5-4, presumably the reactors are or could be similar to those of the Kureha process. Other companies with balanced processes of the type shown in Fig. 5-4 include Société Belge de l'Azote (SBA) (11, 12, 13) and Union Carbide Corp., which uses a Wulff furnace for pyrolysis of the hydrocarbon feedstock.

Storage of Vinyl Chloride   Vinyl chloride is stored as a liquid, generally at less than 60°F (15°C) in order to prevent polymerization or other undesired reactions (67). The factors to be considered during storage will be discussed in Chap. 6.

Processes for Chlorination of Ethylene

Chlorination of ethylene to produce 1,2-dichloroethane is a key step in all modern vinyl chloride plants. Liquid-phase processes are commonly used with hydrocarbon feed streams containing either high or low concentrations of ethylene. Vapor-phase processes are also used in a few commercial units that have feeds with low concentrations of ethylene.

Fundamentals of Liquid-Phase Chlorinations   Liquid-phase chlorinations are operated from about 40°C (104°F) up to perhaps 70°C (158°F), but temperature control is generally not critical (79). Pressures vary up to about 4 to 5 atm, depending mainly on the temperatures used and the purity of the feedstocks. The chlorine and ethylene feed streams are bubbled upward through the liquid, which is predominantly 1,2-dichloroethane. This bubbling, plus some vaporization of dichloroethane, promotes agitation in the liquid.

Ferric chloride is dissolved in the liquid solution in small concentrations, generally from 0.02 to 0.2 weight percent (68). The chlorine and ferric chloride form a complex, $Cl^+FeCl_4^-$, which then reacts with ethylene to produce $CH_2ClCH_2^+FeCl_4^-$. This latter complex reacts with a chlorine molecule, giving 1,2-dichloroethane and regenerating $Cl^+FeCl_4^-$. Ferric chloride can either be fed to the reactor or formed in situ by chlorination of iron if either steel reactors or iron packing is employed.

Overchlorinated ethanes such as 1,1,2-trichloroethane are formed particularly at higher reaction temperatures. Some oxygen in the feed stream, however, minimizes these by-products, which are probably formed by free-radical reactions. Some chlorine radicals are presumably formed at the wall or on the packing (8, 9, 68).

Chlorination occurs primarily in the liquid phase, and ethylene and chlorine are transferred from the gas phase and dissolved in the liquid. Ethylene transfer often is the rate-controlling step. The following analysis is applicable to each reactant, specified here as R:

$$(9) \quad \begin{pmatrix} \text{Net transfer} \\ \text{of R from gas} \\ \text{to liquid phase} \end{pmatrix} = \begin{pmatrix} \text{Net reaction} \\ \text{of R in the} \\ \text{liquid phase} \end{pmatrix} + \begin{pmatrix} \text{Accumulation} \\ \text{of R in} \\ \text{liquid phase} \end{pmatrix}$$

Flow of dissolved reactants from the reactor is small and is ignored in the above equation. For a differential volume dV of the reactor

$$(10) \quad [k_G A (C_R^* - C_R)] \, dV = k[f(C_R, C_{\text{other}})] \, dV + \frac{dC_R}{dt} \, dV$$

where $k_G A$ = mass-transfer coefficient from gas to liquid phase
C_R = concentration of R in liquid phase
C_{other} = concentration of other reactants in liquid phase
C_R^* = equilibrium concentration of R in liquid phase if no reaction were occurring
k = forward rate constant for chlorination

At steady-state conditions in the reactor, the accumulation term is zero. The chlorination reactions are essentially irreversible at reaction conditions; hence only the forward reaction has to be considered. In Eq. (10), the exact form of the kinetic equation is unknown, but it is proportional in some manner to the dissolved concentrations of reactants.

Equation (10) or its integrated form is most useful for analyzing the effect of various operating variables on both the transfer of components between phases and the rate of chlorination in the liquid phase.

1. Temperature affects the forward rate constant k for the chlorination reaction and indirectly the dissolved concentrations of each reactant,

C_R and C_{other}.   The equilibrium concentration of the reactant C_R^* varies somewhat with temperature.   The surface tension and viscosity of the liquid change with the temperature, and as a result the interfacial area A changes with temperature as the gas bubbles upward through the liquid; hence $k_G A$ varies too.

2. Agitation and the type of bubbling affect $k_G A$ and hence indirectly C_R.

3. Total pressure of the system and composition of the reactant gases affect C_R^*.

4. Composition of the liquid phase affects C_R^* and also factors that control mass-transfer resistances.

High dissolved ratios of ethylene to chlorine are desired in order to minimize formation of by-products that are mainly trichloro- and other polychlorinated ethanes (6, 41).   A better understanding of the reaction would be possible if the solubilities of the reactants in the liquid phase were known.   In most cases, the actual solubility C_R is probably much lower than the equilibrium solubility C_R^*; that is, mass-transfer resistances are partially controlling.

Commercial Liquid-Phase Chlorination   Liquid-phase chlorinations are used in numerous plants (14, 15, 38, 43, 44, 61, 68, 79).   The feed gases (ethylene, chlorine, and a small amount of air) are normally introduced through spargers located several feet deep in the liquid.   These spargers should be designed to promote high agitation at the feed point and to minimize high local concentrations of chlorine.   Production capacities per cubic foot of liquid in commercial reactors vary from about 23 to 55 lb/h of dichloroethane.   Temperatures in the tower reactors are controlled either by internal or external heat exchangers or by vaporization of part of the liquid, which is primarily 1,2-dichloroethane. In the latter case (see Fig. 5-7), the overhead vapors are cooled, causing most of the dichloroethane and chlorinated by-products to condense. Part of this condensate is then returned to the reactor as reflux.   Off-gases from the reactor contain unreacted ethylene, chlorine, and air. The airflow to the reactor is adjusted so that the gas stream is not in the explosive range.

When heat exchangers are used for temperature control of the reactor, the reactor can easily be operated at temperatures sufficiently low to ensure that the off-gases from the reactor contain relatively little dichloroethane.   At such temperatures, perhaps $40°C$ or less, fewer by-products are formed.   The product stream from the reactor is liquid dichloroethane containing chlorinated by-products, ferric chloride, and dissolved ethylene and chlorine.

When the product is allowed to vaporize (as shown in Fig. 5-7), however, temperatures in the reactor are usually higher, perhaps 60

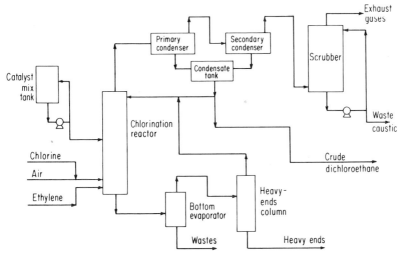

Fig. 5-7 Flowsheet for liquid phase chlorination of ethylene to produce 1,2-dichloroethane.

to 70°C, depending on the pressure, inert gases in feedstock, and the amount of heavy ends that build up in the reactor. Somewhat more overchlorinated materials are produced at these higher temperatures, but an advantage of this technique is the absence of ferric chloride in the overhead product stream. Utility and equipment costs are also somewhat lower with this process modification (68).

Figure 5-7 shows the basic flowsheet commonly used in numerous plants. Two or more condensers condense the dichloroethane and other chlorinated by-products. The first is normally water-cooled, and a refrigerant is used in the second condenser so that essentially all dichloroethane is liquefied and recovered. Uncondensed gases are scrubbed with caustic to remove unreacted chlorine. Part of the condensate is refluxed, and the remainder is fractionated (not shown in Fig. 5-7) in a series of towers to remove light and heavy components and to obtain high-purity 1,2-dichloroethane.

A small liquid stream that is withdrawn from the bottom of the reactor contains mainly dichloroethane contaminated mostly with heavy ends and ferric chloride. The heavy-end content in this liquid can be allowed to build up to a rather significant amount. This bottom stream is separated using a bottom evaporator and a heavy-ends column into a waste stream (containing predominantly heavy ends and ferric chloride), a heavy-ends cut, and dichloroethane (contaminated to some extent with light ends). The dichloroethane stream is recycled to the reactor.

Corrosion is always a problem in the reactor, and feed streams must be dried to less than 30 ppm (68). Steel reactors last 2 to 5 years, and ferric chloride, if used, is then generated in situ. Nickel-based nonferrous alloys, however, are generally used, and significantly longer reactor life can then be expected. In the latter case, makeup ferric chloride is provided from a catalyst mix tank.

Both the Kureha (43, 79) and the Dianor (14, 61) processes employ low-purity ethylenes. Impurities include hydrogen, carbon oxides, methane, and ethane, which although chemically inert act as stripping agents in the reactor. A feed containing only 40 to 45 percent ethylene was reported as being suitable for the Dianor process. In the Kureha process, about 0.96 to 0.98 mol of chlorine is introduced per mole of ethylene (43, 79). The reactor operates at 4 to 5 atm and at 50 to 70°C (122 to 158°F). About 99 percent of the chlorine reacts, compared with 95 to 98 percent of the ethylene. Side reactions are negligible.

A small and now rather obsolete reactor that was packed with iron Raschig rings and cooled using an external heat exchanger has been described (1, 55). The iron packing served two purposes: it maximized mass transfer between the gas and liquid and produced iron chloride catalysts.

Fundamentals of Vapor-Phase Chlorinations   Ethylene is chlorinated in the vapor phase by a free-radical chain mechanism at temperatures from about 90 to 130°C. The initiating step of the overall reaction is probably the rate-controlling one, and chlorine free radicals are produced:

$$(11) \quad Cl_2 \rightleftharpoons 2Cl\cdot$$

Propagation steps involving both ethylene and chlorine molecules are as follows:

$$(12) \quad CH_2{=}CH_2 + Cl\cdot \rightarrow CH_2Cl{-}\dot{C}H_2$$

$$(13) \quad CH_2Cl{-}\dot{C}H_2 + Cl_2 \rightarrow CH_2Cl{-}CH_2Cl + Cl\cdot$$

The free-radical chains terminate when two free radicals combine or when free radicals collide and react at the reactor wall.

The walls of the reactor and/or a catalyst such as iron (probably an iron chloride formed in situ) or calcium chloride have a pronounced effect on the overall reaction (28, 49, 64, 70). Semenov (65) and others have postulated that the walls of the reactor frequently act both to initiate and to terminate the free-radical chains. Bernstein, Fuqua, and Albright (8, 9) confirmed this and found that kinetic equations for chlorination are much more complicated than reported by past investigators because of the reactions occurring both in the gas phase and at the solid surfaces in the reactor. Chlorine free radicals attack many surfaces, especially at higher temperatures of about 200°C. Surfaces which are attacked include steels, Pyrex glass, polytetrafluoroethylene, and even tantalum. Nickel seems to be relatively inert.

In addition to the type of surface, the amount and temperature of the surface are also important. The role of the surface is complicated especially if it is porous or rough; hence the controlling factors may be both mass transfer and heat transfer to it. Rust and Vaughan (64) found that the rate of chlorination is directly proportional to the chlorine concentration and to the square of the ethylene concentration in the gas phase. Increased surface areas caused faster chlorinations.

A pressure of about 7 to 10 atm is generally used to obtain almost complete condensation and recovery of dichloroethane, using water as coolant. Temperature control may be a problem since the reaction is highly exothermic (about 39 kcal per gram mole of product). A large excess of ethylene or an inert gas is generally used for this reason. Excess ethylene also minimizes the formation of overchlorinated products.

Equipment for mixing chlorine and ethylene should provide quick and intimate mixing so as to minimize local concentration gradients. The temperature of each stream should be carefully controlled; otherwise a runaway or even explosive mixture may occur.

When the hydrocarbon stream contains ethylene in rather low concentrations, as in the SBA process (11, 12), the temperature can be controlled without using large, excess amounts of ethylene since adequate inert gases are already present. Some excess of ethylene is normally provided to reduce unreacted chlorine in the process. Propylene and propane in the hydrocarbon feedstream should always be minimized since their chlorinated products are difficult to separate from 1,2-dichloroethane.

Commercial Vapor-Phase Chlorination Figure 5-8 is the flowsheet for a vapor-phase chlorination of ethylene in a dilute hydrocarbon stream, as in the SBA process (1). Liquid chlorine from a storage tank is pumped to a heat exchanger, where it is vaporized and heated

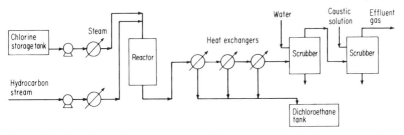

Fig. 5-8 Flowsheet for vapor-phase chlorination of ethylene to produce 1,2-dichloroethane. [*From* (*1*) *with permission of Chemical Engineering.*]

to perhaps 175 to 195°F (80 to 90°C). The hydrocarbon feed stream is also adjusted to a similar temperature in a heat exchanger. The two streams are combined at the inlet of the reactor.

Reactors that are essentially heat exchangers with the catalyst packed inside the tubes have been used. Steel fins inside the tubes may sometimes be used to provide sufficient surface to generate catalyst for the reaction. Such finned surfaces provide good heat transfer and promote turbulence in the fluids. The fins may be coated in some cases with calcium chloride or lead chloride (41). As a result, reaction rates as high as the 0.3 lb mol/(h)(ft³) of dichloroethane or higher seem probable with temperatures of about 255°F (124°C).

Selectivity to dichloroethane of at least 93 percent, based on ethylene, is possible; but the selectivities decrease with increased temperature (70). The reactors are cooled with water, and pressure must be sufficiently low to allow complete vaporization of all products and reactants. The pressures used depend on the amount of inerts in the feed stream.

Exhaust gases from the chlorination reactors are cooled in two or more heat exchangers in series to about −5°F (−20°C) to condense dichloroethane. The uncondensed gases are contacted in two scrubbers, first with water and then with caustic, to remove residual chlorine or any HCl produced. The remaining gases are probably used as fuel.

The condensed product, primarily 1,2-dichloroethane plus small amounts of polychlorinated compounds and ethyl chloride, is stored temporarily. It is then fractionated to produce dichloroethane 99.7 to 99.95 percent pure (68), which is later cracked to produce vinyl chloride and HCl.

Advantages of Liquid-Phase Processes Liquid-phase processes are preferred over gas-phase ones, especially if concentrated ethylene feeds are used, for the following reasons.

1. Better use is made of the heat of chlorination. Relatively cold reactants are introduced into the reactor, and often the heat of reaction provides the energy to vaporize the dichloroethane (10). With this technique, little or no iron chloride catalyst is carried over, and purification steps for the product are simplified.

2. Temperature control of the reactor is better. Large excess of liquid, good agitation of the liquid, and excellent heat transfer between gas and liquid maintain isothermal conditions in the reactor.

3. Safety is increased because chlorine and ethylene do not have to be premixed.

4. Large excesses of ethylene or an inert gas are not needed for adequate temperature control. This advantage is particularly important when essentially pure ethylene and chlorine are used as reactants.

The reactor for a liquid-phase process is generally simple in design and operation.

Oxychlorination Processes

Oxychlorination processes in which ethylene is reacted with hydrogen chloride and oxygen to produce 1,2-dichloroethane were developed in the 1950s and 1960s, and they are now widely used throughout the world (2, 14, 15, 16, 24, 26, 32, 43, 68). Some plants have already been expanded, and additional plants are under construction. Various oxychlorination processes use different ratios of reactants, catalysts, and methods of contacting the reactants with the catalyst. In some processes oxychlorination and chlorination occur in a single reactor. Oxychlorination processes for light hydrocarbons such as ethane have also been announced. These latter processes, which are also useful for the production of vinyl chloride, will be discussed later.

Mechanism of Oxychlorinations Copper chloride catalysts, really a modified Deacon process catalyst, seem to be used exclusively for the oxychlorination of ethylene. The support material for the solid catalysts is of considerable importance and is the subject of numerous patents. The basic chemistry of the process has been reported as follows (32, 34–36, 40):

$$(14) \quad 2CuCl_2 + C_2H_4 \rightarrow CH_2Cl—CH_2Cl + Cu_2Cl_2$$
$$(15) \quad 2Cu_2Cl_2 + \tfrac{1}{2}O_2 \rightarrow CuO \cdot CuCl_2$$
$$(16) \quad CuO \cdot CuCl_2 + 2HCl \rightarrow 2CuCl_2 + H_2O$$

Carrubba and Spencer (18) have presented preliminary data suggesting that ethylene oxide may be an intermediate product. They showed in addition that ethylene oxide reacts to at least some extent with HCl to form 1,2-dichloroethane.

A cupric content of approximately 20 percent or greater is needed in the salt mixture in order to obtain the desired reactions at temperatures as low as $250°C$ $(482°F)$ or even less. Potassium chloride or other alkali-metal chloride is generally added to the catalyst. Fontana et al. (35) have shown that melting points of mixtures of potassium chloride, cupric chloride, and cuprous chloride are as low as $150°C$, whereas the melting point of the eutectic mixture of cupric and cuprous chlorides is $375°C$. At operating conditions, most (if not all) oxychlorination catalysts apparently contain a liquid phase, probably partly adsorbed on the surface. The molar ratio of copper to potassium atoms must be greater than $0.5:1$ in order to obtain a significant reaction. This ratio is probably related to the melting point of the salt mixture.

Oxygen adsorption on the catalyst is of importance. The rate of adsorption increases significantly as the cupric oxide content increases.

Actually, cupric oxide may tend to precipitate from the melt on the surface. When potassium chloride is replaced with sodium, calcium, or lead chloride, the rates of oxygen adsorption decrease. The solid cupric oxide reacts with hydrogen chloride to regenerate the cupric chloride.

Ethylene, oxygen, and HCl must be transferred to the catalyst surface. Although sufficient data are not available to determine with certainty which step in the oxychlorination process is rate-controlling, adsorption (or absorption) or oxygen may be. Such a step seems to be rate-controlling for oxychlorination of methane (36), but oxychlorination of ethylene may involve a different mechanism than that for light paraffins (34).

Temperature has an important effect on the following features of the overall oxychlorination process:

1. True kinetics of the reactions
2. Melting point and viscosity of the surface chloride salts
3. Solubilities or adsorption of reactants on the surface

As low operating temperatures as possible are desired for longer catalyst life, higher yields of dichloroethane (since undesired cracking of the product is minimized), and less corrosive conditions.

Alumina, silica, and other porous materials are apparently preferred supports for the chlorides when solid catalysts are used. These materials form complexes to some extent with the chlorides and may reduce their volatility. In some cases, significant amounts of the chloride salts have been lost from the catalyst by volatilization, thus reducing the catalyst activity. Size and porosity of the support are of considerable importance to the overall performance of the catalyst (30), but little reliable information is available.

Control of Operating Variables   The oxychlorination reaction is highly exothermic, being about 49,835 Btu per pound mole of dichloroethane. Additional exothermicity, perhaps up to 6 percent more, is realized because of side reactions, including oxidation and chlorination. Oxidation reactions usually account for 2 to 8 percent of the ethylene that reacts (68). Heat release is on the catalyst surface, and temperature control of the catalyst is always an important design consideration. Part of the heat of reaction is sometimes used to preheat the reactants, and up to about 47,500 Btu/lb mol can be used to generate steam.

Inert gases in the feed streams to the reactor always help moderate the temperatures in the reactor and act to minimize hot spots on the catalyst surfaces. Such inert gases can be introduced in several ways. Air, which is apparently used in all commercial units for supplying oxygen, of course contains nitrogen. Some hydrogen chloride feed-stocks contain steam, and inert paraffins such as ethane are in some impure ethylene feed stocks. Relatively pure ethylene is required in certain processes (76), however.

Materials with high thermal conductivities, e.g., graphite, silicon carbide, or nickel, are sometimes mixed with the solid catalyst to improve the overall thermal conductivity of packed catalyst beds. Hot spots in the bed are most critical near the feed point. Probably the chloride salts slowly volatilize and migrate away from the inlet as the bed is used. Reversing the direction of flow in the reactor might occasionally be beneficial.

Oxygen has been considered as a substitute for air in order to obtain faster reaction rates and higher conversions for a given reactor, assuming that adequate temperature control could be maintained. Oxygen provides a higher driving force for adsorption (or absorption) on the solid catalyst. Relatively pure oxygen also simplifies the recovery portion of the process. Costs of the oxygen and increased problems in temperature control, however, are major disadvantages. To date, oxygen has apparently not been used commercially.

The exact operating conditions used in commercial reactors are not publicized, but wide variations are probable, depending on the catalyst and feed. Temperatures of 200 to 300°C (392 to 572°F) seem common, but lower ones have been reported in patents. Shell (66) has patented a catalyst containing rare earths and a silica gel having an unusually large surface area (200 m^2/g). This catalyst is reported to be very active at about 250°C. Catalysts that operate at such temperatures generally have a longer life than those operated at higher temperatures because volatilization and subsequent loss of catalyst are a major cause for catalyst deactivation.

Commercial operating pressures are in the range of about 2 to 14 atm (68), the pressures used depending to some extent on whether the catalyst is fluidized or used as a packed bed. Since fluidization depends on the volume of the gas used, lower pressures are preferred. Higher pressures, which facilitate condensation and recovery of the product streams, are more likely with a packed-bed reactor. When inert gases are present, higher pressures may be used for comparable partial pressures of reactants.

Feed streams normally have ratios of ethylene, hydrogen chloride, and oxygen in essentially the stoichiometric ratio of 1:2:0.5. Slight excesses of any one of the three reactants are possible. Excess HCl results in some HCl in the product but in essentially complete conversions of ethylene. In such a case, an aqueous solution of HCl will be obtained upon condensation. When a slight excess of ethylene is employed, however, almost complete conversion of hydrogen chloride generally results. Companies that can use or sell an aqueous solution of hydrogen chloride probably will use a feed stock containing a slight excess of hydrogen chloride.

Conversions per pass of 95 percent or higher are probable in most commercial reactors with essentially stoichiometric ratios of reactants. Overall yields of 1,2-dichloroethane are often about 95 percent, or possibly slightly higher, based on the entering ethylene and HCl. Some losses are caused by incomplete reaction and incomplete recovery of unreacted feeds. By-products include 1,1,2-trichloroethane, ethyl chloride, and other polychloroethanes. Hexachloroethane is produced in small quantities when solid catalysts are used. This compound is highly undesirable in the pyrolysis portion of the vinyl chloride process and, further, is difficult to remove in the distillation part of the process.

The M. W. Kellogg Co. (47) claims that its process, using an aqueous copper chloride catalyst, gives yields of 96 to 98 percent based on the ethylene and HCl, respectively, and only insignificant amounts of hexachloroethane are produced. B. F. Goodrich (26) indicates that it included in its commercial plant a separate distillation unit to remove by-product chlorinated hydrocarbons. The amount of by-product actually produced is so low that this unit will not be incorporated in any future plants. Both the Kellogg process and the Goodrich process (using fluidized beds) have good temperature control, which is important for minimizing by-products.

About 2 to 5 percent of oxygenated products and carbon oxides are present in the product stream of most oxychlorination processes. These compounds must be separated from dichloroethane since they interfere with the pyrolysis operation.

More information is needed before the factors affecting conversions and yields of dichloroethane can be completely understood. Increased conversions occur at higher temperatures, higher partial pressures of reactants, and lower space velocities, i.e., longer residence time in the reactor. High yields are obtained at lower temperatures. Dichloroethane seems to suppress conversions more than would be expected from thermodynamic considerations.

Packed-Bed Reactors Several commercial processes use reactors that are essentially shell-and-tube heat exchangers in which the catalyst is packed in the tubes (15), which have internal diameters of 2 in or less (70). Temperature inside the tubes is controlled to within 230 to 290°C (68) using water as the coolant, and 100 to 150 lb/in^2 gage steam is generated (37). Variables helping to maintain temperature control include high space velocities, inert diluents in the gas stream, and dilution of solid catalyst with inert granular solids (68).

With good temperature control, the catalyst will last about 1 year (75). Since chloride catalysts are relatively volatile, the suggestion has been made of adding some cupric chloride as a vapor in the feed (45). Otherwise, the catalyst migrates toward the exit of the reactor tube, and reactor temperatures slowly vary with catalyst life.

Commercial reactors often have reactor tubes about 1.5 in ID and 20 ft long. A reactor of 7 ft diameter designed to produce 120 million lb/year of dichloroethane has actually been used at rates of 140 million lb/year for rather extended periods. Two such reactors are shown in Fig. 5-9.

At shutdown of a reactor, hydrogen chloride and steam are flushed from the system to minimize corrosion. Nickel, copper-nickel alloys, and high-nickel or high-chromium steels are suitable for the reactor tubes.

Relatively few kinetic data have yet been published for commercial reactors. Shelton et al. (68) report that temperature control limits the productivity to about 250 lb or less of dichloroethane per day per cubic

Fig. 5-9 Vinyl chloride plant showing two packed-bed oxy-chlorination reactors. (*Vulcan Materials Company.*)

foot of reactor volume. Higher rates have been reported in a Vulcan patent (75), however. Reaction rates obtained at 180°C using a "standard commercial *chlorination* catalyst" (18) are considerably lower than the other two reported.

Figure 5-10 is a simplified flowsheet for an oxychlorination process that employs a small excess of hydrogen chloride in the feed stream to a packed-bed reactor. A heat-transfer agent such as Dowtherm is used on the shell side of the reactor, and steam is generated in the reactor cooler.

The product stream from the reactor often contains about 24, 24, and 48 mole percent dichloroethane, steam, and nitrogen, respectively, plus small amounts of other components, including excess hydrogen chloride. Dew-point temperatures of this stream are about 135 or 115°C, respectively, when the total pressure is 200 lb/in² absolute (13.6 atm) or 100 lb/in² absolute (6.8 atm). Excess hydrogen chloride is first recovered from this stream as a 20 weight percent hydrochloric acid solution in the acid absorber. The water used as absorbent must be heated somewhat; otherwise part of the steam condenses, resulting in dilute acid solutions.

The gas stream from the acid absorber is washed first with water and then with a caustic solution to remove the residual hydrogen chloride. The resulting gas mixture is then cooled in primary and secondary condensers that employ normal cooling water and chilled water, respectively. Because of the relatively high pressure of the gas being pro-

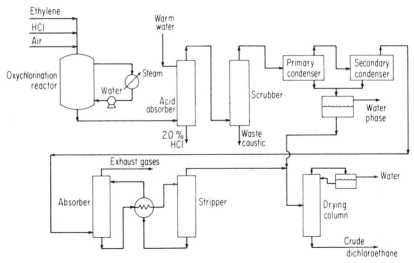

Fig. 5-10 Flowsheet for oxychlorination process using fixed-bed reactor with slight excess of hydrogen chloride.

cessed, most dichloroethane and water are condensed. The resulting condensate forms water and organic layers that are separated by decanting.

Assuming that the gas stream from the secondary condenser is cooled to 15°C (59°F), partial pressures of dichloroethane and water in it are about 1 and 0.2 lb/in² absolute, respectively. Most of the residual dichloroethane is removed in an absorption column, and it is then recovered using a stripping column. The wet dichloroethane streams from the condensers and the stripper are combined and then dried. A heterogeneous azeotropic distillation technique is employed. The crude dichloroethane is further purified by distillation (not shown in Fig. 5-10).

Fluidized-Bed Reactors A fluidized-bed, or perhaps more accurately an expanded-bed, reactor is used in several plants (14, 26, 43), and rather uniform temperatures of 200° to 250°C are maintained throughout. Only a little backmixing of gases occurs as the gases flow upward. Relatively cool gaseous feeds are introduced to the reactor and contacted with the hot fluidized catalyst particles. The exothermic heat of reaction provides the sensible heat for raising the reactant gases to the reaction temperature. High conversions per pass are realized in these reactors, and the recycle of reactants is generally unnecessary.

Details of these fluidized-bed reactors have not been published. PPG Industries, however, has reported some information on an oxychlorination reactor used to produce chlorinated solvents (23). Their reactor may be similar to those used in vinyl chloride plants. The PPG reactor employs 30- to 60-mesh catalyst contained in a vertical bundle of approximately 16-in-ID tubes. Fluidization occurs in these tubes, and boiling water around the tubes removes the exothermic heat of reaction. Makeup catalyst is distributed over the tube bundle as entrained fines are carried away. Precautions are taken using "special feed inlets" to prevent the formation of explosive gas mixtures.

A single bed of catalyst is used in the reactors of some vinyl chloride plants. Possibly in some reactors, several fluidized beds are connected in series. An example of the latter technique is to install the several beds in a single column, with the required number of distribution plates in the column to separate the various compartments. Temperature control when partial conversion has been obtained should not be particularly critical.

Catalyst particles in the fluidized bed erode and fragment with time. Makeup catalyst is occasionally needed to replace the fines carried over with the product. Since the surface of the catalyst is probably semi-liquid, some agglomeration or sticking of catalyst particles may occur. Fines may gradually plug the pores of the larger particles. To obtain reliable design and kinetic data, a rather large pilot plant is probably needed since scale-up techniques for a fluidized bed are not yet well understood.

Fluidized-bed reactors are used in numerous commercial plants, and they may be perhaps more popular than packed-bed reactors. Recovery of the dichloroethane from the product stream is similar to that used with a packed-bed reactor, as shown in Fig. 5-10.

Liquid-Phase Reactor   The oxychlorination process developed by M. W. Kellogg Co. (see Fig. 5-5) uses an aqueous solution of cupric chloride in the reactor (37, 42). The reactant gases (HCl, C_2H_4, and air) are contacted with the solution. Operating conditions for a unit are approximately as follows: temperatures, 170 to 185°C (338 to 365°F); pressures, 16 to 20 atm; and concentrations of cupric chloride solutions, 5 to 7 M. Kinetics of the reaction is first order with respect to cupric chloride concentration. Pilot-plant runs have been made in which about 2.0 to 4.0 g mol of product was obtained per liter of unexpanded solution per hour, that is, 0.125 to 0.25 lb mol/$(ft^3)(h)$.

Mechanically agitating the solution as the gases bubble upward was satisfactory in pilot-plant experiments. As a result, temperature gradients in the reactor were all but eliminated. Temperature control is easy since some water vaporizes and no cooling surfaces are required. Relatively cool feed streams can be introduced into sparging rings or other bubbling devices inside the reactor. These rings quickly disperse the gases in the liquid phase, minimizing high local concentrations of any reactant. The heat of oxychlorination brings the reactants to the reaction temperature.

Tower reactors that are either packed, like the one described earlier for the chlorination of ethylene (55), or contain plates can be employed in the Kellogg oxychlorination process. If plates are installed, cooling devices can be provided between them.

The oxychlorination reaction undoubtedly occurs in the liquid phase. Hence, the reactants must dissolve, or at least contact the liquid. Agitation, caused by the bubbling of the gases through the liquid or a mechanical agitator, creates a large interfacial area between liquid and gas, and hence decreases mass-transfer resistances. Transfer of oxygen is the controlling step in some cases.

The Kellogg reaction system has a big advantage because both oxychlorination and chlorination of ethylene can be performed in the same reactor. 1,2-Dichloroethane is formed in high yields and with high conversions regardless of the ratio of hydrogen chloride to chlorine in the feed stream. Furthermore, aqueous solutions of hydrogen chloride can be used, whereas they cannot with some oxychlorination catalysts. Although liquid-type reactors have probably not yet been commercialized, they appear economically promising.

Pyrolysis of 1,2-Dichloroethane

Processes for the pyrolysis of 1,2-dichloroethane can be divided as follows:

1. Noncatalytic processes with pressures ranging from about 3 to 25 atmospheres (14, 15, 26, 38, 39, 43, 44, 48, 66, 71, 79)
2. Catalytic processes (11, 12, 13, 31, 69, 70, 77)
3. Pyrolysis in the presence of "inert" solids (27, 57)

Chemistry of Pyrolysis The free-radical chain reaction involving the following two propagation steps occurs predominantly if not exclusively in the gas phase, and it is initiated by the formation of a chlorine free radical (7, 46):

$$(17) \quad Cl\cdot + CH_2Cl—CH_2Cl \rightarrow HCl + CH_2Cl—\dot{C}HCl$$
$$(18) \quad CH_2Cl—\dot{C}HCl \rightarrow CH_2{=}CHCl + Cl\cdot$$

There is some evidence that part of the initiation occurs at the wall of the reactor, even for the so-called noncatalytic processes. When a catalyst is used, initiation apparently occurs on the catalyst, possibly as follows:

$$(19) \quad CH_2Cl—CH_2Cl \rightarrow CH_2Cl—\dot{C}H_2 + Cl\cdot$$

Chlorine and oxygen promote the pyrolysis reaction. Presumably, chlorine molecules decompose to produce chlorine free radicals, and oxygen abstracts a hydrogen atom from 1,2-dichloroethane molecules to form the dichloroethyl radicals. Chlorine and oxygen, however, form by-products and are probably not used commercially.

Very pure 1,2-dichloroethane is required to obtain high rates of cracking, to allow high conversions per pass and still obtain high yields of vinyl chloride, to obtain high-purity vinyl chloride, and to minimize coking. Impurities often promote by-products, perhaps because the impurities (or their reaction products) are rather readily adsorbed on the surfaces of the reactor.

Since some reactions occur on the surfaces of the reactor, skin temperatures are important (79). Because of high heat fluxes through the tube walls of the reactors, high skin temperatures are occurring. In unpacked tubular reactors, both tube diameter and flow rates in the tubes affect skin temperatures. As carbonaceous materials are deposited on the tube walls, heat transfer through the walls becomes less efficient. The role of the metal surface compared to carbonaceous deposits relative to surface reactions and adsorptive abilities for various hydrocarbons has not been reported.

With computers, complicated reactions such as the pyrolysis of 1,2-dichloroethane can be modeled for design and operation purposes. A rather preliminary model has been made of this specific reaction (4).

Noncatalytic Pyrolysis Processes   B. F. Goodrich Chemical Co. uses the high-pressure noncatalytic pyrolysis process patented by Krehler (48). The examples in this patent are for a small laboratory unit. Yields of vinyl chloride based on 1,2-dichloroethane are specified from 95 to 98 percent; Goodrich, however, claims commercial yields of 99 percent (26).

British Patent 938,824 (to B. F. Goodrich Co.) describes the design and operation of a pyrolysis furnace for production of commercial quantities of vinyl chloride (39). 1,2-Dichloroethane enters the tubes at about 40°C (104°F) and is heated and vaporized, and the product gases from the tube vary from about 500 to 515°C (930 to 960°F). Pressures of about 23 to 27 atm are probably of most commercial interest, but higher pressure can also be used if desired. Conversions per pass of 50 to 70 percent are obtained for residence times of about 9 to 14 s.

Several reactor tubes are described in the latter patent. One has a diameter of 2.42 cm (0.95 in.) and a length of 102 m (335 ft), and another has a diameter of 8.5 cm (3.35 in.) and a length of 394 m (1,300 ft). Production capacities of vinyl chloride in the larger tubes vary from 70 to 130 million lb/year. Hence only a relatively few tubes are needed even in a large commercial unit. High-nickel alloys such as Inconel are often (if not always) employed as materials of construction.

The inlet portions of each tube, where preheating and vaporization of the dichloroethane occur, are located in the convection zone of the furnace. The pyrolysis sections of the tube are in the radiation zone. In the design shown in Fig. 5-11, two U bends are in the convection zone of the box furnace and three in the radiant zone.

Calculations indicate that the average heat fluxes are as high as 25,000 to 30,000 Btu/(h)(ft^2) for larger tubes. In the pyrolysis section of the tube, even higher fluxes occur, and these furnaces are high-severity ones in the truest sense. Obviously the wall of the tubes in such cases must be considerably hotter than the gaseous mixture inside the tubes. The boundary layer of gases must be quite hot, and a significant portion of the initiation reaction may be occurring there.

By-products of the pyrolysis reaction, which are formed in only small quantities, include acetylene (obtained by pyrolysis of vinyl chloride), benzene (obtained by condensation of acetylene), butadiene, chloroprene, vinylacetylene, and methyl chloride. At periodic intervals, the reactor must be cleaned of coke and tars. The time between decokings depends on the severity of pyrolysis and on the purity of 1,2-dichloro-

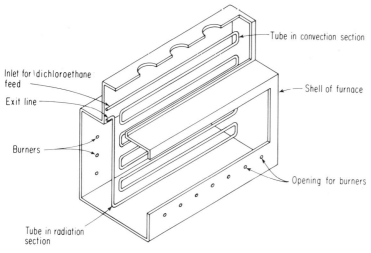

Fig. 5-11   Furnace for pyrolysis of 1,2-dichloroethane.

ethane feedstock. Several months of operation are now possible with high-purity feedstocks at conversions per pass as high as 70 percent (68). Generally steam cannot be used to remove the carbon by means of a shift reaction because of the relatively low temperatures of the unit. Air or oxygen used to burn out the coke (27) causes some metal oxides to form on the tube walls, increasing the kinetics of pyrolysis (68). In the case of propane pyrolysis such metal oxides also result in partial oxidation reactions and coke formation until the surface layer is reduced (29).

Figure 5-12 is a simplified flowsheet for a pyrolysis furnace and auxiliary equipment (39). The feed to the pyrolysis furnace is liquid

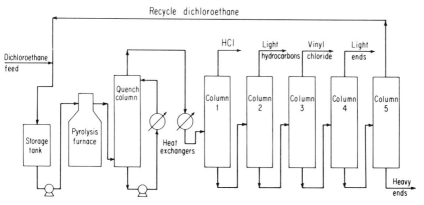

Fig. 5-12   High-pressure pyrolysis process for 1,2-dichloroethane.

1,2-dichloroethane, and the product gases from the furnace are cooled in a quench tower. A stream of crude condensation product from the furnace is continuously sprayed into the quench tower and then recycled. This tower is operated to remove the superheat of the product stream so that the gases leaving the tower are at their dew point. These gases are then condensed, using a heat exchanger. Separation of the condensate to obtain high-purity vinyl chloride, hydrogen chloride, and 1,2-dichloroethane requires at least five columns and will be discussed in detail later.

Other modern pyrolysis units also operate in the range of 7 to 14 atm (68). In the Kureha process, radiant burners fed with off-gas provide heat to the cracker coil (79). Pyrolysis occurs at about 7 atm and 450 to 550°C, to give 60 percent conversion per pass (38). Yield of vinyl chloride is 96 percent.

High-pressure pyrolysis has the advantage over low-pressure (about 2 to 4 atm) pyrolysis that condensation and recovery of vinyl chloride and HCl are cheaper (14). The entire product stream can be liquefied at about 0°C at a pressure of 25 atm. Distillation then separates the product stream into relatively pure components. Yields of the high-pressure processes are as high as those of low-pressure processes or perhaps even higher, and it is doubtful if units operating at less than 6 or 7 atm will be built in the future.

Catalytic Pyrolysis Processes   The SBA process uses a catalyst for cracking 1,2-dichloroethane. Braconier (12) claims that "all secondary reactions and, more particularly, the decomposition into carbon black" are avoided. Details for this process are not known, but patents issued to SBA (69, 70) indicate that a de-ashed active carbon packed in the reactor tubes is used as the catalyst. Operating variables for the process are thought to be in these ranges: temperature 400 to 450°C; pressure, 8 to 10 atm; conversion per pass, 60 to 70 percent. At these relatively low temperatures, selectivities to vinyl chloride of 99 percent or greater are obtained. Temperature gradients obviously occur in the packed tubes.

Wacker-Chemie (78) has reported a carbon catalyst with activity such that temperatures as low as 200 to 350°C can be used. Carbon formation becomes a problem at temperatures approaching 450°C. This latter temperature seems to be the upper limit for carbon-catalyzed pyrolysis. Up to 1 year of operation can be obtained from such catalysts. During this period, the temperature for pyrolysis is slowly raised to maintain the desired conversion. Once the catalyst is deactivated, presumably by carbon and tar formation, it has to be replaced.

At least some catalytic processes operate at slightly lower temperatures than noncatalytic processes, and energy requirements for catalytic pyrolysis are probably somewhat less than those for noncatalytic processes. Yields and conversion levels for both types of process seem to be essentially identical. The operating costs of the catalytic processes are probably in general higher than those of noncatalytic processes, primarily because of catalyst expenses. Available information indicates that catalytic processes generally are not competitive with high-pressure noncatalytic processes.

"Inert" materials, such as gravel, clays, pumice, etc., have been suggested as packing for the tubes of a pyrolysis furnace (27, 57). Advantages of such packing are small at best, and such packings are probably not used industrially.

Purification Steps

Product streams from the pyrolysis furnace, from the chlorination unit, and from the oxychlorination unit are distilled to obtain high-purity vinyl chloride; hydrogen chloride, which is pumped to the oxychlorination unit or to the unit for reaction with acetylene; and 1,2-dichloroethane, which is sent to the pyrolosis furnace. Figure 5-12 is the simplified flowsheet showing how the condensate from the pyrolysis furnace is separated. This condensate contains primarily vinyl chloride, hydrogen chloride, and dichloroethane; however, it contains such a large number of minor components that five towers are generally required to obtain the desired purification of the major components.

The first column of Fig. 5-12, column 1, used for recovery of high-purity (99.8 percent or greater) hydrogen chloride was operated in at least one case at a pressure of about 12 atm (48). At this pressure, the reflux condenser must be cooled to about −40°C (−40°F) to obtain liquefication of the hydrogen chloride. If a higher pressure could be used in the column, a higher temperature would then be allowable; i.e., refrigeration demands for the condenser would be less severe. Unfortunately, at higher pressures, the temperature of the reboiler would also be higher, promoting undesired reactions, including polymerization of the vinyl chloride. Temperatures in the reboiler normally are about 100°C. Precautions must be taken to provide low temperature differences for heat transfer so that skin temperatures of the heat-transfer surfaces are as low as possible.

Column 2 of Fig. 5-12 is for the removal of components boiling between hydrogen chloride and vinyl chloride. Components in this category include butadiene, a highly undesirable impurity since it acts as an inhibitor for vinyl chloride polymerization. More details on the operation of this column are shown in the flowsheet presented by Shelton et al. (68).

Column 3 is the vinyl chloride finishing column, which produces high-purity vinyl chloride as the top product. This column frequently has 30 to 50 trays and is operated at reflux ratios of 2:1 to 3:1 with pressures up to about 5 atm absolute. The bottom product from column 3 is primarily 1,2-dichloroethane containing both light and heavy impurities. As a result, columns 4 and 5 are required for the separation of 1,2-dichloroethane, light ends, and heavy ends. In at least some plants, these last two columns are also used for the purification of the product streams from the chlorination and the oxychlorination units.

Processes Using Ethane as Feedstock

Cost of the hydrocarbon feedstock is a major item in production of vinyl chloride, and a concerted effort has therefore been made to develop processes using ethane rather than ethylene. In 1971, the Lummus Co. announced their Transcat process, which produces vinyl chloride from ethane and chlorine (21, 50). The process integrates chlorination, oxychlorination (using oxygen from air), and dehydrochlorination (or pyrolysis) steps within a single reactor system. A unique molten catalyst, possibly a molten salt (25), is employed, and the overall chemistry is as follows:

$$(20) \quad C_2H_6 + \tfrac{1}{2}Cl_2 + \tfrac{3}{4}O_2 \rightarrow C_2H_3Cl + 1\tfrac{1}{2}H_2O$$

The cost of feedstock for the Transcat process is about 1 cent less per pound of vinyl chloride produced than balanced processes using ethylene (see Table 5-2). Operation costs of the plant have not been announced, but they are apparently comparable to those of current processes. Hence the Transcat process may be most attractive for future vinyl chloride plants.

At least two other balanced processes with the overall chemistry shown above have been reported. In one process, ethane is oxychlorinated (16) using a granular catalyst of phosphate salts of iron (17) to produce mixtures of 1,2-dichloroethane, ethylene, and ethyl chloride. The latter compound can be dehydrochlorinated rather easily to produce ethylene and HCl.

Photochlorination of ethane can also be used to produce 1,2-dichloroethane, ethyl chloride, and vinyl chloride (53, 63). Although vinyl chloride could be produced in high yields, the photochlorination step would probably be too expensive to allow a balanced process to be used commercially.

Selection of Vinyl Chloride Process

With several processes available for the production of vinyl chloride, many engineering economic decisions must be made in selecting the

TABLE 5-2   Reactants Required per Pound of Vinyl Chloride Produced and Cost of Reactants

Process	Reactant, lb					Cost of reactants per pound of VCM
	Eth-ylene*	Acet-ylene†	Ethane‡	Chlo-rine*	HCl¶	
Simple hydrochlorination.....	....	0.44	...	....	0.59	4.7
Balanced:						
Using pure acetylene and pure ethylene (see Fig. 5-1).....................	0.23	0.21	...	0.59	....	4.1
Using oxychlorination (see Fig. 5-2)................	0.49	....	...	0.67	....	3.3
Transcat....................	....	....	0.6	0.58	....	2.3

* At 3 cents/lb.
† At 8 cents/lb.
‡ At 1 cent/lb.
¶ At 2 cents/lb.

"best" process. The major manufacturing cost of vinyl chloride processes is in all cases that of the reactants. These costs vary with plant location and unique situations at each plant. The following prices on a pound basis were probably fairly typical in the early 1970s for areas in the United States where large vinyl chloride plants are in operation: ethylene, 3 cents; chlorine, 3 cents; acetylene, 8 cents; ethane, 1 cent; and hydrogen chloride, 2 cents. Oxygen is also needed in some processes; tonnage oxygen may be used, but air probably is the primary source of oxygen. It will be assumed that air is employed in the following analysis of reactant costs and that the cost of air is included as part of the operating cost for the plant.

Table 5-2 indicates the approximate quantities of reactants required per pound of vinyl chloride for processes using acetylene, acetylene and ethylene, ethylene, and ethane (Transcat process) as hydrocarbon feedstocks. Costs of these feedstocks for the balanced process using an oxychlorination step (Fig. 5-2) are considerably less than those for the simple hydrochlorination process (reaction 1) or those for the balanced process using a combination of pure ethylene and pure acetylene (Fig. 5-1). The Transcat process, using ethane, has by far the lowest feed cost, however, and it or comparable process could become the major vinyl chloride process of the future.

Plant size is highly important relative to the manufacturing costs per given amount of vinyl chloride produced. Spitz (72) has discussed in considerable detail both the manufacturing and capital costs of various sizes of plant using a balanced process involving oxychlorination (Fig. 5-2). The cost of production was about 0.27 cent less per pound of vinyl chloride for a 600 million lb/year unit than a 400 million lb/year unit. Plants have been built with capacities of 1 billion lb/year, and the cost of production of such plants has probably been further decreased somewhat.

Part of the reduced manufacturing cost for larger plants is due to lower labor and overhead costs possible with larger-scale operations. Part is also due to lower depreciation costs, since there are lower capital costs per pound of vinyl chloride capacity. Spitz (72) indicates that the scale-up factor for plants using oxychlorination balanced processes is 0.64. In 1968, a plant with 300 million lb/year capacity cost $8 million, while a 600 million lb/year unit cost $12 million. The cost of a 200 million lb/year unit was predicted in 1965 as being $6 million (32). Operating costs of this latter plant were also presented.

Scale-up of vinyl chloride plants is no problem (72), since the equipment is relatively simple and its operation is quite straightforward. For example, Selas Corp., supplier of commercial pyrolysis furnaces, indicates that modern vinyl chloride plants can be built with only two furnaces. From the capacities reported per tube, four tubes are probably provided in each furnace, as is often the case in furnaces for ethylene plants.

Balanced processes, which generally involve at least three reaction steps, are obviously more expensive to operate than a simpler process such as the one with only a hydrochlorination step (reaction 1). Operating costs (including depreciation) have not been publicized, but some approximate comparisons can be made for large plants with similar capacities. For a plant producting vinyl chloride by reacting acetylene and HCl, the operating costs are estimated to be about 0.5 cent per pound of vinyl chloride. These costs for any oxychlorination type of balanced process (Fig. 5-2) would probably be at least 1.0 cent/lb. A balanced process using pure ethylene and pure acetylene as feeds (Fig. 5-1) would have intermediate operating costs.

A balanced process involving the oxidation of HCl (Fig. 5-3) would be preferred to the popular balanced process with an oxychlorination step (Fig. 5-2) if a sufficiently cheap technique for oxidizing HCl to chlorine is developed. The oxychlorination step is relatively expensive compared to the chlorination of ethylene.

A Deacon type process for oxidation of HCl to chlorine is a portion of the balanced process (Fig. 5-3) of the Shell Development Co. (14, 44). A fluidized-bed reactor is used which contains a catalytic mixture of copper and other (potassium?) chlorides suspended on a silica carrier. An electrolysis route is also used commercially in at least some plants for the oxidation of HCl.

Two more recent developments may lower the cost of HCl oxidation. The Kel-Chlor process of M. W. Kellogg Co. is said to produce chlorine with an operating cost of 0.5 to 0.85 cent per pound of chlorine (19). Nitrogen oxides are used as homogeneous gas-phase catalysts.

The process of Albright and Haug (3) offers a radically new approach to HCl oxidation. In the first step of a three-step process, HCl reacts with a molten mixture of alkali-metal nitrates, such as a 50:50 mixture of sodium and potassium nitrates at 200 to 250°C. Alkali-metal chlorides are formed, and oxides of nitrogen (mainly nitrogen dioxide), steam, and oxygen are evolved from the reactor.

In the second step of the Albright-Haug process, nitric acid is contacted with the molten mixture of alkali-metal nitrates and chlorides. The chlorides are converted back to nitrates, and chlorine, nitrogen dioxide, steam, and oxygen are released. The chlorine and oxides of nitrogen are recovered and separated. The oxides of nitrogen in the third phase of the process are converted back by conventional means to nitric acid, which is recirculated. The overall stoichiometry for the process is

$$(21) \quad 2HCl + \tfrac{1}{2}O_2 \rightarrow Cl_2 + H_2O$$
$$\text{(Air)}$$

Several companies have developed modified vinyl chloride processes. The Dianor process (14, 61), for example, uses a feed stream containing only a relatively small amount of ethylene. It is claimed to be competitive for both small and large vinyl chloride producers, and numerous cost data have been reported. The cost reported to produce vinyl chloride totals almost 6 cents/lb, which seems to make it too expensive for American manufacturers. To date the major interest in the process seems to be in countries with new and developing petrochemical industries. Significant savings in the cost of ethylene are realized in this process since costly separation steps normally employed in ethylene plants are not needed. If the Dianor process is combined with an oxidation step for HCl, a modified process similar to the one shown in Fig. 5-3 will be obtained.

Hydrocarbon pyrolysis processes that produce essentially equal amounts of ethylene and acetylene are uniquely suited to vinyl chloride production, and they are used for balanced processes, as shown in Fig. 5-4. The ethylene and acetylene costs in such processes are considerably less than those requiring pure-ethylene and pure-acetylene feedstock. Capital costs for such a balanced process are relatively high since a high-temperature hydrocarbon pyrolysis unit must also be provided for production of the ethylene and acetylene mixture. A 66 million lb/year plant of this type was reported in 1964 to have an estimated capital cost of $4.150 million (38).

Although detailed and comparable cost data for the balanced processes shown in Figs. 5-2 and 5-4 are not available, the latter process may in some cases be preferred, especially when a cheap source of naphtha feedstock is available.

Combining the chlorination and oxychlorination step in a single reactor seems to offer economic advantages. Up until now, the Kellogg process (37, 42) is the only one for which a significant amount of data have been reported. Such a plant would be likely to save on both capital and operating costs, provide extreme flexibility for using either chlorine or hydrogen chloride as feedstocks, and provide improved control of the operating conditions.

Processes have also been developed for production of chlorinated solvents in addition to 1,2-dichloroethane and vinyl chloride (23, 62, 74). These processes are apparently sufficiently flexible to permit the relative amounts of products to be varied over rather wide limits.

Because of the complexity of the plant, a balanced process requires excellent instrumentation for control purposes. An upset in any portion of the plant tends to upset the remainder. Methods of starting up the unit and maintaining the desired optimum conditions as the catalyst ages and as flows and yields vary require careful planning. Kureha have described in considerable detail their method of handling this problem (38). They were able to save at least 3 percent of the operating costs because of their control procedure.

Several processes for the production of vinyl chloride can be licensed (51). Included in the list are balanced processes involving oxychlorination (Figs. 5-2 and 5-5), oxidation of HCl (Fig. 5-3), special pyrolysis techniques to produce cheap but impure ethylene and mixtures of ethylene and acetylene (Fig. 5-4), and the Transcat process (using ethane). Detailed engineering and economic analyses will be required for selection of the best process for a given situation.

Literature Cited

1. Albright, L. F.: Vinyl Chloride Processes, *Chem. Eng.*, Mar. 27, 1967, pp. 123–130.
2. Albright, L. F.: Manufacture of Vinyl Chloride, *Chem. Eng.*, Apr. 10, 1967, pp. 219–226.
3. Albright, L. F., and H. Haug (to Purdue Research Foundation): Process for Production of Alkali Metal Nitrates and Chlorine, U.S. Pat. 3,348,909 (Oct. 24, 1967).
4. Badguerahanian, L., A. Bellier, and A. Crico: Utilization d'une calculatrice analogique en cinétique chimique appliquée: application à l'étude de la décomposition thermique du dichloroethane-1,2, *Chim. ind.: genie chim.*, 88: 115 (1962).
5. Badische Anilin- und Soda-Fabrik A.G.: Vinyl Chloride from Acetylene and Hydrogen Chloride, British Pat. 769,773 (Mar. 13, 1957).

6. Bahr, H., and H. Zieler: The Interaction of Chlorine on Ethylene, Z. Angew. Chem., **43:** 233 (1930).

7. Barton, D. H. R., and K. E. Howlett: Kinetics of the Dehydrochlorination of Substituted Hydrocarbons, J. Chem. Soc., **1949:** 148–164.

8. Bernstein, L. S., and L. F. Albright: Kinetics of Slow Thermal Chlorination of Hydrogen in Nickel Tubular Flow Reactors, AIChE J., **18:** 141 (1972).

9. Bernstein, L. S., B. B. Fuqua, and L. F. Albright: Thermal Chlorinations of Hydrogen in Tubular Reactors, Symp. Adv. React. Eng., 3d Int. Cong. Chem. Eng. Marianske Lazne, Czechoslovakia, 1969.

10. Bhatnagar, R. K.: Selection of a Process for Manufacture of Ethylene Dichloride, Chem. Age India, September 1966, p. 521.

11. Braconier, F. F.: Manufacture of Vinyl Chloride Starting with Naphtha, Oxygen and Chlorine, Chem. Age India, June 1963, p. 543.

12. Braconier, F. F.: How S. B. A. Makes Vinyl Chloride, Hydrocarbon Process., November 1964, p. 140.

13. Braconier, F. F., and J. A. Godart (to Société Belge de l'Azote): Process for the Manufacture of Vinyl Chloride from 1,2-Dichloroethane and Acetylene, U.S. Pat. 2,779,804 (Jan. 29, 1957).

14. Vinyl Chloride, Br. Chem. Eng. Suppl., November 1967, pp. 77–79, 108.

15. Buckley, J. A.: Vinyl Chloride via Direct Chlorination and Oxychlorination, Chem. Eng., Nov. 21, 1966, p. 102.

16. Burke, D. P., and R. Miller: Oxychlorination, Chem. Week, Aug. 22, 1964, pp. 93–118.

17. Carroll, R. T., and E. J. Dewitt (to B. F. Goodrich Co.): Oxychlorination of Lower Alkanes, U.S. Pat. 3,173,962 (Mar. 16, 1965).

18. Carrubba, R. V., and J. L. Spencer: Kinetics of the Oxychlorination of Ethylene, Ind. Eng. Chem. Process Des. Dev., **9:** 414 (1970).

19. New Process May Reshape Chlorine Industry, Chem. Eng. News, May 5, 1969, pp. 14–15.

20. PVC Producers See Good Business Ahead, Chem. Eng. News, Aug. 4, 1969, pp. 182–183.

21. Vinyl Chloride: New Lummus Process, Chem. Eng. News, Mar. 1, 1971, p. 8.

22. Vinyl Chloride Process, Chem. Eng. News, Aug. 23, 1971, p. 29.

23. Chlorinated Solvents from Ethylene, Chem. Eng., Dec. 1, 1969, pp. 90–91.

24. Vinyl Process: Petrochemical Pacesetter, Chem. Eng., Mar. 27, 1967, pp. 48–49.

25. Vinyl Chloride Monomer Made from Ethane, Chem. Eng., Mar. 8, 1971, p. 51.

26. Bold Stroke at Calvert City, Chem. Week, Aug. 29, 1964, pp. 101–108.

27. Cheney, H. A. (to Shell Development Co.): Halogenated Unsaturated Hydrocarbon Production, U.S. Pat. 2,569,923 (Oct. 2, 1951).

28. Conn, J. B., G. B. Kistiakowsky, and E. A. Smith: Heats of Organic Reactions, VIII: Addition of Halogen to Olefins, J. Am. Chem. Soc., **60:** 2764 (1938).

29. Crynes, B. L., and L. F. Albright: Pyrolysis of Propane and Tubular Flow Reactors, Ind. Eng. Chem. Process Des. Dev., **8:**25 (1969).

30. E. I. du Pont de Nemours & Company, Inc.: Chemical Process and Catalyst, Br. Pat. 941,353 (Nov. 13, 1963); Belg. Pat. 614,580 (Sept. 3, 1962).

31. Eberly, K. C. (to Firestone Tire and Rubber Co.): Catalytic Dehydrochlorination of 1,2-Dichloroethane, U.S. Pat. 2,875,255 (Feb. 24, 1959).

32. Edwards, E. F., and T. Weaver: New Route to Vinyl Chloride, Chem. Eng. Prog., January 1965, p. 21.

33. Fedor, W. S.: Outlook Seventies, Chem. Eng. News, Dec. 15, 1969, p. 94.

34. Fontana, C. M., E. Gorin, G. A. Kidder, and R. E. Kinney: Oxygen Equilibrium Pressures and Oxide Solubility in the Melt, *Ind. Eng. Chem.*, **44:** 369 (1952).
35. Fontana, C. M., E. Gorin, G. A. Kidder, and C. S. Meredith: Ternary System Cuprous Chloride–Cupric Chloride–Potassium Chloride and Its Equilibrium Chlorine Pressures, *Ind. Eng. Chem.*, **44:** 363 (1952).
36. Fontana, C. M., E. Gorin, and C. S. Meredith: Kinetics of Oxygen Absorption by the Melt, *Ind. Eng. Chem.*, **44:** 373 (1952).
37. Friend, L., L. Wender, and J. C. Yarze: Liquid-Phase Oxychlorination Provides High Selectivity Route to Vinyl Chloride, *Pap. Div. Pet. Chem. Am. Chem. Soc. Meet., New York, Sept.* 11–16, 1966.
38. Gomi, S.: Japan's New Vinyl Chloride Process, *Hydrocarbon Process.*, November 1964, p. 165.
39. B. F. Goodrich Co.: Preparation of Vinyl Chloride, British Pat. 938,824 (Oct. 9, 1963).
40. Gorin, E., C. M. Fontana, and G. A. Kidder: Chlorination of Methane with Copper Chloride Melts, *Ind. Eng. Chem.*, **40:** 2128–2138 (1948).
41. Groll, H. P. A., G. Hearne, F. F. Rust, and W. E. Vaughan: Chlorination of Olefins and Olefin-Paraffin Mixtures at Moderate Temperatures: Induced Substitution, *Ind. Eng. Chem.*, **31:** 1239 (1939).
42. Heinemann, H., K. D. Miller, and M. L. Spector: Olefin Chlorination in Homogeneous Aqueous Copper Chloride Solutions, *Pap. Div. Pet. Chem. Am. Chem. Soc. Meet., New York, Sept.* 11–16, 1966.
43. Vinyl Chloride, *Hydrocarbon Process.*, November 1965, p. 290; November 1967, pp. 239–242; November 1969, p. 248.
44. Dichloroethane and Vinyl Chloride: Shell Development Co., *Hydrocarbon Process.*, November 1967, p. 166.
45. Imperial Chemical Industries, Ltd.: Chlorohydrocarbons, Belg. Pat. 632,044 (Nov. 18, 1963).
46. Kapralova, G. A., and N. N. Semenov: Study of the Mechanism of 1,2-Dichloroethane Decomposition by the Calorimetric Method, *Zh. Fiz. Khim.*, **37:** 73 (1963).
47. M. W. Kellogg Co.: private communications, 1966–1968.
48. Krehler, H. (to Farbwerke Hoechst): Process of Preparing Vinyl Chloride, U.S. Pat. 2,724,006 (Nov. 15, 1955).
49. Kuhlmann, E.: Production of Vinyl Chloride, French Pat. 1,139,124 (June 25, 1957).
50. Zasloff, H. B., and others (The Lummus Co.): personal communications, 1971.
51. Miller, R. L.: List of Processes for License or Sale, *Chem. Eng.*, Apr. 20, 1970, pp. 122–144.
52. Miller, S. A.: "Acetylene: Its Properties, Manufacture, and Uses," vols. 1 and 2, Academic, New York, 1965 and 1966.
53. Mottern, H. O., and J. P. Russell (to Air Reduction Company): Processes for Producing Chlorination Ethanes, U.S. Pat. 3,505,193 (Apr. 7, 1970); U.S. Pat. 3,506,553 (Apr. 14, 1970).
54. Nair, K. S.: Commercial Manufacture of Vinyl Chloride by Low-Pressure Synthesis, *Chem. Age India*, January 1963, p. 80.
55. Nair, K. S.: Notes of the Manufacture of Commercially Important Products, *Chem. Age India*, May 1965, p. 382.
56. Nesmeyanov, A. N.: Quasi-Complex Organometallic Compounds, *Bull. acad. sci. USSR, Cl. sci. chem.*, **1945:** 239.
57. N.V. de Bataafsche Petroleum Maataschappij: A Process for the Manufacture of Vinyl Chloride, British Pat. 663,221 (Dec. 12, 1949).

58. Othmer, D. F.: Make 3- to 5-cent Acetylene, *Hydrocarbon Process.*, March 1965, p. 145.

59. Ogunye, A. F., and W. H. Ray: Optimization of Vinyl Chloride Monomer Reactor, *Ind. Eng. Chem. Process Des. Dev.*, 9: 619 (1970).

60. Peters, E. H.: Ethylene: Organic Chemical Building Block, *Chem. Eng. Prog.*, June 1965, p. 87.

61. Remirez, R.: Dilute Ethylene Opens Door to Vinyl Chloride Production in Small Plants, *Chem. Eng.*, Apr. 22, 1968, pp. 142–144.

62. Rosenzweig, M. D.: Vinyl Chloride Process Has Wide Range of By-Products, *Chem. Eng.*, Oct. 18, 1971, pp. 105–107.

63. Russell, J. P. (to Air Reduction Company): Preparation of Vinyl Chloride from Ethane Using Chlorine with Actinic Light, U.S. Pat. 3,506,552 (Apr. 14, 1970).

64. Rust, F. F., and W. E. Vaughan: The High-Temperature Chlorination of Olefin Hydrocarbons, *J. Org. Chem.*, 5: 472 (1940).

65. Semenov, N. N.: "Some Problems in Chemical Kinetics and Reactivity," vol. 1, pp. 211–227, Eng. trans. by M. Boudart, Princeton University Press, Princeton, N.J., 1958.

66. Shell International Research: Halogenation of Hydrocarbons, Br. Pat. 907,435 (Oct. 3, 1962).

67. Shelley, P. G., and E. J. Sills: Monomer Storage and Protection, *Chem. Eng. Prog.*, 65(4): 29 (1969).

68. Shelton, L. G., D. E. Hamilton, and R. H. Fisackerly: Vinyl and Vinylidene Chloride, in E. C. Leonard (ed.), "Vinyl and Diene Monomers," pt. 3, pp. 1206–1289, Wiley-Interscience, New York, 1971.

69. Société Belge de l'Azote: Vinyl Chloride, French Pat. 1,290,953 (Mar. 12, 1962).

70. Société Belge de l'Azote: Process for the Preparation of Vinyl Chloride, British Pat. 954,791 (Apr. 8, 1964).

71. Solvay et Cie: Process for Producing Vinyl Chloride, Br. Pat. 605,277 (July 20, 1948).

72. Spitz, P. H.: Vinyl Chloride Economics: Effects of Plant Size and Changing Technology, *Chem. Eng. Prog.*, 64(3):19 (1968).

73. Stobaugh, R. B.: Acetylene: How, Where, Who—Future, *Hydrocarbon Process.*, August 1966, p. 125.

74. Tsuda, S.: Tri- and Perchloroethylene Made by Simple New Route, *Chem. Eng.*, May 4, 1970, pp. 74–76.

75. Vulcan Materials Co.: Oxychlorination Process, Br. Pat. 980,983 (Jan. 20, 1965).

76. Vulcan Materials Co.: 1,2-Dichloroethane by Oxychlorination Reaction, *Rep.*, 1966.

77. Wacker-Chemie GMBH: Process for the Manufacture of Vinyl Chloride by Catalytic Splitting of 1,2-Dichloroethane, British Pat. 979,309 (Jan. 1, 1965).

78. Wacker-Chemie GMBH: Prevention of Vinyl Chloride Decomposition during the Catalytic Dissociation of Dichloroethane, Belgian Pat. 610,498 (May 21, 1962); German Pat. 1,135,451 (Nov. 25, 1960).

79. Washimi, K., and M. Asakura: Computer Control of a Vinyl Chloride Plant Provides Process Optimization, *Chem. Eng.*, Oct. 24, 1966; pp. 133–138; Nov. 21, 1966, pp. 121–126.

80. Wesselhoft, R. D., J. M. Woods, and J. M. Smith: Vinyl Chloride from Acetylene and Hydrogen Chloride: Catalytic-Rate Studies, *AIChE J.*, 5: 361 (1959).

CHAPTER SIX

Production of Polyvinyl Chloride Polymers

1. INTRODUCTION AND POLYMERIZATION FUNDAMENTALS

Polyvinyl chloride (PVC) polymers are the second largest family of high polymers in the United States, based on the amount produced; this was over 3 billion lb by 1970 (21) and accounted for approximately 16 percent of all plastics produced. The term PVC polymers as used here includes polymers produced from pure vinyl chloride or mixtures of comonomers that are predominantly vinyl chloride. PVC polymers are by far the most important member of the family of vinyl polymers,* which also includes polyvinyl acetate, polyvinyl alcohol, polyvinyl acetals, and polyvinylidene chloride polymers.

Production of PVC polymers in the United States increased in the 1960s on an average of over 12 percent per year. The rates of increase will probably not be as high in the 1970s but should be substantial. It has been predicted that by 1980 over 6 billion lb/year will be produced

---

* Several authors use the term vinyl polymers differently. Billmeyer (6), for example, uses it to refer to all polymers in which the repeating unit of the polymer has the formula —CH_2—CHX—, where X is any atom or group of atoms except hydrogen.

(21). In many parts of the world, production rates have increased at a relatively more rapid pace than in the United States. In 1946, the United States produced 80 percent of all PVC but presently produces only about 25 percent (1). Production in Common Market countries now surpasses that in the United States, and Japanese production is over 75 percent of that of the United States. The major increase of PVC production in the rest of the world is caused in part by delays in achieving adequate production of other thermoplastic polymers, including polyethylene. When these other polymers are produced abroad in larger quantities, they may displace PVC for certain uses. Nevertheless, the growth rate for PVC polymers probably will remain substantial.

Increased production of PVC during the last few years has been due in part to a significant decrease in the selling price. The price for general purpose PVC resins dropped to as low as 8.5 cents/lb in 1968 even though the nominal list price was 12 cents/lb (11). In 1970, the list price ranged from about 10.5 to 11 cents/lb, but some actual prices may have been lower. Since vinyl chloride costs are approximately 4.5 cents/lb, direct polymerization costs are about 2 to 3 cents/lb, and expenses for packaging, warehousing, and freight are perhaps 2 cents/lb, prices like those listed above do not provide much margin for sales costs and profits. Much further price reduction seems doubtful. Only companies with rather large sales volumes can profit in such a market, and currently over 20 companies produce PVC resins in the United States.

Uses for PVC are many and varied (32, 46), as indicated by Table 6-1. PVC polymers can vary from soft and rubberlike materials to

TABLE 6-1 Diversity of Polyvinyl Chloride Markets

Market	Consumption, %
Calendering	18
Extruded products	14
Wire and cable	12
Calendered flooring	10
Film and sheet	6
Paper and textile coating	5
Sound records	5
Plastisols	5
Protective coatings and adhesives	4
Injection and blow molding	3
Coated flooring	2
Other domestic uses	11
Exports	5

tough, horny ones. They can be produced to combine high strength, relatively low density, and inertness to many different media. Good processing and color characteristics, toughness, and electrical insulation are features of many PVC polymers. The physical properties can be varied significantly by plasticization of the PVC polymers or by copolymerization. Flexible or at least semiflexible PVC polymers are the most common varieties, but rigid PVC finds large uses in tubing, fittings, bottles, records, and sheeting. PVC producers are spending large sums of money in an effort to find new uses and to develop improved products.

PVC containing dioctyltin stabilizers is used in large quantities to produce blow-molded bottles that are frequently clear and are employed for packaging foods and pharmaceuticals, including mineral water, wine, beer, distilled spirits, cooking oils, vinegar, shampoos, and mouthwashes. Techniques for production of large PVC containers are available. PVC film is widely used for packaging meats, vegetables, and fruits. A heat-shrinkable film is effective for irregularly shaped foods.

Increased uses of PVC polymers can be expected in construction operations. Polymers have been developed for home siding, window frames, conduits, gutters, and tiling, but it is often necessary to have outmoded building codes changed to allow the use of these plastics. In some cases, the PVC polymers are easier and cheaper to install than presently used materials. Some building unions have actively opposed legislation that would allow the use of PVC or similar plastics.

Chemistry of Polymerization

Vinyl chloride and mixtures of comonomers that are predominantly vinyl chloride are polymerized commercially by free-radical chain reactions (1, 6, 23) similar to those of the high-pressure polymerization of ethylene and the commercial polymerization of styrene.

Initiation is accomplished with an initiator to produce free radicals at relatively low temperatures. Initiators soluble in vinyl chloride are used for suspension, bulk, and solution polymerizations. Such initiators include isopropyl peroxydicarbonate (IPP), lauryl peroxide, and azobis(isobutyro)nitrile (9). Percarbonate-type initiators can also be generated in situ (41, 42). About 30 percent of all peroxide initiators used for polymerization of various polymers are used for vinyl polymers, which include polyvinyl acetate; total consumption of peroxide initiators was about 13 to 14 million lb in 1963 and has grown since then. Polymerization data obtained using numerous established initiators and several newer ones have been published by the Lucidol Division of Pennwalt Corp. (35). Some of the newer initiators result in more uniform polymerization rates, more controllable reactions, and shorter batch reactions.

Water-soluble initiators are used for emulsion polymerizations. Redox initiators (often hydroperoxides) and persulfates are common examples (9).

Propagation involves the addition of vinyl chloride to the free radical at the end of a growing chain. R in the following equation represents the fragment of the initiator:

$$R—(CH_2—CHCl)_n—CH_2—\dot{C}HCl + CH_2{=}CHCl \xrightarrow{k_p}$$
$$R—(CH_2—CHCl)_{n+1}—CH_2—\dot{C}HCl$$

The reaction rate constant k_p for propagation is essentially independent of the length of the growing chain.

Termination steps, in which the concentration of free radicals decreases, involve the reaction between two growing chains. Coupling is the combination of two long-chain free radicals to form a single high-polymer molecule. Disproportionation, another type of termination, results in two polymer molecules; one molecule is saturated, and the other has a double bond at one end.

Equations have been developed to represent the kinetics of polymerization, especially at the beginning of a batch run (6, 23). These reaction rates are frequently proportional to the initiator concentration of the $\frac{1}{2}$ power and to the vinyl chloride concentration to either the first or $1\frac{1}{2}$ power. Complexities of the polymerization system often make such kinetic equations only semitheoretical at best. In particular, more than one phase occurs in the initial stages of some commercial polymerizations, and more than one reaction site may be present.

Chain transfer terminates the growth of a free-radical polymeric chain but produces a new free radical to start another chain. Such a transfer does not change the concentration of free radicals, as indicated by the transfer with a vinyl chloride monomer:

$$R—(CH_2—CHCl)_n—CH_2\dot{C}HCl + CH_2{=}CHCl \rightarrow$$
$$R—(CH_2—CHCl)_n—CH_2—CH_2Cl + CHCl{=}\dot{C}H$$
$$\text{or } CH_2{=}\dot{C}Cl$$

Vinyl chloride molecules then add to the new radical to form another growing chain.

Chain transfer also occurs with the initiator or PVC polymer molecules. In the latter case, a long-chain branch results. Transfer steps do not affect the overall rate of polymerization. The molecular weight of the polymer is decreased, however, when transfer is with vinyl chloride or with the initiator but not when transfer is with a polymer molecule.

The operating conditions used during polymerization affect the chain-transfer steps and consequently the characteristics of the PVC molecules. Increased temperature and in some cases increased concentrations of initiator decrease the molecular weight of the PVC polymer. Chain branching is most important when the concentration of polymer is high, i.e., at high level of polymerization. Comonomers, solvents (if used), and additives like those used in suspension and emulsion polymerization also result in chain-transfer steps.

Side reactions occur during the polymerization of vinyl chloride. Some $CH_2\!\!=\!\!\dot{C}H$ radicals formed by chain-transfer steps combine to form butadiene:

$$2CH_2\!\!=\!\!\dot{C}H \rightarrow CH_2\!\!=\!\!CH\!\!-\!\!CH\!\!=\!\!CH_2$$

Butadiene is a chain terminator that adds to a growing polymer chain, forming an unreactive resonance-stabilized allyl radical. Small amounts of HCl are evolved as vinyl chloride is polymerized (18). If suspension or emulsion polymerization is employed, the pH of the water may decrease as polymerization progresses.

Multiphase Polymerization

Polymerizations of vinyl chloride always involve more than one phase during at least part of the polymerization. This fact is particularly obvious for suspension or emulsion polymerization, two common methods. At least two phases are present then, namely the discontinuous organic phase and the continuous water phase. Since PVC is only slightly soluble in vinyl chloride and in most common solvents, a solid-polymer phase forms after only a relatively small amount of polymerization for all types of polymerization. This polymer phase is porous and a good adsorbent for vinyl chloride. At least part of the monomer, however, may actually be trapped in the pores as a liquid and not be absorbed.

Suspension Polymerization Winslow and Matreyek (49) have presented a schematic diagram of the states of dispersion of vinyl chloride for suspension polymerization as shown in Fig. 6-1. Dispersed vinyl chloride droplets are suspended throughout the water phase in the initial stages of polymerization, and they are subjected to shear resulting from mechanical agitation. The larger droplets are broken into smaller droplets, which tend to coalesce, reforming larger droplets. A dynamic equilibrium of dispersion and coalescence occurs that affects the porosity and bulk density of the final PVC product.

Protective colloids, such as starch, proteinaceous materials, methyl cellulose, and polyvinyl alcohol, as indicated in Fig. 6-1, are added to stabilize the vinyl chloride droplets and help prevent agglomeration of the PVC droplets. These colloids are water-soluble but relatively insoluble in vinyl chloride. They act to increase the viscosity of the water phase and hence delay coalescence (40). Inorganic materials are occasionally added to regulate coalescence. Materials such as kaolin, barium sulfate, magnesium carbonate, talcum, neutral phosphates, and bentonite clay are effective. They are generally quite insoluble in both phases and concentrate at the interface between the water and organic phases.

The initiator used to start polymerization is soluble in vinyl chloride. The viscosity of the organic phase increases as polymerization occurs, and polymer molecules form throughout the droplets, which become syrupy as a PVC phase forms there. Agglomeration (or sticking) of particles is a problem during this phase of polymerization. When the PVC phase forms, autoacceleration is often noted and the rate of polymerization increases even though the vinyl chloride concentration is decreasing.

Autoacceleration has been explained by a relative decrease in the rates of termination steps for the overall reaction. Evidence has been obtained to indicate that polymerization occurs in both the vinyl chloride and the PVC phases. Polymerization in the PVC phase results as vinyl chloride diffuses to the active sites of the phase (30). These active sites tend to be located close to the surface of the semisolid PVC phase, and polymerization causes the PVC phase in the droplet to grow in size. Because of the limited mobility of the active sites in the PVC phase, these active sites cannot easily react with each other to cause coupling or disproportionation reactions that destroy free radicals. When autoacceleration occurs, the degree of polymerization (or the average molecular weight of the polymer molecules) also increases,

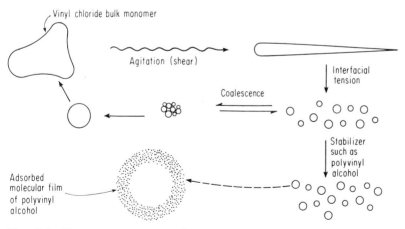

Fig. 6-1 Phenomena occurring during suspension polymerization of vinyl chloride. Type of coalescence affects density and porosity of final PVC product. [*From (1) with permission of Chemical Engineering.*]

which supports the hypothesis concerning termination steps. Because of the complexity of the reactions and because of the several phases present in the system, modeling the kinetics is not easy and requires several approximations. Several investigators (15, 18, 19, 30) have reported results clarifying the basic mechanism.

Emulsion Polymerization   The theory of emulsion polymerization is not completely understood. Although the mechanism proposed by Harkins (27) has been widely accepted, it does not explain all details of the emulsion polymerization of vinyl chloride (25, 34, 38).

At least four components are employed in emulsion polymerization— water as the continuous phase, vinyl chloride as the discontinuous phase, water-soluble initiators, and emulsifying agents to stabilize the agitated emulsions. The emulsifiers, either ionic or nonionic in character, dissolve in the water at low concentrations. When the emulsifier concentration is increased, the surface tension decreases between the phases and the electric conductivity of the mixture increases. Eventually a critical micelle concentration is reached, above which both the surface tension and conductivity change less rapidly as the concentration varies. The emulsifier, which was previously distributed throughout the water phase, begins to agglomerate into groups of emulsifier molecules, or micelles, that contain 20 to 30 molecules (27).

The initiator in the water phase decomposes, forming free radicals. Some investigators (34) believe that propagation reactions start in the water phase with dissolved vinyl chloride; this growing radical then transfers to a micelle. Others believe that the initial radicals migrate to the micelle, where propagation starts. In any case, most propagation steps presumably occur in micelles which gradually are transformed into dispersed PVC particles; these are stabilized with emulsifier. Figure 6-2 summarizes the mechanism proposed by Harkins (27).

Vinyl chloride molecules diffuse from the dispersed droplets of vinyl chloride through the water to the growing polymer chain. As polymerization progresses, the dispersed vinyl chloride droplets decrease in size and the polymer particles grow. Emulsifier molecules also transfer from the monomer droplets to the polymer particles. Obviously the level of agitation in the reaction vessel is important in affecting the kinetics of these transfer steps. When 12 to 20 percent conversion is reached, the micelles have often disappeared and all the emulsifier is located at the surface of the particles (27). At higher conversions, perhaps 60 percent in at least some cases, the monomer droplets have disappeared, and all the monomer is in the polymer phase.

The number of polymer particles formed (and hence the size of the particles eventually formed) is controlled by the early stages of polymerization. Not all micelles give birth to polymer particles. If a growing polymer chain occurs in most micelles, there will be many particles. In at least some cases, the number of polymer particles remains essentially constant after the micelles disappear (34). Apparently polymerization cannot easily be initiated in the dispersed vinyl chloride droplets.

Autoacceleration sometimes occurs in emulsion polymerization (34), and both monomer and polymer phases occur in the polymer particles. After all the emulsifier has been transferred to the polymer interface and as the polymer particles grow, less emulsifier is available for a given area of surface, in which case the stability of the emulsion is decreased. The amount of emulsifier used affects the stability of the emulsion, the number of micelles, and hence the number of polymer particles produced; in some cases it affects the rate of polymerization.

PVC polymers produced by emulsion polymerization tend to have high molecular weights since termination steps involving coupling or disproportionation are rare because the number of polymer chains in each dispersed polymer particle is small. Occasionally a second free radical enters the polymer particle and results in termination or initiates another polymer chain.

The kinetics of emulsion polymerization of vinyl chloride is complicated since polymerization reactions presumably occur both in the micelles and in the water phase. Peggion et al. (34) indicate that the polymerization rate is proportional to the $6/10$ or $7/10$ power of the ini-

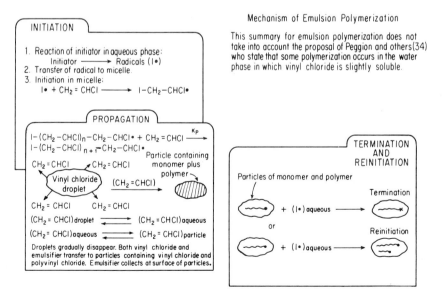

Fig. 6-2 Simplified mechanism of emulsion polymerization. [*From (51) with permission of Chemical Engineering.*]

tiator concentration, whereas Smith and Ewart (45) predict that it should be proportional to the $\frac{4}{10}$ power. Although several investigations (14, 22, 27) present valuable kinetic information concerning emulsion polymerization, more is needed. There are obviously many operating variables of importance, and each variable has complex effects on the final reaction and on the character of the emulsion formed (3).

Bulk Polymerization   The chronology of events for bulk polymerization has been summarized by Thomas (48) and Berger (5). First very fine (perhaps 0.1 μm in diameter) elementary grains of PVC are formed at polymerization levels less than 1 percent vinyl chloride. The number of grains does not increase with further polymerization. At polymerization levels of about 2 percent, the elementary grains agglomerate to form a highly porous structure. Agglomeration is apparently shifted to higher levels of polymerization as the temperature is decreased (47), but viscosity may play a role in this step. The basic structure and character of the elementary grains are retained throughout the remainder of polymerization.

As polymerization progresses, the elementary grains in the granules grow in size, and the viscosity of the mixture increases, since considerable amounts of vinyl chloride are absorbed or trapped inside the porous particles. Slurries containing up to 15 percent PVC can be agitated and pumped (48). A rather sticky intermediate stage occurs at polymerization levels from about 15 to 20 percent PVC. At 20 percent and higher levels, almost all free vinyl chloride is in the PVC granules.

The elementary grains in the PVC granules eventually reach a size of 0.5 to 1.5 μm at polymerization levels of 85 to 90 percent. The granules are of course much bigger than this, often ranging from 50 to 280 μm. Operating conditions employed up to about 7 to 10 percent polymerization levels are critical in setting the size and degree of agglomeration of the elementary grains. Higher degrees of agitation produce smaller grains and also smaller granular particles.

The kinetics of bulk polymerization is not yet well understood. The rate of polymerization is generally not proportional to the square root of the initiator concentration, as is often true with other types of polymerizations. Perhaps the free radicals on the end of the polymeric chain become less accessible to reaction when PVC precipitates (48). Autoacceleration is also sometimes noted when the solid phase is formed.

The molecular weight of the bulk-polymerization product is controlled primarily by temperature and not by the level of initiator (48). Chain transfer from the polymer chain to vinyl chloride is apparently critical and is temperature-dependent. Although much information concerning the mechanism of bulk polymerization is known (2, 4, 15, 30, 48), more is needed.

Solution Polymerization   Solution polymerization as practiced commercially is not a true solution polymerization since PVC precipitates from the solution. It is sometimes referred to as *precipitation polymerization* and is related to bulk polymerization, in which the polymers also precipitate. The solubility of the polymers depends on the solvent, the concentration of vinyl chloride in the solution, and the composition and molecular weight of PVC or copolymer (30). With poorer solvents such as cyclohexane, a precipitate occurs from the start of the run. With better solvents, such as tetrahydrofuran, precipitation does not start immediately. Temperature control is good because the solvent acts as a heat sink.

When the polymer begins to precipitate from solution, autoacceleration is often noted since the precipitate contains solvated vinyl chloride (plus possible comonomer). Polymerization continues in the precipitate, yielding PVC with a higher molecular weight than that produced in solution. The latter PVC may have a fairly low molecular weight because of chain-transfer reactions with the solvent. At low temperatures, chain-transfer steps may not be important, and high-molecular-weight PVC is often produced. A thorough discussion of the complex mechanism of solution (and precipitation) polymerization is presented by Mickley, Michaels, and Moore (30).

Vapor-Phase Polymerization   Preliminary information on the vapor-phase polymerization of vinyl chloride is reported in a patent (39) and patent applications. Vinyl chloride and an initiator are adsorbed (or solvated) on solid PVC particles that are fluidized with vinyl chloride gas. Subsequent polymerization causes the size of particles to grow. The adsorbed concentrations of reactants on the particles are key factors affecting the PVC quality, desired porosity of particles, and temperature control. Transfer of vinyl chloride to the particles is by two methods. First, some particles are often sprayed or contacted with liquid vinyl chloride. Second, some vinyl chloride is always transferred from the gas phase. The initiators used are however quite nonvolatile, and solutions containing initiator are generally sprayed so as to contact the PVC particles. Frequently these solutions are sprayed on seed granules just before they are added to the reactor.

In a fluidized bed, the solids are classified to a considerable extent, and the fines are carried over with the exit-gas stream. Recovery of the fines for seed granules is generally an important feature of the process. Spraying only the fines with initiator solution ensures that only the fines grow in size in the reactor. When the initiator content of a particle is depleted, growth of that particle stops. Other physical steps that are important in this polymerization process include agglomeration and fragmentation of the PVC particles.

Characteristics of PVC Molecules

Vinyl chloride monomers react during the propagation reaction to form primarily a head-to-tail arrangement of the repeating units, $-(CH_2-CHCl)-$, in the chain (6, 23):

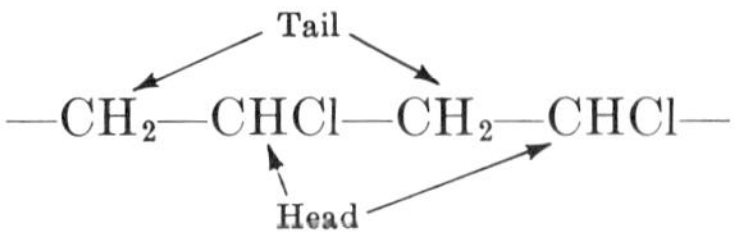

$$-CH_2-CHCl-CH_2-CHCl-$$

with "Tail" labeled above and "Head" labeled below.

Stereospecificity   The carbon atom to which the chlorine atom is attached is asymmetric, and as a result PVC polymers are capable of occurring in various stereospecific arrangements, namely atactic, syndiotactic, and isotactic (7, 43, 44). PVC is generally considered to contain primarily mixtures of syndiotactic and atactic regions (6). Conventional PVC (that has been polymerized at 40 to 60°C) has a lower level of syndiotacticity than PVC produced at −60 to −40°C (36). Conventional PVC, often considered to be primarily atactic, probably has at most 5 to 10 percent crystallinity, whereas as much as 20 percent is obtained in PVC with higher levels of syndiotacticity.

There is some question whether isotactic PVC has ever been produced. Such a polymer might not be commercially of much interest because of an expected high softening temperature (1).

Molecular-Weight Considerations   Commercial PVC polymers have weight-average molecular weights that vary from about 50,000 to 150,000 (43, 44). Routine laboratory tests for PVC generally involve the measurement of the intrinsic viscosity of the polymer rather than the direct measurement of the molecular weight.

The higher-molecular-weight PVC resins are often plasticized (as will be discussed later) and used for flexible tubing, welting, electrical components, garden hose, and calendered film. These products are obtained by extrusion, and the material is subjected to elevated process conditions for only a short time (17). Intermediate-molecular-weight resins are used in film and sheet, coated fabrics, and rigid applications. Low-molecular-weight resins are used in fluidized-bed coatings, phonograph records, and injection-molded parts. Tables 6-2 and 6-3 indicate the effect of molecular weight on both unplasticized and plasticized PVC polymers. The molecular weight has a significant effect on the impact resistance for unplasticized polymers and on the elongation plus tensile strength at 100 percent elongation for the plasticized products.

Chemical Factors Affecting Stability   PVC polymers are relatively unstable at higher temperatures (especially 200°F or more) and in light (13); HCl is evolved, forming double bonds. These bonds are attacked

**TABLE 6-2   Effect of Molecular Weight on Unplasticized PVC
and Blends of PVC and ABS Polymers (17)**

Property	Vygen 65*	Vygen 85*	Vygen 120*	70 parts Vygen 85: 30 parts ABS	70 parts Vygen 120: 30 parts ABS
Molecular weight of PVC........	62,000	74,000	107,000	74,000	107,000
Intrinsic viscosity..............	0.70	0.80	1.18		
Tensile strength, lb/in².........	7,750	7,775	7,850	6,150	6,250
Flexural strength, lb/in²........	11,750	12,000	12,500	9,225	9,600
Flexural modulus, lb/in² × 10⁻⁵..	4.2	4.2	4.4	3.3	3.6
Notched Izod at 77°F, ft-lb/in...	0.44	0.50	0.80	14.0	18.0
Heat distortion (264° lb/in²), °C:					
10-ml deflection..............	69	69	75	66	72
60-ml deflection..............	75	76	80	74	78

* Product of General Tire and Rubber Co.

TABLE 6-3   Effect of Molecular Weight on Plasticized PVC * (17)

Property	Vygen† 85	Vygen† 105	Vygen† 110	Vygen† 120
Molecular weight..............	74,000	83,000	93,000	107,200
Intrinsic viscosity.............	0.80	0.93	1.03	1.13
Tensile strength, lb/in².........	2,160	2,490	2,730	2,890
Ultimate elongation, %........	230	300	340	350
Tensile strength at 100% elongation, lb/in²..................	1,290	1,400	1,420	1,460

* Formulation: 100 parts resin, 50 parts plasticizer (dioctyl phthalate), and 2 parts barium-cadmium stabilizer.
† Trade name of General Tire and Rubber Co.

by oxygen or enter into cross-linking reactions.   Severe degradation of the physical properties and appearance may result, depending to some extent on the plasticizer used.   Although not all the factors affecting stability are known, several features of the polymer molecule contribute to this instability.   Hence manufacturing techniques that minimize these features are of importance.

Unsaturated end groups of the PVC molecules such as $-CH=CHCl$ or $-CH=CH_2$ are "weak" points at which decomposition reactions begin.   Such groups are formed by disproportionation reactions during termination or because of chain-transfer reactions involving vinyl chloride.   These groups weaken the carbon-hydrogen bond which is allylic to the double bond.   HCl is then evolved from the carbon atoms adjacent to the initial double bond.   The new double bond then allows further dehydrochlorination to continue down the polymer chain. This phenomenon is referred to as *zipper dehydrochlorination.* The rate of dehydrochlorination increases as the molecular weight of the polymer decreases and is essentially proportional to the number of end groups. Chlorination of the double bonds minimizes dehydrochlorination.

Initiator fragments on the end of PVC polymer molecules may also be the starting point for decomposition reactions. Chain branching, which occurs to only a relatively small extent, also decreases the stability of the polymer, since tertiary carbon atoms occur with chlorine atoms attached. Such chlorine atoms are much easier to abstract than chlorine atoms attached to secondary carbon atoms. If any head-to-head arrangement of the repeating units occurs in the polymer chain, the polymer molecule is relatively unstable at the point where the two chlorine atoms are adjacent.

Quality Control Tests Tests generally conducted before the PVC is sent to the consumer include intrinsic viscosity (to determine the average molecular weight), bulk density, plasticizer take-up, irreversible plasticizer take-up (blotter resins), moisture content, mill stability, clarity, fisheyes (gel particles), foreign particles such as dirt, particle-size distribution, and press stability. For electrical-grade PVC, the conductivity and pH are also measured.

Plasticizer take-up is of interest since it indicates the rate at which the plasticizer and PVC resin mix and hence the rate at which extrusion or injection molding can be done. Careful control of quality is obviously an important consideration.

Compounding PVC Polymers

PVC are often compounded, i.e., blended, with plasticizers and always with stabilizers. Lubricants and pigments are also sometimes added.

Plasticizers Hard, horny PVC is converted into a softer and rather flexible material by compounding it with a plasticizer. Plasticizers decrease the tensile strength of the polymer, decrease the processing time for extrusion or molding operations, increase the allowable elongation of the polymer, and increase the impact strength and low-temperature flexibility (20, 33). The liquid plasticizer acts to partially solvate the polymer chains, causing them to be partly separated. The intermolecular forces between the chains are decreased, and any crystallinity originally present in the PVC polymers is destroyed. Table 6-4 indicates one example of the relationship between various degrees of plasticization and important physical properties (17).

Several classifications have been given to plasticizers. External plasticizers are mixed physically with the PVC resin, whereas internal plasticizers react chemically with the polymer chain. The following discus-

TABLE 6-4   Effect of Plasticizing PVC with
Dioctyl Phthalate (DOP) (17)

Physical property	Parts of DOP per 100 parts PVC*					
	0	30	40	50	60	70
Tensile strength, lb/in².	7,750	3,550	3,200	2,850	2,500	2,100
Elongation, %.	5–25†	265	295	345	370	410
Shore A hardness, 10-s.	115	98	92	86	79	72

* Vygen 120 (marketed by General Tire and Rubber Co.) has intrinsic viscosity of
1.18 dl/g and a molecular weight of 107,000.
† Estimated from plastic properties chart, in "*Modern Plastics* Encyclopedia 1965,"
McGraw-Hill, Inc., New York, 1964.

sion pertains to external plasticizers, which are divided into primary and secondary plasticizers. Primary plasticizers are highly compatible with the resin, but secondary ones are of intermediate compatibility.

With high compatibility, a true solution is present at all temperatures. With a lower compatibility, a true solution (or single phase) is present only at high temperatures, as used in extrusion or molding. In such a case, as the mixture is cooled to ambient temperatures, PVC and plasticizer phases form. The degree of compatibility has an important effect on the properties of the final product. In addition, especially with low compatibility, the plasticizer may slowly diffuse out of the final product, changing the physical properties significantly with time.

Plasticizers should also have low volalities to minimize odor and loss of plasticizer. Other desired features include good stability relative to heat, light, or various types of radiation; nonflammability; nontoxicity; satisfactory low-temperature properties; and reasonable cost. No plasticizer meets all the above requirements, and thus the selection of the plasticizer or mixture of plasticizers involves a compromise.

Plasticizers are commonly organic esters with a rather high molecular weight, about 300 to 1,500; 679 million lb of phthalic anhydride esters was produced as plasticizers in 1965, of which di(2-ethylhexyl) phthalate was the major one (10). Other phthalate esters used include various C_4 to C_{10} alkyl phthalates. Esters of sebacic, adipic, azelaic, and phosphoric acid are also good plasticizers. Alcohols for these esters are frequently obtained using the Oxo process. Straight-chain alcohols, e.g., those produced by Continental Oil's Alfol process, may have certain advantages in some cases over branched-chain alcohols (20). Epoxidized oils and esters and polymeric esters with molecular weights from about 2,000 to 5,000 have also been used commercially.

Mechanical Blends with Other Polymers   Excellent impact strength is obtained by blending PVC that contains little or no plasticizer with nitrile rubbers or ABS polymers (produced from acrylonitrile, butadiene, and styrene). As indicated in Table 6-2, 70 percent PVC and 30 percent ABS results in impact strengths up to 20 times greater than those of pure PVC, but the tensile and flexural strengths of the blends are somewhat less.

Copolymers of ethylene and vinyl acetate and of methyl methacrylate, butadiene, and styrene and chlorinated polyethylenes are also used in mechanical blends with PVC to improve impact strength and processability.

Stabilizers Certain stabilizers have been found useful in minimizing decomposition reactions caused by heat, light, or ultraviolet radiation. Heat is always a factor in the extrusion or molding operations of the polymer. Most stabilizers are metal salts, such as lead, tin, or combinations of barium and cadmium or of calcium and zinc (37). By making an opaque plastic, stabilization against ultraviolet radiation or light is accomplished. With transparent PVC, several stabilizers have now been found to be quite successful. In general, the stabilizers are used in relatively low concentrations, frequently less than 1 to 2 percent, in the final polymeric material.

A detailed analysis of the complicated problems of stabilization and available stabilizers is given by Chevassus and deBroutelles (13).

Other Additives Fillers, pigments or dyes, lubricants, and fungicides or pesticides are sometimes added, depending on the final use of the compound. Fillers are generally cheap extenders, such as clays, used for certain electrical-grade PVC resins. The clays adsorb free acids and other polar compounds. Asbestos fillers are used in certain floor-tile applications.

Copolymers

Copolymers containing 60 percent or greater vinyl chloride and the remainder primarily vinyl acetate are of commercial importance (24, 26). Copolymers perhaps account for 25 percent of the production capacity of vinyl chloride polymers. Copolymers increase two important properties, namely flexibility and limited solubility in solvents. External plasticizers are one method of increasing flexibility, but plasticizers are not always adequate for solubility requirements.

Copolymers in one sense are internal plasticizers, and certain comonomers, e.g., vinyl stearate or long-chain esters of maleic anhydride, are sometimes copolymerized into the final product. In such a case, the plasticizer is chemically bonded in the polymer chain and hence is a true internal plasticizer.

The properties of copolymers of vinyl chloride and vinyl acetate depend of course on the relative ratio of the two comonomers that have reacted, assuming the same molecular weights of the final polymers. Copolymers containing less than 10 percent vinyl acetate have physical properties quite similar to those of the homopolymer of vinyl chloride except that lower temperatures are required for compounding. Differences in the mechanical properties increase rather rapidly as the concentrations of vinyl acetate increase above 10 percent. Table 6-5 indicates properties of two copolymers containing 3 and 15 percent vinyl acetate, respectively.

Copolymers containing 13 percent or more vinyl acetate and with relatively low molecular weights are used for protective and decorative coatings, flexible film, floor tiles, and compression moldings where exceptionally good flow characteristics are required (26). Phonograph records are one example in which precise duplication and hence excellent flow characteristics are needed. An 87:13 vinyl chloride-vinyl acetate copolymer has been used, but homopolymers of a relatively low molecular weight offer an alternative solution if the heat-stability problem can be solved (43). Copolymers containing less than 13 percent vinyl acetate and with relatively high molecular weights are used for rigid sheeting, extruded rods, and calendered articles.

A small amount of maleic anhydride or other polymerizable carboxyl hydrocarbon is sometimes copolymerized with the vinyl chloride and vinyl acetate when the copolymer is to be used for protective or decorative coatings (41). The presence of carboxyl or epoxy groups improves the adhesion properties of the coatings. In other cases, a portion of the acetate groups on the polymer chain is removed by hydrolysis to produce hydroxyl groups, which also improve adhesion.

Since vinyl chloride–vinyl acetate copolymers are produced by batch polymerizations, the relative ratios of vinyl chloride and vinyl acetate that react tend to vary with the extent of polymerization unless vinyl

TABLE 6-5   Properties of Rigid Press-polished Sheets of Vinyl Chloride–Vinyl Acetate Copolymers (26)

Property	Acetate, %	
	3	15
Mill-roll temperature for softening, °F	340	230
Hardness, Rockwell M	60	50
Notched impact, ft-lb/in	0.65	0.20
Abrasion loss in 2,000 cycles, %	0.02	0.13
Heat distortion (66 lb/in²), °C	67	57
Tensile strength, lb/in²	8,200	8,500
Yield stress in flexure, lb/in²	12,800	12,200

chloride (the more reactive comonomer) is added in order to maintain a constant ratio of comonomers for the reaction (26).

Vinyl chloride is also copolymerized commercially in significant quantities with acrylonitrile and vinylidene chloride (39). Copolymers of vinyl chloride and propylene, which were announced several years ago (8, 10, 28, 29), are now finding important uses. These latter copolymers are extruded at temperatures ranging about 30 to 50°C lower than those used for PVC homopolymers. As a result, thermal decomposition of the modified PVC polymers is decreased. Furthermore zipper dehydrochlorination is less of a factor since the repeating unit resulting from propylene tends to stop the degradation reaction along the chain. These copolymers normally contain 3 to 10 percent propylene groups and the remainder vinyl chloride. They have received clearance from the Federal Drug Administration for food and drug applications.

II. PROCESS DETAILS

Vinyl chloride is polymerized commercially by suspension, emulsion, bulk (or mass), and solvent (or precipitate) polymerizations. Major improvements have recently been realized in several of these processes, resulting in improved products and/or reduced operating costs. Each process offers unique advantages in producing certain of the large number of PVC products currently marketed. In addition, a vapor-phase polymerization technique has recently received considerable attention and may shortly be commercialized.

Suspension Polymerization

Suspension-polymerization processes since the 1950s have been the major method used for polymerization of vinyl chloride (50, 55, 78, 80, 81, 43, 44). Although attempts have been made to develop continuous-flow variations of suspension polymerization, only one type of commercial flow process has apparently been developed. Various modified batch processes will be described later; Fig. 6-3 is the flowsheet used by one major producer.*

Monomer Storage   Vinyl chloride at ambient temperatures and pressures is a colorless gas with a faint, sweet odor. Mixed with air, it is explosive (94); a large plant was severely damaged by an explosion that occurred after a sight glass on a storage tank broke (57). High

---

* Considerable information on this process was presented at a seminar sponsored by the AICE for chemical engineering instructors in October 1964. The information on polymerization was presented by D. J. Clark, W. J. Hanlon, and J. L. Weaver, of General Tire and Rubber Company. The author wishes to thank that company for permission to use this information.

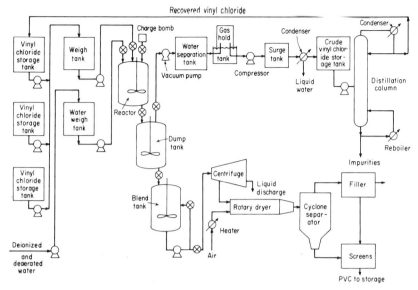

Fig. 6-3 Suspension polymerization of vinyl chloride for production of granular PVC. [*From* (50) *with permission of Chemical Engineering.*]

concentrations of vinyl chloride in the air are toxic to man. In storing vinyl chloride, precautions are necessary to protect plant personnel, to prevent fires and explosions, and to maintain high-quality monomer.

In the past vinyl chloride to be shipped contained a small amount of phenol inhibitor to minimize polymerization. This inhibitor was removed before polymerization by washing the vinyl chloride with caustic and then with water. Inhibitors, however, are not necessary in highly purified vinyl chloride and are generally not used now. The vinyl chloride is frequently stored in a wet state, as is also the recycle vinyl chloride. The temperature of the vinyl chloride in the insulated storage tanks is often maintained at 60°F or less. Vapor recompression or external cooling is preferred in these tanks to internal coils. At such temperatures, hydrolysis reactions or peroxide formation is effectively suppressed.

Formerly, both glass-lined and stainless-steel storage tanks were used. In one plant, 316 stainless steel was preferred to 304 stainless steel (50). Trace quantities of water and HCl (formed by hydrolysis of vinyl chloride) led to corrosion and iron pickup. With the use of lower temperatures for storage and for high-purity vinyl chloride, low-carbon steel is suitable (89). Care should be taken to exclude copper or brass from the system because trace amounts of acetylene may react to form copper acetylides.

One company uses several storage tanks of 15,000 gal capacity, and each tank can hold up to 110,000 lb of liquid vinyl chloride. At 60°F, the pressure in the tank is approximately 27 lb/in^2 gage. The number of storage vessels depends of course on the plant size, size of vessels, and the availability of vinyl chloride. When vinyl chloride is purchased, more storage capacity is generally needed than when vinyl chloride is produced at the plant.

Shelley and Sills (89) describe in considerable detail the equipment and techniques to be used in storing vinyl chloride.

Batch Reactors   Batch reactors provided with one or more agitators are used for the polymerization step of the suspension process. Figure 6-4a shows the tops of several reactors in which the agitator shaft enters through the top. Figure 6-4b is a sketch showing the main features of a reactor.

Temperature control of the exothermic polymerization is a key factor in designing and operating the jacketed reactors. Water at the desired temperature is rapidly pumped through the jacket. Part of this water is recirculated, and part is discarded. Normal cooling water, refrigerated water, steam, or blends of the three provide the necessary makeup for the jacket (52). The reactors are jacketed generally both on the side walls and on the bottom. Jacketing the top head is also employed on occasion. Water and especially vinyl chloride vaporize during polymerization, condense on the cold top, and then reflux back to the reaction mixture. Additional heat-transfer surfaces are sometimes provided by cooled baffles, external heat exchangers, and external reflux condensers. Cooled baffles may account for an additional 10 percent of heat-transfer surface. In the past, however, external heat-transfer surfaces have had to be cleaned at rather frequent intervals. Temperature control tends to be more difficult for larger reactors because of the lower ratios of heat-transfer surface to volume of reaction mixture.

Glass-lined reactors were the preferred type of reactor up to about 1970. These reactors varied in size from about 2,000 to 6,000 gal. Bigger glass-lined reactors were in general impractical because the rather low overall heat-transfer coefficients, about 30 to 50 Btu/(h)(ft^2)(°F), did not allow adequate temperature control. Glass-lined vessels were once thought to be necessary in order to minimize polymer buildup (or sticking) on the walls of the reactor. Such buildup is most undesirable, since it decreases the heat-transfer coefficients and also causes "fish eyes" in the final polymer product. These reactors can be operated for several batch runs before they have to be cleaned.

Stainless-steel-clad or all-stainless-steel reactors have now been developed; in these, polymer deposition is not a critical problem. Such reactors have been used in many, if not most, recent installations. Their

(a)

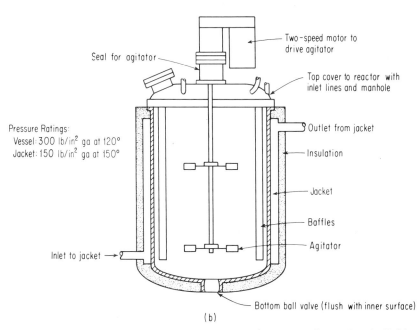

Two-speed motor to drive agitator

Seal for agitator

Top cover to reactor with inlet lines and manhole

Pressure Ratings:
Vessel: 300 lb/in² ga at 120°
Jacket: 150 lb/in² ga at 150°

Outlet from jacket

Insulation

Jacket

Baffles

Agitator

Inlet to jacket →

Bottom ball valve (flush with inner surface)

(b)

Fig. 6-4   (a) Top section of suspension PVC reactors (*Goodyear Tire & Rubber Co.*); (b) schematic sketch of a suspension PVC reactor.

inner surfaces must be highly polished, and The Brighton Corp., for example, used an electropolishing technique to produce an almost mirror-like finish on 304 stainless steel.   Overall heat-transfer coefficients are much higher in these vessels, being about 75 to 130 Btu/(h)(ft²)(°F). As a result, much larger reactors are practical, and/or the time for a run can be significantly shortened. Hence significantly lower polymerization costs have resulted.

Two very large stainless-steel reactors have recently been reported. The Shinetsu Chemical Co. of Japan (60) uses a 35,000-gal (131-m^3) reactor that is operated for at least 100 batches without cleaning. Treating the inner walls of these reactors and the accompanying reflux condenser with various polar compounds, dyes, or pigments before a polymerization run reduces polymer depositions, often by factors of 50 or more (72a). Such treatments presumably reduce vinyl chloride adsorption on the surfaces. In Germany, a 52,000-gal reactor is being employed.

The type of jacket and the type of steel used for construction of the reactor are both important design considerations. The water in the jacket should flow with a relatively high velocity to promote high heat-transfer rates. Dimpled, half-pipe coil, and conventional (but baffled) jackets are available (76). Dimpled and half-pipe coil jackets also promote structural strength of the vessel. Since stainless steel has a lower thermal conductivity than low-carbon steel, there is not only an economic but also a heat-transfer advantage in using plain steel clad with stainless steel for the construction of the vessel.

Several methods are available for cleaning the reactor. In some modern reactors, an automatic hydraulic reactor cleaner that provides high-velocity water streams is installed inside the reactor. A reactor can usually be cleaned in this manner within 20 to 30 min, and some companies clean their reactors after each polymerization. Solvent cleaning with tetrahydrofuran, N-methyl-2-pyrrolidone, or other solvents is used by some companies after perhaps every four to six runs. The solvents are expensive, and they must be recovered and recycled. Manual cleaning was widely used at one time; the reactor is opened after every several runs, and the polymer on the wall is mechanically chipped and scraped away. This type of cleaning, however, tends to cause scratches so that subsequent polymer deposition occurs. The cleaning procedure must be carefully standardized to ensure safe operating conditions and to minimize entry of air (or oxygen) into the reactor.

Polymer buildup or skins on the reactor walls often occur throughout the entire reactor. In the top portion of the reactor, the depositions may be caused by higher skin temperatures of the surface or because of oxygen that entered during cleaning operations. Hydrogen chloride formed by hydrolysis of vinyl chloride may also promote skin formation.

The degree of agitation in the reactor affects the average size of the vinyl chloride droplets formed, the porosity of the final PVC product, and the heat-transfer coefficients obtained. The following factors are important: number of impellers, impeller size and design, speed of agitator, number and type of baffles, and materials of construction. In most reactors, the impeller shaft enters through the top of the reactor, but in some cases it enters through the bottom. A popular impeller is a three-retreating-blade agitator, but use of axial-flow impellers is increasing. One company that uses a rather small autoclave operates their agitator at about 120 r/min. An ammeter is normally employed to indicate if the power requirements are changing. At least two impellers are often provided in large reactors.

Tubular-finger, single-blade, and flattened-coil baffles have been employed in the reactors. The position of the baffle can often be varied, and a trial-and-error procedure is used to find the best position. Scale-up and design of larger reactors involve at best only a semitheoretical approach. There is definitely a need for additional theoretical information on the scale-up of this two-phase system.

Factors Affecting Reactor Size   The size of the reactors depends on at least the following factors:

1. Average temperature of coolant used in the jacket of the reactor
2. Maximum rate of polymerization during the batch run
3. Temperature of polymerization
4. Ratio of water to vinyl chloride
5. Heat-transfer coefficients

The temperature for the coolant fluid has been calculated (50) for glass-lined reactors with overall heat-transfer coefficients of 46 Btu/(h)(ft^2)($^\circ$F) and containing suspensions of 182 weight parts of water per 100 parts of vinyl chloride. The heat-transfer surfaces in typical 2,000-, 3,000-, 4,000-, and 5,000-gal autoclaves are about 180, 245, 295, and 340 ft^2, respectively. The heat of polymerization is 65 Btu per pound of vinyl chloride polymerized. When the polymerization occurs at 130°F (54°C), and when the maximum rate of polymerization during the batch run is 16 percent vinyl chloride per hour, energy balances indicate that the average temperature of the water (or brine) in the jacket must be 60, 53, 43, and 35°F, respectively, for the 2,000-, 3,000-, 4,000-, and 5,000-gal reactors. The temperature of the feedwater (or brine) to the jacket must of course be less than that of the water being recirculated, and hence refrigeration is necessary. Basel and Papp (52) have also made similar calculations with a 3,700-gal reactor. In stainless-steel reactors, which have significantly higher heat-transfer coefficients, the jacket temperatures are substantially higher.

The large Shinetsu 35,000-gal reactors require a special combination of initiators in order to obtain an essentially flat or uniform heat release (or rate of polymerization) during the entire batch polymerization. These reactors are also provided with a variable-speed drive for the agitator. Possibly the speed of agitation is changed during the batch run.

If a higher or lower ratio of water to monomer is employed than in the above example, the coolant temperature can be higher or lower, respectively, since the amount of PVC formed per batch is changed.

Polymerization Initiators The type and amount of the initiator employed have a large effect on the rate of polymerization, as shown in Table 6-6. Lauroyl peroxide has been a popular initiator for many years, and a typical recipe on a weight basis is as follows: vinyl chloride, 100; deionized water, 200; lauroyl peroxide, 0.1; and polyvinyl alcohol, 0.1. The last material is the protective colloid used to stabilize the monomer particles in the two-phase system. The amount of stabilizer affects the size of monomer droplets formed plus the eventual size of the polymer particles. Further, it affects the degree of coalescence of the particles, the porosity (or bulk density) of the PVC product, and apparently the tendency of the polymer to deposit on the reactor walls. Some companies have changed their recipes slightly when they switched from glass-lined to stainless-steel reactors. Differences in the skin temperatures of the reactor wall and the nature of the surface apparently affect the deposition of polymers.

A typical relationship of conversion vs. time for polymerizations at 122°F (50°C) using lauroyl peroxide is shown in Figure 6-5. About

TABLE 6-6 Suspension Polymerization of Vinyl Chloride*

Initiator	Amount, parts	t, °C	Time for 45 lb/in² pressure drop, h
Lauroyl peroxide..........	0.125	54	15
IPP....................	0.035	54	6
	0.035	51	8
	0.035	43	16
	0.025	60	5.5
	0.025	54	8.0
	0.025	51	10.5
	0.020	54	9.3
	0.015	54	12.3

Reactor size: 20 gal
Formula:
  Vinyl chloride, 100 parts
  Water, 200 parts
  Polyvinyl alcohol, 0.065 part

* Data furnished by PPG Industries.

20 to 21 h is a typical time required to obtain 95 percent polymerization of vinyl chloride, and the maximum rate of polymerization is about 8 percent conversion per hour after 11 to 15 h when autoacceleration occurs.   An induction period of 1 h is often experienced when lauroyl peroxide is used.   It is probably caused by polymerization poisons, e.g., oxygen, that leak in during cleaning of the reactor.   In addition, time is required for the buildup of the free-radical concentration.

Diisopropyl peroxydicarbonate, more commonly called isopropyl percarbonate (IPP), has been widely accepted as an initiator within the last few years.   It is generally added to the reaction vessel in a solution using an inert solvent such as hexane.   IPP has a half-life of about 2.0 h at 122°F, compared with 50 h for lauroyl peroxide.   IPP has several important advantages over lauroyl peroxide (11, 93).

1. Time for polymerization is reduced.   Batch runs are often made within 4.5 to 12 h, depending on the amount of IPP employed and on the temperature.   Figure 6-5 indicates a run with IPP which was completed in 10 to 11 h and in which the maximum rate of polymerization was 16 percent conversion.

2. Little or no induction period is necessary, since an appreciable concentration of free radicals is formed quickly.

3. Fewer initiator fragments and less unreacted initiator are incorporated in the polymer.   Improved polymer quality is the result.

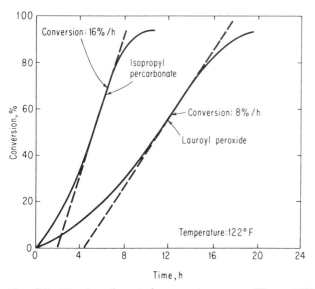

Fig. 6-5   Kinetics of typical suspension runs.   [*From (50) with permission of Chemical Engineering.*]

4. Polymerization costs are reduced.   Although lauroyl peroxide costs less on a pound basis than IPP, considerably less IPP is used, perhaps as little as one-tenth as much.   The biggest saving however, results, from the shorter times required for batch runs.   The reaction rate furthermore is more linear throughout the run when IPP is used.   Total savings of nearly 0.75 cent per pound of PVC produced are claimed (11).

5. There is less long chain branching.

Other peroxydicarbonate initiators now available include di-*n*-propyl, di-*sec*-butyl, di-2-ethylhexyl peroxydicarbonates.

Smith and Rogers (41, 42) have developed a method to make diethyl peroxydicarbonate in situ from ethyl chloroformate and sodium peroxide as follows:

$$2C_2H_5—O—\overset{\overset{\displaystyle O}{\|}}{C}—Cl + Na_2O_2 \rightarrow 2NaCl$$

$$+ C_2H_5—O—\overset{\overset{\displaystyle O}{\|}}{C}—O—O—\overset{\overset{\displaystyle O}{\|}}{C}—O—C_2H_5$$

The ethyl chloroformate is mixed with the vinyl chloride; hydrogen peroxide and sodium bicarbonate are added to the water.   The pH of the water phase is greater than 7.0.   The diethyl peroxydicarbonate is formed during the polymerization run, presumably at or close to the interface of the water and organic phases.   Advantages claimed for this technique are as follows:

1. Shipping and handling of unstable initiators are eliminated.
2. Uniform and controllable reaction rates are realized during the entire batch run.
3. Polymerization costs are reduced because of lower material costs for the initiator.   Cost savings of almost 0.09 cent/lb are claimed as compared with those for IPP (41).
4. Safety in operation is improved.   If a power failure should occur so that agitation stops, there is less probability of a fast runaway reaction, since the formation of the initiator almost stops.

Polymerization times as low as 4.5 h can be realized in 10,000-gal reactors using the above technique (91).

The relationship between polymerization temperature and molecular weight is essentially identical for both conventional and in situ systems, as shown in Fig. 6-6.   PVC polymers with intrinsic viscosities varying from 0.8 to 1.3 dl/g have been produced commercially by in situ systems.

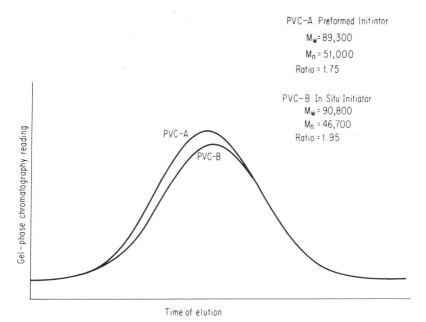

PVC–A Preformed Initiator
$M_w = 89,300$
$M_n = 51,000$
Ratio = 1.75

PVC–B In Situ Initiator
$M_w = 90,800$
$M_n = 46,700$
Ratio = 1.95

PVC–A

PVC–B

Gel–phase chromatography reading

Time of elution

Fig. 6-6 Comparison of molecular weights of PVCs produced with conventional percarbonate initiator and initiator formed in situ. (*Data from Goodyear Tire & Rubber Co.*)

Limited experience has also been obtained in producing resins with intrinsic viscosities down to 0.6.

Considerable effort has been devoted to developing still faster or more active initiators. More uniform rates of polymerization are frequently obtained if two or more initiators are mixed. The Shinetsu process, for example, is reported to have an almost uniform rate, but the specific initiators and recipe have not yet been reported.

Factors Affecting Suspension The amount of water used in the suspension polymerization varies commercially from about 1.3:1 to 4:1 (on a weight basis). Factors of importance in setting this ratio include the following:

1. Temperature control. With increased amounts of water, faster polymerizations can be made and still maintain adequate temperature control.

2. Amount of polymer formed per batch. At lower ratios, more monomer is added to the reactor in each batch.

3. Sufficient water to maintain a free-flowing two-phase system. The size of the monomer droplets (and hence the size of polymer particles formed) will be affected by the water-hydrocarbon ratio. High-porosity PVC particles soak up significant quantities of water. A higher ratio

may be necessary with such a polymer, whereas a lower ratio is adequate with nonporous PVC. Factors affecting porosity will be discussed later.

The degree of agitation of course affects the characteristics of the suspension, and it may change somewhat as the run progresses. The volume of the organic phase shrinks during polymerization by about 35 percent since the density of the polymer is greater than that of the monomer. The slurry level decreases, primarily in the final stages of polymerization after the sizes of the particles are relatively well established. The suggestion has been made to introduce additional water to compensate for volume changes during the reaction. This method is claimed to reduce the amount of suspension agent required and to minimize the formation of polymer skins. It is doubtful whether the method is widely used, however.

Deionized and deaerated water is needed for suspension polymerization. The water helps maintain good temperature control and provides a substrate in which the suspending agents control the surface properties of the polymer particles formed.

Failure of the agitator in the reactor may lead to serious trouble. In such a case, the organic phase separates and coalesces, which may lead to a runaway reaction because of inadequate temperature control. If the following action is quickly taken, the danger can be minimized. A shortstopping agent such as styrene, isoprene, or butadiene is added to destroy most of the free radicals, and vinyl chloride is vented to obtain sufficient evaporative cooling to prevent excess temperatures.

The gross physical properties of the resin produced, such as the size and porosity of the polymer particles, depend on a complex relationship of the type and amount of the suspending agent, presence (or absence) of salts, agitation, temperature, and ratio of water to monomer. Smith (44) has presented three basic recipes for good-quality general-purpose resins, all of which use lauroyl peroxide as initiator but employ different suspending agents. The three agents are polyvinyl alcohol, gelatin (type B), and methyl cellulose. The bulk density, particle size, and the time required for dry blending of the three polymers vary significantly. The ability to blend quickly with the plasticizer is most important to the fabricator of the final plastic product since it affects the rate of production in his equipment.

Operational Details   Start-up of a batch run varies with the companies. One company adds the charge (vinyl chloride, water, initiator, suspension agent, etc.) to the reactor at essentially ambient temperature. Weigh tanks are used for measuring each. Agitation is then started, and steam is used as a heating agent in the jacket to heat the mixture to the desired reaction temperature, usually from 113 to 158°F (45 to 70°C). At least $\frac{1}{2}$ h is required for this heating. As the polymerization starts, cooling water or brine is substituted for steam in the jacket. Sometimes the temperature is allowed to overshoot by 9°F during the initial portion of the run.

The temperature is then controlled closely during the run. Frequently a master-slave cascade instrument arrangement is used to measure both the internal reactor and jacket temperatures (44), and steam or cooling fluid is automatically supplied to the jacket as needed. In some cases, the temperature is also allowed to rise somewhat during the final stages of a run. The latter technique produces a lower-molecular-weight PVC during this time, and the resulting polymer can be processed somewhat more easily probably because of its lower melt viscosities.

Increased temperature increases the rate of polymerization, and in addition, as already indicated, it results in lower-molecular-weight polymers. The stability of the polymer is reduced at higher temperatures or because of hot spots resulting at these temperatures. Polymers produced at higher temperatures have lower polymerization costs since the rapid rate of polymerization results in increased capacity of the reactor. When PVC of rather low molecular weight is desired, chain-transfer agents such as chlorinated hydrocarbons and isobutylene can be added.

Figure 6-7 indicates the total pressure vs. conversion for a run at about 130°F (50). The pressure at the beginning of the run increases to about 115 lb/in² gage (130 lb/in² absolute) as the temperature of the reaction mixture rises to 130°F. This pressure is the sum of the

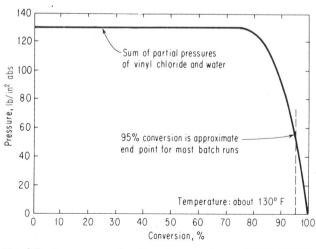

Fig. 6-7   Pressure as a function of conversion at 130°F (54°C) [*From (50) with permission of Chemical Engineering.*]

vapor pressures of water and vinyl chloride. Up to about 70 percent conversion, both a vinyl chloride and PVC phase are present in the suspended droplets in the suspension. At about this conversion, the vinyl chloride phase is depleted since the remaining vinyl chloride is essentially adsorbed in the PVC phase. The pressure then begins to drop; vinyl chloride in the gas phase condenses and enters the liquid suspension. Polymerization beyond 90 to 95 percent is slow, and degradation of product (increased long-chain branching, decreased resin porosity, and more color) occurs.

One company uses a *hot-charge* technique for starting the polymerization run. The cold monomer is added first, and then preheated water containing the required amount of stabilizer is pumped into the reactor. Agitation is started, and the initiator, generally in a solution, is added, using a charge bomb located on top of the reactor. The reaction mixture is at a sufficiently high temperature to eliminate additional preheating of the mixture, and the polymerization starts essentially immediately.

The following hot-charge technique has been tested. Hot water is added to the reactor first, followed by a relatively cold vinyl chloride containing dissolved initiator. The temperature of the mixture is sufficiently high for polymerization to start but the resulting polymer is of relatively poor quality. Analysis of the above technique indicates that the first portion of the vinyl chloride that enters the reactor vaporizes but probably only a small part of the initiator does. Hence during the initial stages of the run, some of the vinyl chloride droplets have high concentrations of initiator, resulting in product with reduced quality.

The following modification might increase the quality of PVC formed in this latter type of hot charge. Pure vinyl chloride (containing no initiator) should be added to the reactor until the desired pressure has been obtained, as shown in Fig. 6-7. Then and only then should vinyl chloride containing initiator be added. This technique would be likely to minimize concentration gradients of initiator in the organic-phase droplets.

Many, if not all, reactors are provided with two-speed agitators. The higher rate of agitation is employed during the polymerization portion of the batch run. The decreased rate is needed as the liquid level drops when the reactor is drained; otherwise, a vortex forms, slowing the draining rate.

High-Porosity PVC   At least three techniques can be employed to obtain high-porosity PVC resins. First, the batch polymerization is stopped while a significant amount of vinyl chloride is still present (54, 44). The unreacted monomer is vaporized, leaving highly porous granular solids. This technique can be used to produce resins with bulk densities as low as 0.3 g/cm³, but it is relatively expensive since the production capacity of the reactor is reduced significantly because only partial polymerization is realized.

In a second and popular method, after partial conversion has been obtained, the pressure on the reaction vessel is suddenly released for a short time. Part of the unreacted vinyl chloride is rapidly vaporized, causing the granular PVC particles to swell (54). Following this, polymerization is then often continued to high conversion levels. In general, the bulk density of the PVC is lowered as the pressure drop is increased. The overall amount of polymer formation in a reactor is greater by this method than by the first one.

The porosity of the PVC particles can also be increased by adding secondary emulsifiers, e.g., sulfonated oils or acids, condensation products of ethylene oxide, and other surfactants, to the reaction mixture. PVC products that have relatively large particles (principally 80 to 100 mesh), flow readily in dry blends, and have relatively good adsorptivity can be produced in this way. The secondary emulsifier encourages formation of rough, porous particles since it promotes the desired degree of partial coalescence of the dispersed particles. Inorganic buffers are sometimes added with gelatin systems to maintain the proper isoelectric relationship in the suspension and maintain its colloidal efficiency. The exact formulas used in suspension polymerization are closely guarded trade secrets.

Cycle for Batch Run   A complete cycle for a batch reaction involves several steps. The times required for each step of the cycle vary of course with the specific unit, but the following times are typical for a larger reactor when IPP is the initiator:

	Hours
Charging reactor with reactants	0.5
Reaction time	10.0
Drop and flush time	0.5
Scheduling and turnaround	2.0
Total	13.0

Although the reactor does not have to be cleaned after each run, approximately 2 h is required on the average for cleaning and maintenance per batch. Assuming 350 operating days per year and 5 additional days in which the equipment has operational difficulties, approximately 550 batches per year are possible for each reactor. For each 1,000 gal of reactor volume, the production capacity is about 1.3 million lb/year of PVC, which is equivalent to about 0.15 lb/h per gallon of reactor space. One company uses 17 reactors, each with 3,500 gal capacity, for production of 75 million lb/year of product. The time required for essentially complete polymerization of a batch is sometimes as low as 5 h with a fast initiator. In such a case, the production capacity is increased by about 150 percent.

Direct digital control of 80 batch reactors and the necessary mixing tanks has proved successful for improved plant operation (81). The computer was able to charge over 100 different formulations to the reactors.

Recovery System for Vinyl Chloride   After the reaction is completed, the suspension is emptied from the reactor into a stainless-steel dump (or drop) tank.   About 10 percent of the unreacted monomer is often solvated in the PVC particles, and the remainder is present either as a dispersed phase in the water or in the vapor phase above the suspension.   Separation of the solvated monomer is difficult since it must diffuse through the particles.   Diffusion is a function of time, temperature, and pressure (or vacuum).

When the suspension is in the dump tank, simple flashing and stripping are employed.   The rate at which the monomer plus water is removed depends on the vacuum obtainable and also on the temperature maintained in the tank.   Heat is provided, otherwise vaporization of the monomer would cool the system.   Live steam is usually injected directly into the suspension.   Care must be exercised in this step to prevent degradation of the product.   The steam should be injected so that quick and intimate mixing occurs and none of the polymer is overheated, as would occur if a hot heat-transfer surface were used.

The total time for a batch in the dump tank is about 2 h: about $\frac{1}{2}$ h to empty a charge from the reactor into the dump tank, 1 h for stripping vinyl chloride, and another $\frac{1}{2}$ h to unload the tank.   These tanks are agitated to prevent phase separation of the suspension, and they are at least 30 percent larger than the reactor because of foaming while the monomer vaporizes and is removed.   One company uses a dump tank large enough to hold three batches from a reactor (75).   A small amount of polymerization may occur in the dump tank.

One dump tank is often provided for three to six reactors.   Polymerization runs must be carefully scheduled so that a dump tank is ready for each batch of polymer.   The remainder of the recovery system, including the compressors and condensers, should be operated as continuously as possible.   The suction tank, probably a gasometer, before the compressor provides storage of crude vinyl chloride gas for at least several hours.

A number of positive-displacement compressors, e.g., water-sealed ones, are often used in parallel.   All the compressors are used during peak periods, but fewer compressors are employed otherwise.   Vinyl chloride plus water vapor is compressed to about 5 to 6 atm pressure, and normal cooling water is used for condensation. The condensed water and vinyl chloride are essentially immiscible, and the water phase is often discarded. Crude vinyl chloride liquid is pumped into a storage tank that can be constructed of plain carbon steel since the vinyl chloride is generally purified by distillation before reuse (44). Recovery procedures for the unreacted vinyl chloride are generally sufficiently effective to realize 95 to 97 percent or even higher yields of polymer (where yield is defined as the percent of feed monomers converted in the overall process to recovered polymer). Yields have tended to increase somewhat in the last few years because of improved recovery techniques and operation.

Recovery of Polymer The slurry from the dump tank is pumped to a blend tank, which is a slurry storage tank (see Fig. 6-3). Tanks with a capacity of 15,000 gal constructed of 304 stainless steel are used in one installation. Continuous agitation is provided to prevent separation of the polymer particles. In addition, the slurry is continuously recirculated through the exit pump and line to prevent plugging there. Often about five batches from a dump tank can be blended in this way, perhaps from 20,000 to 35,000 lb of polymer in about 105,000 lb of total slurry. Quality-control tests are performed on the batch before it is processed further.

The slurry from the blend tank is sometimes pumped to a centrifuge for the separation into a water solution and wet PVC solids containing 23 to 25 percent water. The slurry enters a horizontal feed pipe positioned at the axis of the centrifuge. The centrifuge is conical in shape with the bowl rotating at 500 r/min for larger units and with a plow mechanism rotating in the same direction as the bowl but at reduced speeds. The solids are transported to the small end of the bowl, where they are discharged. The liquid is discharged at the large end of the bowl.

The wet solids in the centrifuge can be washed with deionized water, if desired. This technique is particularly applicable for electrical-grade resins since washing reduces the impurities in the polymer. The amount and temperature of the wash water affect the removal of the soluble ions. Up to about 60 percent of the ions are removed with 120°F water added at an 8:1 ratio of wash water to the water originally retained in the wet solids.

Some companies separate the suspension by filtration or screening instead of centrifuging. They can then wash the residue with water. The centrifuges or filtration equipment are normally operated as continuously as possible.

Drying Polymer with Rotating Tubular Dryer Drying the polymer is a critical phase of the PVC process. At about 150°F (65°C) and above, degradation of the polymer occurs. A rotating tubular dryer is often used, as shown in Fig. 6-3. One producer provides two centrifuges and one dryer for each blend tank. In other cases, up to three blend tanks may be used for a single dryer.

A dryer rated to produce 4,000 lb/h of dry PVC is constructed from stainless steel and is 10 ft in diameter and 30 ft long. The hot clean air and wet solids flow concurrently through the dryer. This particular dryer vaporizes 1,300 to 1,500 lb/h of water. Latent-heat requirements for this amount of water approximate 1.4 to 1.6 million Btu/h. Entering air is first heated to about 300°F in a finned-tube heat exchanger that can transfer 3 million Btu/h. The flow of air is about 55,000 lb/h with a velocity approximating 200 ft/s in the dryer proper. The time required to dry a batch of polymer of the blend tank is from about 5 to 8 h.

Air entering the dryer at 300°F causes a polymer surface temperature approximating 120°F (the wet-bulb temperature). As the air and solids pass through the dryer, their temperatures approach 146 and 136°F, respectively. The PVC particles at the end of the dryer contain about 0.25 weight percent moisture.

The cross-sectional area at the exit end of this specific dryer is constricted to raise the air velocity sufficiently to entrain (or fluidize) the dry PVC particles. A cyclone separator is highly effective in removing coarser particles (99.93 percent) and even fines (99.4 percent). Bag filters clean the exit-air stream, and they are cleaned by regulating pulses of air that strike the neck of the bag. Such pulses set up sinusoidal patterns sufficient to loosen adhering particles. This system has a high mechanical efficiency without appreciable operating difficulties.

In one plant, the solid PVC particles recovered from the cyclonic separators and bag filters are sized using triple-deck screens. Oversize particles are designated off-specification material.

The screened PVC particles are air-conveyed to storage bins or silos and finally are packaged. Standard 50-lb bags are still common, and at least 20 bags per minute can be filled by the loading apparatus. Bagging is being replaced to a considerable extent by packaging techniques involving much larger quantities. The most economical method is to fill a railroad hopper car.

Flash and Other Drying Methods Several flash-drying techniques are used for drying wet PVC granules. Allied Chemical Company uses a combination of flash and fluidized-bed driers in their plant with a nominal product capacity of 100 million lb/year (79). The PVC-water slurry is first screened, yielding a solid with 30 to 35 percent moisture. A continuous centrifuge reduces the water content to 20 percent, and these solids are stored in a large hopper. Each of three dryers is fed from a hopper; two vibrators on each hopper facilitate smooth flow of the solids.

In the Allied installation, the flash dryer is a 3-in duct about 50 ft long. Filtered and heated air sweeps the wet solids through this duct, reducing their water content to about 1 to 3 percent. The exit solids then pass downward through two fluidized-bed chambers; hot air flows upward through these chambers, which are operated batchwise. The dried (0.2 percent water) PVC polymer is reported to be of excellent quality. A picture of these dryers has been published (79). Three flash or fluidized-bed dryers can process up to 36,000 lb/h of product.

Firestone Plastics Co. selects the preferred drying method to a large extent according to the size of the PVC particles, as indicated below (75):

Drying Method	Particle Size, Mesh
Spray drying	<325
Flash-rotary	200
Rotary	142
Two-stage flash	100

Firestone uses a solid-bowl centrifuge to dewater the slurry for products to be dried by the last three methods. In their flash-rotary method, the wet solids are air-conveyed using 300°F air in a 30-in duct; the resin is dried to 8 to 10 percent water. The polymer is then separated from the air in a cyclone separator 9 ft in diameter. The second stage of drying is in a rotary dryer, as already described.

In the two-stage flash-drying method, the wet polymer from the first stage is dried to the desired level in the second step. Flash drying results in highly porous, adsorptive resins.

When product with a very fine particle size is to be spray-dried, the slurry is fed directly to the spray-drying column without dewatering. Firestone uses a spray dryer that is three stories high, 18 ft in diameter at the top, and conical at the bottom. Spray drying is used primarily for the products formed by emulsion polymerization (to be discussed later). The resin produced by this method is a very fine, white flour. This drying method is used when separation of the water-polymer mixture by either centrifuging or filtering is difficult. It is expensive since large quantities of water are vaporized.

Economics of Batch Suspension Processes Labor costs for operating a plant using batch suspension-type reactors are always a significant operating expense. For most American plants producing about 100 to 150 million lb/year of polyvinyl chloride, the following labor is necessary: 20 to 30 operators, 8 to 10 maintenance personnel, and 10 to 14 supervisory and support personnel. Operating costs due to labor then total about 0.6 to 0.9 cents per pound of PVC. These costs of course depend to a significant extent on the number and size of reactors used.

Using larger and hence fewer reactors that are well instrumented has been done to minimize labor costs. One company now indicates that they have been able to reduce their labor cost to about half that mentioned above. Furthermore, yields of PVC based on feed vinyl chloride are above 97 percent. Whereas total operating costs (excluding the costs of vinyl chloride) for many plants are in the range of about 2.0 to 2.5 cents per pound of PVC, costs for one company are said to be about 1.5 cents/lb.

Copolymers of Vinyl Chloride   Suspension-polymerization processes are used for producing several copolymers in which vinyl chloride is the predominant comonomer. The same equipment and similar operating conditions are often applicable for the copolymerizations as for homopolymerizations. Copolymers containing up to 10 percent of reacted vinyl acetate are prepared by suspension processes, but for higher amounts, agglomeration during polymerization becomes a problem (44). As a result, these latter copolymers are often prepared by a solution (or precipitate) process, described later.

Copolymers containing several weight percent of bound light olefins, especially propylene, appear to have a bright future for injection-molded, extruded, or blow-molded products and for food-grade packaging containers (68, 69). These copolymers, introduced in the late 1960s, can be fabricated at fairly low temperatures, 300 to 400°F (150 to 200°C). They are also thermally more stable than homopolymers of vinyl chloride for several reasons.

1. The $-C_3H_6-$ groups in the polymer chain minimize or interupt zipper-dehydrochlorination reactions that occur during decomposition of conventional PVCs.

2. When propylene is used as a comonomer, the chain-transfer steps during polymerization are modified, causing a high percentage of stable $-CH_2-CH_2Cl$ and $-CH_2-CH=CH_2$ end groups to be formed, compared with less stable ones for conventional PVC polymers.

3. Chain-transfer activity of the propylene reduces the amount of branching that normally occurs during the homopolymerization of vinyl chloride. Branch points are relatively unstable.

The results of pilot-plant runs for production of copolymers containing up to about 8 percent propylene have been reported (8). Vinyl chloride and propylene form ideal-liquid mixtures; i.e., they follow Raoult's law. Vinyl chloride is more reactive, and as it reacts and hence disappears, the partial pressure of propylene rises. The copolymer composition tends to become richer in propylene groups as the batch run progresses unless additional vinyl chloride is added. Reaction conditions used varied from temperatures of about 40 to 60°C and maximum pressures of 100 to 200 lb/in² gage. Up to 10 to 15 percent propylene was required in the hydrocarbon feed to incorporate up to about 8 percent propylene in the copolymer. Batch runs of 10 to 12 h often gave conversion levels of 80 percent or greater when *tert*-butyl peroxypivalate was used as the initiator. IPP and lauroyl peroxide can also be used as initiators.

Continuous-Flow Reactor Systems Most companies that produce PVC have given at least some attention to continuous-flow systems for suspension polymerization. Problems experienced in a flow reactor include obtaining adequate product quality (fisheyes or polymer skin often form) and avoiding plugging of the transfer lines.

The patented process of General Tire and Rubber Co. utilizes three continuous-flow stirred-tank reactors (53). Two of the reactors are operated in parallel at a 70 percent conversion level, which is the region of the maximum rate of conversion. The exit product of these two reactors is fed to the third vessel, where the reaction is completed, i.e., to 95 percent conversion. The quality of the PVC is good, but problems of line plugging have not yet been solved.

A semicontinuous suspension process is used by one company in the United States to produce a PVC paste resin (81). Vinyl chloride containing dissolved initiator is first homogenized with water containing the dissolved stabilizing agent. To do this, the two liquid phases are combined and forced at high velocities through jets or heads containing numerous orifices (64). The resulting mixture is sometimes not completely stable, and part of the vinyl chloride may separate. The remaining mixture, containing about 0.8 to 1.0 part by weight vinyl chloride per part water, is stable, however. This stable mixture is then fed continuously without agitation into the top of a tower reactor.

In one plant, the column reactor is 2.5 ft in diameter and 65 ft high. When lauroyl peroxide is used as initiator, a residence time of about 30 h is given as an example to obtain 90 percent polymerization (64). Calculations indicate that in such a case the column would produce about 2.3 million lb of PVC per year. If a more active initiator were used, reducing the residence time to 10 h, the production capacity would increase by a factor of 3. Reactor temperatures of about 45 to 50°C have been reported.

The two-phase product from the tower reactor is pressure-filtered to obtain the PVC granules, which are then washed. The viscosity of plastisol eventually produced is more sensitive than that of a plastisol produced by emulsion polymerization. Hence a slower casting speed is required with the suspension product.

Flow processes will probably be at best only marginally cheaper to operate than the current batch processes. Since in most plants numerous varieties of PVC must be produced—probably over 20 at least—a batch process is quite practical, particularly for specialty-grade PVC polymers produced in relatively small quantities. A special problem of a flow system would be to schedule switches from one type of PVC to another without producing off-grade materials. Some such materials can be blended with other PVC materials to produce desired products, but such blending operations are relatively expensive.

A flow process would probably be most suitable for PVC products prepared in large quantities. Such a process might then result in significant savings in labor costs and in increased plant capacities. The recovery systems for both the unreacted vinyl chloride and the PVC polymer would need to be modified to take advantage of the new reactor system.

Processing Suspension PVC Polymers Suspension polymers are highly suitable for most uses requiring plasticized PVC materials. Suspension polymers, but not emulsion polymers, are easily mixed or dry-blended with plasticizers. The resin, plasticizer, stabilizer, lubricant, filler, and colors are blended, often in a ribbon-type blender. The resulting material is a "dry," free-flowing powder, since the suspension PVC resin particles have the ability to adsorb the liquid plasticizer and yet exhibit minimum wetness. PVC resins that can be dry-blended can be easily and rapidly processed in several types of fabrication equipment.

The ability of a polymer to dry-blend depends on the uniformity, size, shape, and porosity of the granular particle. These properties of the particle depend to a considerable extent on the conditions used for polymerization. Uniformity of particle size is desired to obtain consistency in blending. Increased porosity and hence lower bulk densities are usually required when large amounts of plasticizer are to be used. Bulk densities of commercial suspension polymers vary from about 0.25 to 0.70 g/cm³, and the size of the polymer particles is generally between 40 and 200 mesh.

The molecular weight of the suspension PVC resin is important to the application. High-molecular-weight resins are used to make flexible tubing, garden hose, calendered film, welting, and insulation for wiring. Medium- and low-molecular-weight resins are used in film and sheeting and as coating for fabrics.

Low-molecular-weight resins are being used to an increasing extent for the so-called rigid PVC materials. Little or no plasticizer is used, but in some cases the resins are blended with nitrile-type synthetic elastomers. The low-molecular-weight polymers permit more rapid throughput in some processing equipment because of more rapid fusion. Hence the polymer can be processed more rapidly, and further the polymer is subjected to high temperatures (at which degradation reactions occur) for much shorter periods of time. Rigid PVC is used for pipes, skylights, valves, tank linings, and flume ducts.

Emulsion Polymerization

Emulsion polymerization is the oldest technique for the commercial production of PVC. It was the main process till about 1950, when suspension-polymerization processes were developed. Recently some of the unique PVC products produced by emulsion processes have found expanded uses (51, 43, 44).

PVC resins produced by emulsion polymerization generally retain all or at least most of the emulsifier added to the reaction system, often about 1 to 5 percent. Suspension polymers, however, contain less surfactant or stabilizing agent.

For emulsion polymers, the emulsifier is concentrated especially at the surface of the granular particles, which are beadlike and relatively smooth. When the plasticizer is mixed with these PVC particles, it wets them but penetrates only slowly into the beads. Such mixtures of PVC resin and plasticizer often form a relatively fluid mixture, depending on the amount of plasticizer used. They are called plastisols, and when an organic solvent is also included in the formulation, the liquid mixture is termed an organosol.

Plastisols and organosols are used in dip coatings, slush moldings, rotational moldings, and foam applications. In the final steps of the fabrication operation, heat is applied to raise the temperature to 160°C for a short time. The heat causes total or at least partial solvation and uniform mixing of PVC and the plasticizer. Upon cooling of these mixtures, the plasticizer may be rather uniformly mixed throughout the plastic. In other cases, the final plastic has a PVC continuous phase and a plasticizer discontinuous phase. Such a plastic obviously has different physical and chemical properties from one containing only one phase.

Factors Affecting Emulsion Polymerization Many recipes are employed commercially in emulsion polymerization, also sometimes called dispersion polymerization. The following materials are always employed, plus others in some cases. The approximate range at which each is added is given on a weight basis (14, 22, 43, 44):

	Parts
Vinyl chloride	100
Water	150–250
Emulsifying agent	2.0–5.0
Water-soluble initiator	0.1–0.4

The vinyl chloride and water should both be highly purified, as in the suspension-polymerization process.

A wide variety of soaps or surfactants are employed as emulsifying agents, including sodium lauryl sulfate, other sulfated esters, ammonium stearate, sulfonated castor oil, sulfonated higher-alkyl alcohols, alkyl naphthalene sulfonic acids, many synthetic detergents, and just plain soaps (salts of lauric, myristic, palmitic, and stearic acids). In addition, nonionic emulsifiers are sometimes used, especially for polymers for electrical applications.

Hopff and Fakla (71) investigated the effect of various emulsifiers on the rates of polymerization and the characteristics of the PVC produced. Saturated emulsifiers in general resulted in faster kinetics than unsaturated ones. The results obtained with the sodium salts of the sulfuric acid esters of C_{10} to C_{18} fatty alcohols are interesting. High-molecular-weight PVCs were produced in all cases, but the kinetics with sodium capryl sulfonate was considerably lower than with the others. Paraffin sulfonates, sulfates, and salts of fatty acid all caused similar results. Nonionic emulsifiers tended to be less active than anionic ones. Although the theoretical basis of the complete role of emulsifiers is not well understood as yet, major changes are obviously obtained when different emulsifiers are used or when amounts are varied (38).

PVC resins for electrical applications, such as wire coatings, must be prepared either with nonionic emulsifiers or with decomposable emulsifiers. The latter are quite stable during polymerization, but they can be saponified later. Residues of the emulsifier are removed by washing the granular PVC product. Some emulsifiers in this category include hydroxyalkylsulfonic acid–fatty acid esters and sulfonic and phosphonic acid esters.

The initiators used are water-soluble; important ones include ammonium persulfate, potassium persulfate, hydrogen peroxide, and various redox systems such as chlorate-bisulfite combinations. The initiators are of course soluble to some extent in the vinyl chloride phase. As a result, a small amount of polymerization may occur in that phase.

Temperatures used commercially vary from about 40 to 55°C. Temperature affects not only the rate of decomposition of the initiator and other chemical reactions but also the nature of the emulsion and the transfer of reactant to the growing polymer particle.

Batch Processes for Emulsion Polymerization   In the United States, batch processes are used predominantly if not exclusively. At least one flow process is apparently in use in Europe, however (81). Batch processes have at least two advantages:

1. Increased ease of producing numerous distinct PVC resins without producing significant quantities of off-specification-grade polymers. In a continuous-flow process, relatively poor-quality materials may be produced, especially during start-up or while switching from one product to another.

2. Better control of operating variables during polymerization and hence higher-quality polymers. Difficulty is probably still experienced in obtaining the highest-quality resins in continuous-flow processes.

Much of the process equipment used for emulsion polymerization is the same as that used for suspension polymerization. The reactors used for suspension polymerization are suitable for emulsion polymerization if the level of agitation can be varied over rather wide limits. Many companies design their overall plant so that both emulsion and suspension polymerization can be practiced.

In the Firestone plant, the only major equipment difference seems to be in the drying phase of the process (75). Since much of the equipment is used interchangeably, the operation steps are also very similar.

Several variations of the method of adding the reactants to the autoclave are probable. The emulsifier, initiator, and possibly buffer are generally dissolved first in the deionized water. Relatively cool water is normally used since the initiator begins to decompose as soon as the temperature is raised. The water solution is added to the autoclave, and care is taken to exclude air (or oxygen) from the system. The desired amount of vinyl chloride is added, and agitation is started to form a relatively stable emulsion.

When the emulsion is heated, the reaction begins. Temperature control is often critical during the most rapid phases of polymerization, just as for suspension polymerization. Adequate temperature control is maintained by cold brine or water in the jacket. Baum (3) outlines some of the methods employed to increase the overall heat-transfer coefficients:

1. Introduction of the brine through nozzles to produce tangential, high-velocity, swirling brine or water.

2. Baffled design for the jacket. Proper installation is relatively expensive, and this method is apparently not widely used.

3. Rapid convection or pumping of brine through the jacket.

When possible, cold vinyl chloride is added to the polymerization vessel in order to moderate the temperature there. Baum (3) indicates that a reflux condenser is also sometimes employed for the emulsion polymerization of some monomers. As indicated earlier (51), such a technique is not always applicable for production of PVC resins.

Often 90 to 95 percent of the vinyl chloride is polymerized in each batch run. Polymerization rates decrease rapidly above 90 percent conversions, and 10 to 15 h is usually provided for polymerization. The pressure drops as the run nears completion, similar to that of suspension-polymerization runs.

Separation of the unreacted vinyl chloride from the emulsion and its subsequent recovery, purification, and recycling involve most of the equipment used in the suspension process. The remaining emulsion or latex is stored in blend tanks until sufficient quantity, i.e., several batches, is collected. The most popular method for recovering the solid PVC particles is by spray drying since it is relatively hard to separate the emulsion by centrifuging or filtering. Rysavy (88) has indicated some of the factors of importance regarding spray drying, and Labine (75) describes the spray dryer used by Firestone. Drum driers have also been employed, and in addition the emulsion is sometimes coagulated, separated by filtration or centrifugation, and then dried (63).

Modified Emulsion Polymerization Batch   In some modified emulsion methods, few if any micelles are present. The conventional initiation of polymer particles is bypassed, and instead the reactor is seeded with tiny polymer particles. The polymerization then proceeds in the conventional manner; namely polymerization occurs on (or in) the latex particles (or seed particles). Several modifications of seeding probably are employed, but the patented process by Powers (82) is an important one.

Powers explains the important relationship between the PVC particle size and the amount of dispersing agent, such as soap, necessary to maintain a stable emulsion, i.e., prevention of coalescence of the PVC particles or formation of additional dispersed particles in the emulsion. The area A covered to a thickness of one molecular layer by a mole of dispersing agent is given by

$$(1) \quad A = N_0 A_0$$

where N_0 = Avogadro's number
A_0 = cross-sectional area of molecule of the dispersing agent (such values are often available in handbooks)

Since the individual PVC particles are essentially spherical in shape, the weight W of the dispersing agent of molecular weight M required to cover the surface of a particle of diameter d is

$$(2) \quad W = \frac{\pi d^2 M}{N_0 A_0}$$

The weight W_d of polymer with density ρ in each particle is

$$(3) \quad W_d = \frac{\pi d^3 \rho}{6}$$

Then the weight percentage P of the dispersing agent (based on the weight of the polymer) which is needed to cover the surface of a particle is obtained by dividing Eq. (2) by Eq. (3) and then multiplying by 100:

$$(4) \quad P = \frac{W}{W_d} = \frac{600M}{N_0 A_0 \rho d}$$

Of course, the particles in an emulsion have a range of diameters. The surface area is proportional to the diameter squared, whereas the volume of the particles is proportional to the diameter cubed. In general, the diameters of the particles present in an emulsion follow a normal probability-distribution function. Corrections must then be made to Eqs. (2) to (4) when P is being calculated. A semitheoretical correction technique has been presented by Powers, who reports that when 20 to 60 percent of P is used, preferably 25 to 40 percent of P, significant coalescence and formation of particles do not occur.

The actual fraction of P required is undoubtedly to some extent a function of temperature, the amounts of the reactants and other additives, plus the degree of agitation. By implication, Powers apparently believes that 25 to 40 percent of P is good value for the usual range of operating conditions used in emulsion-polymerization processes.

In the Powers process, additional vinyl chloride and dispersing agent are added as the batch reaction progresses. The additional vinyl chloride provides additional monomer for polymerization, and the additional dispersing agent is required to maintain a stable emulsion as the PVC particles grow in size. The vinyl chloride would undoubtedly be added cold since it would help control the reactor temperature. A higher rate of polymerization is hence possible.

In an example of the Powers process, the average diameter of the final PVC latex is chosen to be 2500 Å, and sodium stearate is the dispersing agent. The following charge is added at the start of a batch run in a 15-gal reaction vessel:

	Pounds
Vinyl chloride	9.0
Water	36.7
Potassium persulfate	0.27
Ammonium hydroxide (28% solution)	0.27
Seed latex	0.80
Sodium stearate	0.013

The seed latex contained PVC particles with an average diameter of 342 Å, as indicated by electron-microscope measurement. Table 6-7 shows how the vinyl chloride and dispersing agent were added during the run plus the formation of polymer. The dispersing agent was added in a 5 percent water solution, which was added more rapidly at the beginning of the run than at the end. Vinyl chloride was added, however, at an approximate rate of 2.4 lb/h for the first 10 h of the run and at about 3.2 lb/h for the next 6 h. Then the rate of vinyl chloride addition was decreased for the next 4 h and finally stopped at the end of 20 h. The latex particle produced at the end of the run (23.2 h and 95 percent conversion) had an average particle size of 2530 Å, which was essentially identical to that predicted based on the calculation procedure reported in the patent.

Adding both the vinyl chloride and the dispersing agent at a variable rate during the batch run is certainly a complicating factor, since variable-flow feeding devices are required. Several analyses of the emulsion during the run are probably required in order to check the course of the polymerization as a function of time. Highly trained operators and/or sophisticated process controls are needed to ensure high-quality PVC.

The polymer formation lags behind the amount of vinyl chloride added to the reactor by 4 to 13 lb, as indicated in Table 6-7. The biggest lag occurred at about 16 h. In the range from 16 to 23 h, the rate of polymerization was most rapid, except possibly for the rate during the first hour of the run. An autoacceleration reaction is apparently occurring during these periods.

TABLE 6-7   Example of Batch Run by Powers Process (82)

Time, h	Total vinyl chloride added to reactor, lb	Total dispersing agent added to reactor, lb	Theoretical amount P of dispersing agent in reactor, %	PVC polymer in reactor, lb
0.0	9.0	0.013	35	0.15
1.0	14.4	0.164	22	4.2
2	14.4	0.164	30	8.7
4	18.9	0.20	26	14.1
6	23.2	0.21	23	18.0
8	27.8	0.37	35	22.8
10	33.3	0.37	32	26.4
16	52.4	0.53	35	39.6
20	60.0	0.60	34	49.1
23.2	60.0	0.60	31	57.0

With the Powers process, PVC particles varying in size from 500 to 20,000 Å can be produced, and the emulsions formed are highly stable. Latices in which the PVC particles are as small as 100 to 200 Å can be used for seeding, but these emulsions are quite unstable and must be used within a few hours. If the seed latices have larger particles, they will be stable up to several days or even weeks.

The volume percent of PVC particles that can be stabilized in an emulsion varies up to about 65 percent. In most commercial applications, the volume percent employed is probably reasonably high in order to minimize the problems of removing the water from the polymer.

The Powers emulsion technique has important advantages over regular emulsion processes. The size of PVC particles that can be produced uniformly is much larger. In addition, the amount of dispersing agent can be minimized since it is added only as needed to stabilize the particles. In the above example, the final PVC polymer contains about 1.0 percent emulsifying agent. An even smaller quantity would be present if larger PVC particles were produced. The modified method of emulsion polymerization is related to suspension polymerization in that essentially only polymer-monomer particles occur in the dispersed phase of the water.

The method of preparing the seed polymer for this modified method is not reported. Presumably it would be prepared in a relatively small reactor.

Continuous-Flow Emulsion Polymerization   An old German process has been described by Dunlop and Reese (63), Herte (70), and Krehler and Wick (74). The reactor vessels had a relatively large height-to-diameter ratio. One vessel was described as being 23 ft tall and about 5.4 ft in diameter with a capacity of 3,600 gal (63). It was glass-lined and jacketed, being cooled with brine or cooling water. A simple blade or paddle agitator was provided near the top of the vessel. This paddle was specified as being 4 ft in diameter and 1.5 ft high. A variable-speed motor to turn the agitator up to 50 r/min was provided.

The water, containing dissolved emulsifiers, initiator, and buffer, was fed continuously to the top of the reactor in one feed stream, and vinyl chloride was metered to the reactor through another inlet line. The maximum degree of shear in the emulsion was apparently needed near the feed point, but Dunlop and Reese (63) characterized it as gentle swirling. They further indicated that the level of agitation was most important. Certainly relatively little agitation was provided in the bottom of the reactor. As PVC particles form, they are heavier than vinyl chloride droplets in the emulsion because of the greater density of PVC. It is doubtful, however, that any significant phase separation occurred in the reactor since the emulsions normally obtained were stable.

Variations in the arrangement of reactors occurred in German plants. In one case, multiple reactors were arranged in parallel-flow arrangements (70). In another case, the relatively large reactor described above was connected in series with a much smaller reactor (63). In this case, about 88 percent polymerization of the entering vinyl chloride occurred in the first (and larger) reactor, and an additional 4 percent of the vinyl chloride reacted in the second unit. The theoretical residence time of vinyl chloride in the two reactors was 28 h. Blue dye injected in the feed stream did not appear in the exit stream for 20 h. The blue color persisted in the exit stream, however, for 100 h. Some backmixing was apparently occurring at the top of the reactor but with plug flow in the bottom portions.

Arranging the multiple reactors in series minimized the number of feed pumps and lines required and helped make temperature control easier. Although the reason for using reactors in series was not specified, such an arrangement provided more flexibility for producing many types of PVC resins.

Temperature control was reported as being the critical factor governing reactor size. Fairly uniform temperatures varying from about 38 to 50°C, depending on the specific PVC polymer desired, were needed throughout the reactor, but with the laminar flows that undoubtedly occurred in the bottom of the reactor the overall heat-transfer coefficients certainly tended to be low. Some convection currents obviously occurred because of temperature gradients resulting from the exothermic heat of polymerization. Operators of the German plants indicated that a "delicate" temperature balance occurred (63). Increased initiator concentrations raised the temperature and the rate of reaction. The temperature of the brine or water in the jacket was also most important for good temperature control.

Occasionally the reactors had to be cleaned out. The reactors were shut down, and workmen entered each reactor through a manhole. Such cleanouts were performed every few weeks or in some cases could be delayed several months.

Starting up the reactor required several hours' time in order to minimize the production of off-grade polymer. The reactor was filled with cool water, emulsifier solution, initiator, and vinyl chloride (63). Warm water was provided in the jacket of the reactor until the reaction mixture was sufficiently heated to start polymerization. Gradually, cooling water was substituted in the jacket, and the density of the emulsion was measured frequently until it reached a desired level. At this time the flows of vinyl chloride and the water solution were started. Start-up was then considered to be complete.

Processing the emulsified product from the reactor was similar to that in the batch process except that a continuous-flow system could be used. The German plant using this process is supposedly still in operation (95) and produces a paste resin of broad particle-size distribution (0.3 to 1.0 μm).

Bulk Polymerization

Bulk polymerization of vinyl chloride has become an important process within the last few years (5, 72, 95). World capacity for the Pechiney–St. Gobain two-step process was over 800 million lb/year in 1970 (96), and it is expected to more than double in the next several years. Three companies in the United States were reported by 1971 to be using the process.

Figure 6-8 is the flowsheet for the process, including the polymerization steps, recycle of unreacted vinyl chloride, and sieving and recovery of the PVC product. Details of the process have been reported (5, 61, 73, 83–86, 95), and the more modern version involves a two-step batch polymerization.

Prepolymerizer The first reactor (commonly called the *prepolymerizer*) is a vertical stainless-steel-clad vessel with a flat-blade turbine and baffles to provide turbulent agitation (95). The operating conditions, especially agitation and temperature, are important in regulating the characteristics of the product PVC granules obtained from the second reactor (commonly called the *autoclave*). More than half the vinyl chloride feed is added to the prepolymerizer, and sufficient initiator is also added to obtain the desired level of polymerization, usually in the range of 7 to 12 percent (5, 86, 95).

Each prepolymerizer is used in conjunction with four autoclaves. Two sizes of prepolymerizers are used commercially (86). In a 20,000 metric tons/year (44 million lb) plant, a single prepolymerizer of 8 m³ (280 ft³) capacity is sufficient, but a 13-m³ (460-ft³) unit is used in a 30,000 tons/year plant. Typical batch information for the smaller prepolymerizer is reported as follows:

Vinyl chloride added, metric tons (lb)..... 5.5 (12,600)
Reaction time, min..................... 30
Total time for cycle, h................. 2.5

The recommended initiators have not been reported, but obviously highly active ones are required. Two recipes found in the literature (83) are shown on page 259.

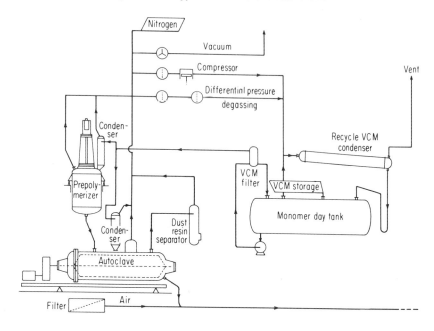

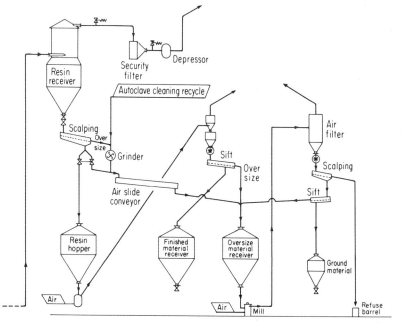

Fig. 6-8 Mass, or bulk, polymerization of vinyl chloride by Pechiney–St. Gobain two-step process. (*Pechiney–St. Gobain.*)

	Parts by weight	
	Recipe 1	Recipe 2
Vinyl chloride..................	100	100
Azobis(isobutyro)nitrile..........	0.016	0.02
Diisopropyl percarbonate.........	....	0.00005

All or most of the initiator is consumed in the prepolymerizer (95). The time for this phase of the run depends of course on the level of initiator used and on the temperature, which is in the range from 40 to 70°C. The conversion level in the prepolymerizer is determined by measuring the amount of heat evolved. The temperature is controlled by transferring heat to the jacket and by refluxing part of the vinyl chloride. The pressure is essentially the vapor pressure of vinyl chloride, which varies from about 70 to 170 lb/in^2.

The product (a slurry of PVC and vinyl chloride) from the prepolymerizer is generally transferred to the second reaction vessel. Since polymerization has essentially stopped (or can be stopped by cooling), the slurry can be stored, however, before being polymerized further. Occasionally delays may be experienced in having a second reactor (autoclave) available.

The prepolymerizer and the autoclave are often operated by automatic controls (86). Some units are operated by an electromechanical sequential program whereas computer controls are used in a few cases. The prepolymerizers are cleaned every five to eight batches.

Autoclave The slurry from the prepolymerizer, the remainder of the vinyl chloride feed, and additional initiator are added to the second reactor (or autoclave) for the second stage of polymerization. Each autoclave is a horizontal cylindrical vessel provided with incurvated (or helicoidal) agitator blades or ribbon blenders (86, 95). Two different autoclaves are used, depending on the size of the prepolymerizer and the desired plant capacity:

	Smaller autoclave	Larger autoclave
Volume, m^3..........	16	25
Diameter, m.........	1.8	2.2
Length, m...........	6.8	7.3

Details of the ribbon blender used are not known, but a patented version has two or three ribbons wound on whorls with different diameters (61). One of the ribbons turns in an opposite direction to the other(s). The autoclave blades are continuously cleaned by the stirrer blades, and good heat exchange is maintained (84). The ribbon blenders are designed to prevent sudden jerks as they penetrate the polymer powder. They rotate with minimum wall clearances at several revolutions per minute.

Figure 6-9 shows the 16-m^3 (560-ft^3) autoclave, the bottom of a pre-polymerizer, and the connecting line between the vessels. This autoclave is often loaded with a total of 8.3 metric tons (the batch from a prepolymerizer plus additional vinyl chloride and initiator). After the autoclave has been loaded, a small amount of vinyl chloride is vented to remove any oxygen from the gas phase. The mixture is then heated rapidly to reaction temperature by hot water in the jacket of the autoclave. As polymerization starts, the heat of reaction is removed through the autoclave jacket, a cooled agitator shaft, and a reflux condenser. Heat evolved during the reaction is measured in order to determine the conversion level. When the desired level of about 80 to 85 percent is reached, unreacted vinyl chloride is removed with the aid of a compressor able to pull a vacuum; the recovered vinyl chloride is stored temporarily in the monomer day tank.

Fig. 6-9 Autoclave and bottom of prepolymerizer. (*Pechiney–St. Gobain.*)

Calculations indicate that the autoclave is approximately 60 to 65 percent filled on a volume basis at the beginning of a run. At the end of the run, when it contains approximately 7 metric tons (15,400 lb) of PVC having a bulk density often about 0.6 g/cm³, the autoclave is about 70 to 75 percent filled. The level of loading is undoubtedly of importance to obtain the desired agitation throughout the run.

The exact operating conditions used in the autoclave are not specified. Reaction temperatures are probably in the range of 45 to 60°C. The pressure in the autoclave is essentially the vapor pressure of vinyl chloride. Apparently conversions are not obtained at high enough levels so that the pressure begins to fall, as shown in Fig. 6-7 for suspension polymerization. The characteristics, including porosity of the final product, are undoubtedly affected to some extent by the conversion level obtained.

Deposition of polymer on the reactor surfaces is at best only a partially solved problem since the autoclave must be cleaned after each run. The most acute periods of run for deposition probably occur with "wet" PVC solids. Such a condition is of course prevalent as the thick slurry is converted to rather dry solids, i.e., about 15 to 20 percent polymerization levels. Refluxing of liquid vinyl chloride, however, also maintains, to some extent at least, wet solids during all or most of the run. The method of returning the liquid vinyl chloride to the autoclave may be a critical one.

A reaction time of 6 h has been reported (86) as typical for a run in the smaller of the two autoclaves. Cleaning the autoclave after each run takes several hours and considerable labor. The total time for a cycle is approximately 11 h.

Recovery Steps for Bulk Polymerization   Recovery of the unreacted vinyl chloride is much simpler for bulk polymerization than it is for suspension or other polymerization processes. The vinyl chloride is quickly removed from the PVC granular material at the end of the polymerization. Providing heat to the jacket of the autoclave may be necessary to speed up the vaporization step. Most of the vinyl chloride is then recovered by condensation in a recycle condenser cooled with water. A final condenser cools the tail gases to about −35°C to separate trace amounts of vinyl chloride (86).

Condensed vinyl chloride does not need to be purified by distillation, as is sometimes done in the suspension process. The liquid vinyl chloride, however, is passed through a column packed with solid caustic soda to remove any moisture, HCl (formed by decomposition of vinyl chloride), or materials formed by a decomposition of the initiators.

After most of the vinyl chloride has been removed from the PVC granules in the autoclave, the autoclave is purged with an inert gas.

Hence explosive mixtures of air and vinyl chloride are not formed when the polymer is discharged from the autoclave by air blowing (86). The particles are screened to remove oversize particles, which are milled. Transfer to storage, bagging, loading to bulk transportation devices, etc., complete the operation.

Modified Bulk Polymerizations   Attempts have been made to perform the complete polymerization in the same vessel. In one case (83), the initial prepolymerization was conducted using a very high rate of agitation. Less vigorous agitation and a different agitator were employed for later stages of the polymerization. Probably this method, sometimes called a *two-speed* method, is not used commercially, and it would have an important disadvantage in that a large vessel would be needed for both stages of the polymerization whereas a small vessel is adequate for prepolymerization in the two-stage process.

Autoclaves built in 1954 were filled with steel balls and rotated (95). They are currently used in the second step of the bulk process.

Increased interest has recently been shown in low-temperature bulk polymerization of vinyl chloride since the resulting polymer has a relatively high level of syndiotacticity and makes a most attractive fiber. Furthermore, this product has a low dehydrochlorination rate, high resistance to solvents, and improved mechanical properties compared with the more conventional product. Polymerization temperatures from about −20 to −60°C are often used, but finding a suitable and cheap initiator or catalyst is a problem. Organometallic compounds such as boron, cadmium, aluminum, lead, or silver alkyls have been tried but apparently with only moderate success (77); organic hydroperoxides with sulfur dioxide have been claimed to be effective (77). Applicazioni Chimichi Societa per Azioni (ACSA) is reported to have a low-temperature unit for production of 11 million lb/year of a narrow-molecular-weight-range PVC (58). The initiator system and details of their process have not been published.

Bulk PVC Product   The type of PVC produced by the Pechiney–St. Gobain process can be varied significantly by the operating variables employed (56). The molecular weight can be controlled over a wide range by varying the temperature during the second step. Particle-size distribution is set primarily by the agitation of the first step. Bulk density and porosity of the final PVC particles are closely linked to temperatures in both steps and to the final degree of conversion in the second step.

Although the bulk process is apparently used only for the production of the homopolymers of vinyl chloride, pilot-plant results indicate that many copolymers employing olefins (such as propylene and vinyl acetate) can be produced without major difficulties. For copolymers containing large amounts (about 15 percent) of vinyl acetate, crusting may be a problem.

Since the bulk product contains no surfactants or additives (like those in the suspension process), it is extremely clear and finds good acceptance for bottles, sheets, film, etc. The bulk product is highly porous and free-flowing. It is reported (5) to have higher bulk densities, yet great porosities, and faster plasticizer take-up rates than PVC produced by the suspension process.

Economics of Bulk Process   A Pechiney–St. Gobain plant to produce 88 million lb/year (40,000 metric tons) of PVC was estimated at $3.25 million (14.7 million French francs) in 1970 (86). The polymerization, sieving, and PVC-recovery portions of the plant are included in the estimate. The cost of such a plant appears to be less than that for a plant employing the suspension process.

Simple recovery steps for vinyl chloride and (especially) PVC product result in significant savings over the other processes (81). About 97 to 98 percent of the vinyl chloride feed is converted into PVC product. The remainder probably is lost as PVC deposits on the reactor walls and because of handling problems. Such a high yield is somewhat higher than that in most suspension plants. Labor costs, however, are higher than those of suspension processes. The total operating costs for the bulk process are perhaps less by 0.1 to 0.2 cent/lb than for many suspension processes.

Large producers of PVC in the United States are known to be interested in this process, which probably will be used to an increasing extent. Whether it will eventually displace the suspension process will depend on the performances and expected process improvements of the current bulk process.

Solution Polymerization

Solution polymerization is used mainly for the manufacture of solution coating resins which are often copolymers of vinyl chloride and vinyl acetate. Polymerization costs are relatively high, but the high-quality polymers obtained and the convenience of preparing solutions of polymers often justify the higher costs. The process is particularly applicable for copolymers since they are considerably more soluble in the solvents used than the homopolymers of vinyl chloride are.

The basic process, probably used for many years by Union Carbide Corp., is described by two patents (16, 87). The process is employed to prepare copolymers containing 75 to 90 percent vinyl chloride and the remainder vinyl acetate. The mechanically stirred autoclave maintained at approximately 40°C contains a slurry consisting of about 80 percent solvent, designated as n-butane. The remaining 20 percent is the comonomers plus some copolymers dissolved in the solvent and the solid copolymer present as a precipitate. The slurry is pumped to a filter press to remove the precipitate, and the filtrate is returned to the autoclave. Makeup comonomers and initiator are continuously added to the autoclave to replace those which react.

This process, as described in the 1933 and 1937 patents, has undoubtedly been modified as newer equipment has become available. Two batch filter presses were originally specified. One filter press was emptied and cleaned while the other was being used. Such filters would require considerable labor, and there is the danger of air leaking into the system, since the filter presses are opened and exposed to air during the cleaning cycle. Either centrifugal separators or continuous-flow filters are probably used in modern commercial processes. Lead-lined autoclaves were specified in the patents, but glass-lined or stainless-steel reactors (not readily available 30 years ago) are probably used now.

Benzoyl peroxide was specified in the patent as the initiator, used in concentrations of about 0.5 weight percent. Popular initiators such as lauroyl peroxide and isopropyl percarbonate probably would be satisfactory in the present commercial process. 2,2-Azobis(isobutyro)nitrile was used in the recent experimental program of the precipitation polymerization of vinyl chloride (30).

The separated solid polymer (or copolymer) particles contain considerable amounts of dissolved vinyl chloride plus some solvent (30). Polymerization reactions will continue to occur until the vinyl chloride is separated from the solid particles. The temperature and pressure to which the separated particles are subjected are important relative to flashing of the vinyl chloride.

Solvents other than n-butane used in at least experimental investigations (24, 74) include benzene, chlorinated aliphatics, cyclohexane, and tetrahydrofuran. Normal paraffins such as n-butane contain only primary and secondary carbon atoms, and they are less active as chain-transfer agents than isoparaffins, e.g., isobutane, which contain a tertiary carbon atom.

An important advantage of solvent polymerization is that water is not used and hence the recovery of the final product is simplified. The unreacted vinyl chloride and solvent are generally relatively easy to remove, especially if a fairly volatile solvent such as n-butane is used. The recovered vinyl chloride and solvent are probably recirculated as a mixture. The final polymer product contains few impurities since presumably no emulsifiers or surfactants are employed.

Solvent polymerization of vinyl chloride is also one of the techniques to produce an improved polymer used in 100% PVC fibers. Operating temperatures are generally low, probably less than $-20°C$, and hence the resulting polymers are highly syndiotactic.

In France, PVC fibers have captured the following percentages of the market (58): flags, 60; curtains, 35; underwear, 30; ski and mountain clothing, 25; sweaters and blankets, 20; and auto and airplane interiors, yarns, and artificial furs, 10. Several American companies are interested in these PVC fibers, which are cheaper than nylons, polyesters, and acrylic fibers. Société Rhovyl has produced fibers with tensile strengths up to 4.2 g/denier, elongation at break of 23 percent, and shrinkage in boiling water of 6.6 percent (67). Shrinkage in trichloroethylene (a dry-cleaning fluid) is often low, less than 10 percent. Adhesive fibers can be prepared if a copolymer of vinyl chloride and vinyl acetate is produced or dissolved in various water-soluble solvents before spinning (92). Suitable solvents include acetone, tetrahydrofuran, dimethylformamide (DMF), and dimethylacetamide (DMA).

Copolymerization results in interesting mass-transfer features. The relative rates of transfer of the comonomers from the solvent phase to the solid copolymer particle probably are quite different. If so, the composition of the copolymer would presumably change as polymerization progresses.

Vapor-Phase Polymerization

Several companies have shown considerable interest in the so-called vapor-phase process for polymerization of vinyl chloride, and the process has been developed to at least a pilot-plant stage. Both batch and continuous-flow modifications have been considered. Moberly and Kahle (31) describe the process assigned to Phillips Petroleum Co., which in several respects is similar to the vapor-phase processes being used for production of polyethylene and polypropylene.

Batch operation requires a two-step process. In the first stage, which may be similar to the prepolymerization step of the Pechiney–St. Gobain process, small granules of PVC are produced, separated, and used as the seed particles in the fluidized-bed reactor, the second step of the process. Such a first step, however, is not required in the continuous-flow version of the process.

The fluidized-bed reactor is operated at about 100 to 165°F (38 to 74°C) and at absolute pressures of 4 to 12 atm. These temperatures are similar to those used in other processes for vinyl chloride. Higher temperatures of course promote thermal instabilities of the polymer and may also lead to excessive agglomeration of the solid particles in the fluidized bed.

The two examples of the Phillips patent (31) report on the operation of the fluidized-bed reactor in considerable detail.   Continuous-flow operation was provided at 140 F and 7.8 atm pressure.   About 28.5 percent of the vinyl chloride feed was introduced as a liquid, probably through a spray nozzle located in the fluidized solids, to provide adequate temperature control.   This liquid vinyl chloride wet the fluidized solids; part of it was polymerized, and the remainder was vaporized.   The heat of vaporization was essentially equal to the heat of polymerization. About 70 percent of the feed vinyl chloride was added to the reactor as the fluidizing gas.   Upward flow velocities of about 0.1 to 0.4 f/s were used, depending on the size of the PVC particles desired in the product.

The seed PVC granules, ranging in size from 100 to 400 mesh, are pretreated before a run by spraying them with solutions containing an initiator and a stabilizer.   Pentane solutions of IPP and dibutylin laurate have been used.   The pretreated seed granules were transferred to the reactor by means of gaseous vinyl chloride (about 1.5 percent of the feed).

Approximately 5 percent of the entering vinyl chloride is polymerized in Phillips fluidized-bed reactors.   The overhead product of reactor contains unreacted vinyl chloride and PVC fines that are separated by a cyclonic separator.   The vinyl chloride is recycled, and the fines, after suitable screening and grinding, are used as seed granules.   Larger (and heavier) PVC particles are withdrawn from the bottom of the reactor through a star valve.   These particles, containing adsorbed vinyl chloride, are first flashed to separate vinyl chloride, which is recycled.   Then the PVC product is screened; oversized particles are ground; fines are used as seed particles; and the remaining intermediate-sized particles constitute the desired PVC product.

The properties of the product have not been reported, but presumably it would have a high density and be relatively nonporous.   If so, it would not be suitable for dry-blending applications.   Since the seed particles are recycled from the previous batch, portions of the product have been exposed to reaction temperatures for extended periods of time.   The thermal stability of the product might then be rather poor.

Sticking or coating of PVC polymer on the wall of the reactor apparently is not a problem.   Pretreating the seed granules so that a uniform reaction can be obtained on each particle is important, however.   As indicated, about 95 percent of the vinyl chloride has to be recovered and recycled.   Perhaps a reactor with a suitable agitator, e.g., the autoclave employed in bulk polymerization, can be developed so that recycle of vinyl chloride can be significantly reduced.   Ease in recovering the PVC product is a major advantage of this process, which may find commercial applications.

Chlorination of PVC Resins

The high-temperature stability of PVC polymers is improved if the unsaturated end groups of the polymer molecules are chlorinated. As a result, zipper dehydrochlorination is significantly suppressed.

In the process developed by B. F. Goodrich Co. (62), granular PVC is suspended in water, and a small amount of chloroform or other chlorinated hydrocarbon is added to promote swelling of the PVC solids. Chlorine gas is bubbled through the resulting slurry in the presence of light. The chlorinated PVC produced is separated from the water phase, and it is then washed, first with water and then with a dilute solution of sodium bicarbonate. The PVC solids are separated by filtering or centrifuging, and they are often washed with methanol and then dried. Various additives (65, 66, 90) such as silica are often added to the reaction mixture to permit a high PVC content in the slurry, to minimize coagulation, and/or to promote chlorination.

Although the amount of chlorinated PVC produced is relatively small, significantly increased production is expected in the near future. This chlorination step adds significantly to the cost of the final product.

Future of PVC Polymers

The future for PVC polymers both in the United States and in the rest of the world is a bright one. Recent process improvements for producing vinyl chloride and for polymerizing it have ensured that the polymer will be highly competitive for years to come. Uses and production of various PVC polymers will certainly expand in the future.

One factor of concern is how to dispose of PVC polymers. Incineration is not a satisfactory solution since HCl gas is released. It is hoped that methods to recycle (reprocess and then reuse) PVC products can be developed.

Several companies are currently offering to sell or license PVC knowhow. Although considerable information on the polymerization techniques is available, details on the polymerization fundamentals which might be helpful in devising improved processes are still needed.

Literature Cited

1. Albright, L. F.: Polymerization of Vinyl Chloride, *Chem. Eng.*, **74**(10): 151 (1967).
2. Arlman, E. J., and Wagner, W. M.: Some Experiments on Bulk Polymerization of Vinyl Chloride, *J. Polym. Sci.*, **9**: 581 (1952).
3. Baum, S. J.: Engineering Aspects of Emulsion Polymerization, *Ind. Eng. Chem.*, **49**: 1797 (1957).
4. Bengough, W. I., and R. G. W. Norrish: Mechanism and Kinetics of Heterogeneous Polymerization of Vinyl Monomers: Benzoyl Peroxide–catalyzed Polymerization of Vinyl Chloride, *Proc. R. Soc.* (*Lond.*), **A200**: 301 (1950).

5. Berger, F.: The Polymerization Structure and Properties of Bulk PVC in Comparison with Suspension PVC, *Interreg. Petrochem. Symp. Petrochem. Ind. Dev. Countries, UN Ind. Dev. Organ., Baku, USSR,* October 1969.

6. Billmeyer, F. W.: "Textbook of Polymer Science," Interscience, New York, 1962.

7. Bockman, D. C.: Stereoregular Crystalline PVC, *Br. Plast.,* 38(6): 361 (1965).

8. Cantow, M. J. R., C. W. Cline, C. A. Heiberger, D. T. A. Huibers, and R. Phillips: Vinyl Chloride/Propylene Copolymers, *Mod. Plast.,* June 1969, pp. 126–138.

9. Competitive Growth for Peroxide-Catalysts, *Chem. Eng. News,* Feb. 24, 1964, p. 25.

10. Plasticizer Sales Pass One Billion Pounds, *Chem. Eng. News,* Aug. 1, 1966, p. 20.

11. Polyvinyl Chloride: Makers Switch Catalyst, *Chem. Eng. News,* Aug. 26, 1968, p. 22.

12. PVC Producers See Good Business Ahead, *Chem. Eng. News,* Aug. 4, 1969, p. 182.

13. Chevassus, F., and R. deBroutelles, "The Stabilization of Polyvinyl Chloride," trans. by C. J. P. Eichkorn and E. E. Sarmiento, St. Martin's, New York, 1963.

14. Corso, C., and E. Ferrari: Emulsion Polymerization of Vinyl Chloride, *Mater. Plast.,* 28: 10 (1962); 27(8): 781 (1961).

15. Cotman, J. D., M. F. Gonzalez, and G. C. Claver: Studies of Poly (Vinyl Chloride): The Role of the Precipitated Polymer in the Kinetics of the Polymerization of Vinyl Chloride, *J. Polym. Sci.,* (A-1)5: 1137 (1967).

16. Douglas, S. D. (to Union Carbide and Carbon Corp.): Process for Producing Vinyl Resins, U.S. Pat. 2,075,429 (Mar. 30, 1937).

17. Douglas, W. C., J. M. Gyenge, G. Hackin, and A. J. Hanley: PVC Fabrications and Applications, *Am. Inst. Chem. Eng. Sem. General Tire and Rubber Co.,* October 1964.

18. Enomoto, S.: Polymerization of Vinyl Chloride, *J. Polym. Sci.,* (A-1)7: 1255 (1969).

19. Farbar, E., and M. Korrar: Suspension Polymerization Kinetics in PVC, *Polym. Eng. Sci.,* January 1968, p. 11.

20. Fedor, W. S.: Plasticizers, *Chem. Eng. News,* Nov. 13, 1961, pp. 118–138.

21. Fedor, W. S.: Outlook, *Chem. Eng.,* Dec. 15, 1969, pp. 94–100.

22. Fitch, R. M.: The Theory of Emulsion Polymerization, *Off. Dig. J. Paint Technol. Eng.,* 37(489, pt. II): 32–48 (1965).

23. Flory, P. J.: "Principles of Polymer Chemistry," Cornell University Press, Ithaca, N.Y., 1958.

24. Gabbett, J. F., and W. M. Smith: Copolymerizations Employing Vinyl Chloride or Vinylidene Chloride as Principal Components, chap. 10, in G. E. Ham (ed.), "Copolymerization," Interscience, New York, 1964.

25. Gardon, J. L.: Emulsion Polymerization: Review of Theory and Data, *Am. Inst. Chem. Eng. 65th Natl. Meet., Cleveland,* 1969.

26. Ham, G. E.: "Copolymerization," pp. 587–637, Interscience, New York, 1964.

27. Harkins, W. D.: A General Theory of the Mechanism of Emulsion Polymerization, *J. Am. Chem. Soc.,* 69: 1428–1444 (1947).

28. Heiberger, C. A., and L. Fishbein: Polymers Based on Vinyl Chloride, Belgian Pat. 668,473 (Dec. 31, 1964).

29. Heiberger, C. A., and R. Phillips: New Propylene-modified PVC Resins for

Rigid Applications, *VIPAG Newsl.*, 3(1): 22–23 (Sept. 30, 1966); also presented at 22d *ANYEC, Montreal, March* 7–10, 1966.
30. Mickley, H. S., A. S. Michaels, and A. L. Moore: Kinetics of Precipitation Polymerization of Vinyl Chloride, *J. Polym. Sci.*, **60**: 121 (1962).
31. Moberly, C. W., and G. R. Kahle (to Phillips Petroleum Co.): Vinyl Monomer Polymerization Process, U.S. Pat. 3,578,646 (May 11, 1971).
32. Olivier, G.: What's the Future for PVC, *Hydrocarbon Process.*, **45**(9): 281 (1966).
33. Park, R. M.: Plasticizers: Versatile, Necessary Resin Modifier, *Hydrocarbon Process.*, **41**(3): 120 (1962).
34. Peggion, E., F. Testa, and G. Talamin: A Kinetic Study of the Emulsion Polymerization of Vinyl Chloride, *Makromol. Chem.*, **71**: 173 (1964).
35. Pennwalt Corp., Lucidol Division: Free Radical Initiators for the Suspension Polymerization of Vinyl Chloride, *Tech. Bull.* 30.90, Buffalo, N.Y., 1971.
36. Pezzin, G.: Structure and Properties of Crystalline Polyvinyl Chloride, *Plast. Polym.*, **37**: 295 (1969).
37. Richard, W. R.: Stabilizers: Key to Ageless Plastics, *Hydrocarbon Process.*, **41**(3): 123 (1961).
38. Roe, C. P.: Surface Chemistry Aspects of Emulsion Polymerization, *Ind. Eng. Chem.*, **66**(9): 20 (1968).
39. Schildknecht, C. E.: "Vinyl and Related Polymers," Wiley, New York, 1952.
40. Schildknecht, C. E.: "Polymer Processes," Interscience, New York, 1956.
41. Smith, E. S., and T. H. Rogers: In Situ Peroxydicarbonate Initiator System, *Meet. Am. Inst. Chem. Eng., New Orleans, March* 1969.
42. Smith, E. S.: Polymerization Process with Peroxydicarbonate Initiator Formed in Situ, U.S. Pat. 3,022,281, Feb. 20, 1962.
43. Smith, W. M.: "Vinyl Resins," Reinhold, New York, 1958.
44. Smith, W. M.: "Manufacture of Plastics," vol. 1, pp. 303–343, Reinhold, New York, 1964.
45. Smith W. V., and R. H. Ewart: Kinetics of Emulsion Polymerization, *J. Chem. Phys.*, **16**: 592 (1948).
46. Spencer, F. J.: Progress in Polymers Today, *Hydrocarbon Process.*, **45**(7): 83 (1966).
47. Talamini, G., and G. Vidotto: Polymerization of Vinyl Chloride Initiated by Tri-*n*-butylborane–Oxygen Systems, *Makromol. Chem.*, **50**: 129 (1961).
48. Thomas, J. C.: New Improved Bulk PVC Process, *Hydrocarbon Process.*, **47**(11): 192 (1968).
49. Winslow, F. H., and W. Matreyek: Particle Size in Suspension Polymerization, *Ind. Eng. Chem.*, **43**: 1108 (1951).
50. Albright, L. F.: Vinyl Chloride Polymerization by Suspension Processes Yields Polyvinyl Chloride Resins, *Chem. Eng.*, **74**(12): 145 (1967).
51. Albright, L. F.: Vinyl Chloride Polymerization by Emulsion, Bulk and Solution Processes, *Chem. Eng.*, **74**(14): 85 (1967).
52. Basel, L., and L. Papp: Polymerization Procedures, Industrial, in N. Bikales (ed.), "Encyclopedia of Polymer Science and Technology," vol. 11, p. 280, Interscience, New York, 1969.
53. Bingham, R. E. (to General Tire and Rubber Co.): Continuous Polymerization of Vinyl Monomers, U.S. Pat. 3,125,553 (Mar. 17, 1964).
54. Bingham, R. E., H. D. Forest, W. J. Hanlon, and J. L. Hutson (to The General Tire and Rubber Co.): Process of Producing Blotter-Type Vinyl Halide Resin and Product Obtained Thereby, U.S. Pat. 3,062,759 (Nov. 6, 1962).

55. Suspension or Emulsion Type PVC, *Br. Chem. Eng. Proc. Supp.*, November 1967, p. 69.
56. Chatelain, J.: Two-Step Bulk Polymerization Process of Vinyl Chloride, *163d Am. Chem. Soc. Meet., Boston, Mass.*, 1972.
57. PVC Plant Goes the Limit for Safe Operation, *Chem. Eng.*, Sept. 13, 1965, pp. 124–126.
58. Europe Leads the Way in 100% PVC Fibers, *Chem. Eng. News*, Mar. 13, 1967, pp. 36–37.
59. Systems Ease PVC Reactor Cleanup, *Chem. Eng. News*, Feb. 1, 1971, p. 49.
60. Japanese Tell of New Route to PVC, *Chem. Week*, Aug. 5, 1970, 43–44.
61. Compagnie de Saint-Gobain: Continuous Bulk Polymerization, French Pat. 1,261,921 (Apr. 17, 1962).
62. Dannis, M. L., and F. L. Ramp (to B. F. Goodrich Co.): Resin Composition with Improved Heat Resistance, U.S. Pat. 2,996,489 (Aug. 15, 1961).
63. Dunlop, R. D., and F. E. Reese: Continuous Polymerization in Germany, *Ind. Eng. Chem.*, **40:** 654 (1948).
64. Enk, E., and H. Reincke (to Wacker-Chemie): Process for Production of Vinyl Chloride Polymers by Dispersion Process in Aqueous Media, U.S. Pat. 2,981,722 (Apr. 25, 1961).
65. Gateff, G. (to B. F. Goodrich Co.): Process for Chlorination of Polyvinyl Chloride, U.S. Pat. 3,334,078 (Aug. 1, 1967).
66. Gateff, G., and H. H. Bowerman (to B. F. Goodrich Co.): Process for Halogenation of Synthetic Resins, U.S. Pat. 3,167,535 (Jan. 26, 1965).
67. Gord, L. J. (to Société Rhovyl): Polyvinyl Chloride Fibers and Process for Producing Same, U.S. Pat. 3,236,825 (Feb. 22, 1966).
68. Heiberger, C. A., and L. Fishbein (to Air Reduction Co.): Vinyl Chloride–Ethylene Copolymers, U.S. Pat. 3,468, 840 (Sept. 23, 1969).
69. Heiberger, C. A., and L. Fishbein (to Air Reduction Co.): Vinyl Chloride–Propylene Copolymers, U. S. Pat. 3,468,858 (Sept. 23, 1969).
70. Herte, P.: Herstellung von Polyvinylchlorid (Production of PVC), *Chem. Tech.*, **4:** 327 (1952).
71. Hopff, H., and I. Fakla: The Influence of the Composition of the Emulsifier on the Polymerization of Vinyl Chloride, *Makromol. Chem.*, **88:** 54–74 (1965).
72. Polyvinylchloride: Produits Chimiques Pechiney–St. Gobain, *Hydrocarbon Process.*, November 1969, p. 231.
72a. Koyanagi, S., S. Tajima, and K. Kurimota (to Shinetsu Chemical Co.): Method for Preparing Polyvinyl Chloridet by Suspension Polymerization, U.S. Pat. 3,669,946 (June 13, 1972).
73. Krause, A.: Mass Polymerization for PVC Resins, *Chem. Eng.*, Dec. 20, 1965, pp. 72–74.
74. Krehler, K., and G. Wick: "Kunststoff-Handbuch," Bd. II, "Polyvinylchlorid," Pt. I, pp. 41–72, Hanser, Munich, 1963.
75. Labine, R. A.: Drying Tricks Tailor Resin Properties, *Chem. Eng.*, Nov. 16, 1959, pp. 166–169.
76. Markovitz, R. E.: Picking the Best Reactor Jacket, *Chem. Eng.*, Nov. 15, 1971, pp. 156–162.
77. Mazzolini, C., L. Patron, A. Moretti, and M. Campanelli: Catalytic System for Low Temperature Polymerization of Vinyl Chloride, *Ind. Eng. Chem. Prod. Res. Dev.*, **9:** 504 (1970).
78. Meinhold, T. F., and M. W. Smith: Produces Dust-free PVC Resin, *Chem. Process.*, July 1959, pp. 61–62.

79. Meinhold, T. F., R. Williams, and C. Rehfuss: Flash Fluidized Bed Dryers for PVC, *Chem. Process.*, November 1968, pp. 21–22.

80. Nass, L. I.: "Chemistry and Technology of Poly (Vinyl Chloride) and Related Compositions," Dekker, New York, to be published.

81. Platzer, N.: Design of Continuous and Batch Polymerization Processes, *Ind. Eng. Chem.*, 62(1): 6 (1970).

82. Powers, J. R. (to B. F. Goodrich Co.): Polymerization of Vinyl Compounds, U.S. Pat. 2,520,959 (Sept. 5, 1950).

83. Produits Chimiques Pechiney–St. Gobain: Preparation Process for Mass Polymerization of Polymers and Copolymers of Vinyl Chloride, French Pat. 1,357,736 plus first, second, and third additions to patent (Apr. 10, 1964).

84. Produits Chimiques Pechiney–St. Gobain (by E. Berjot and J. C. Thomas): Autoclave for Bulk Polymerization, French Pat. 1,360,251 (May 8 ,1964); also French additions 83,327 (July 24, 1965).

85. Produits Chimiques Pechiney–St. Gobain (by J. C. Thomas): Bulk Polymerization of Vinyl Chloride, French Pat. 1,382,072 (Dec. 18, 1964).

86. Produits Chimiques Pechiney–St. Gobain: *Brochure* MT/GC 3.1970, 1970.

87. Reid, E. W. (to Carbide and Carbon Chemicals Corp.): Vinyl Resins, U.S. Pat. 1,935,577 (Nov. 14, 1933).

88. Rysavy, D.: Preparation of Poly (Vinyl Chloride) Paste-Types in Spray Installations, Kunststoffe, 48(3): 108 (1958).

89. Shelley, P. G., and E. J. Sills: Monomer Storage and Protection, *Chem. Eng. Prog.*, 65(4): 29 (1969).

90. Shockney, J. C. (to B. F. Goodrich Co.): Post-Halogenating of Halogen Containing Resins, U.S. Pat. 3,100,762 (Aug. 13, 1962).

91. Smith, E. S. (Goodyear Tire and Rubber Co., Akron, Ohio): Personal Communication, 1971.

92. Société Rhovyl (by L. J. Gord and F. Michel): Adhesive Fibers, French Pat. 1,414,948 (Oct. 22, 1965).

93. Strong, W. A.: Organic Peroxides: Diisopropyl Peroxydicarbonate, *Ind. Eng. Chem.*, 56(12): 33 (1964)

94. Suga, M.: Studies of Explosive Reaction of Vinyl Chloride Mixed with Oxygen, *Rev. Phys. Chem. Jap.*, 29(2): 73 (1960).

95. Thomas, J. C.: New Improved Bulk PVC Process, *Hydrocarbon Process.*, 47(11): 192 (1968).

96. Toulas, M. (Produits Chimiques Pechiney–St. Gobain): personal communications, 1970.

CHAPTER SEVEN

Processes for Styrene Production

Production of styrene has grown at a substantial rate since it was first produced commercially in the 1930s, and by 1970 about 5 billion lb/year were produced in the United States. With a predicted growth rate of about 8 percent per year, production of at least 8 billion lb/year is expected by 1980 (13, 15, 51).

In the United States, about 62 percent of the styrene is used for the production of homopolymers of styrene (called straight or regular polystyrenes); 17 percent for modified polystyrenes, which are copolymers requiring major fractions of styrene; 16 percent for butadiene-styrene elastomers and 5 percent as a comonomer for the production of numerous polymeric materials, including polyester resins and alkyds. The fraction of styrene used for the production of straight polystyrenes is expected to grow within the next few years to 68 percent (51).

Styrene production was started almost simultaneously in the 1930s by the Dow Chemical Company in the United States and by Badische Anilin- und Soda-Fabrik (BASF) in Germany (8). In the late 1940s, most production was in the United States, but European Common Market countries, Japan, and Russia are now all large producers. The relative growth rate for production generally will be faster in these countries than in the United States (32, 51).

Since 1955, the selling price for styrene has decreased from 17 cents/lb until currently it is about 7 cents/lb. Some American producers fear that prices may even drop to 6 cents/lb especially if too much production capacity is built. Some decreases in manufacturing costs have resulted in the last few years for the following reasons:

1. Costs of feedstocks (ethylene and benzene) have decreased somewhat.

2. Operating costs per pound of styrene produced are lower because of improved technology and the large size of the newer plants.

More modern plants are huge. Dow's newest unit in Texas has a capacity of 1 billion lb/year, and that of Monsanto Co. at Texas City, Texas, has a capacity of 1.3 billion lb (13, 35). Plants smaller than 0.5 billion lb may soon become uncompetitive, at least in the United States.

All styrene is currently produced commercially by the catalytic dehydrogenation of ethylbenzene. The product stream from dehydrogenation is fractionated to obtain high-purity styrene, and the unreacted ethylbenzene is recycled. In the United States, about 90 percent of the ethylbenzene is obtained by alkylation of benzene with ethylene, using several available processes. The remaining 10 percent is recovered by superfractionation of mixed xylene streams of refineries. The sales price of ethylbenzene has recently decreased to 4 cents/lb (12). Figure 7-1 is a simplified flowsheet of a styrene plant that includes ethylbenzene production from ethylene and benzene and then dehydrogenation of ethylbenzene. Commercial processes will be discussed in detail later.

Some styrene is produced in ethylene-propylene plants during the pyrolysis of heavier hydrocarbons such as naphthas. The styrene can be recovered by extractive distillation of the C_8 cut of the product stream. As much as 70 million lb can perhaps be recovered per billion pounds of ethylene. Although such a process is probably not yet used commercially, it appears promising.

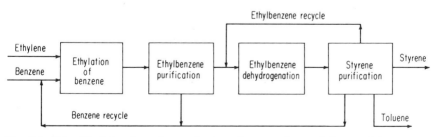

Fig. 7-1 Simplified flowsheet for production of styrene from ethylene and benzene.

Production of Ethylbenzene from Benzene and Ethylene

Benzene is catalytically alkylated with ethylene, producing ethylbenzene:

$$C_6H_6 + C_2H_4 \rightarrow C_6H_5C_2H_5$$

Ethylbenzene unfortunately is also ethylated, resulting in diethyl and higher polyethylbenzenes that can be reacted with benzene under suitable conditions to produce ethylbenzene. These latter reactions, frequently referred to as dealkylation, transalkylation, or disproportionation, are employed in all commercial processes to obtain high yields of ethylbenzene on both a benzene and ethylene base.

Liquid-phase and vapor-phase processes and several catalysts are available for production of ethylbenzene. Liquid-phase processes using catalysts of the aluminum chloride, $AlCl_3$, and boron trifluoride, BF_3, types predominate.

Ethylbenzene Production Using $AlCl_3$-Type Catalysts   Processes for ethylation of benzene using $AlCl_3$-type catalysts are relatively old, and many details have been described in the literature (8, 46, 48, 49). The catalysts are effective for both the ethylation of benzene and the dealkylation of overalkylated products. The chemistry and engineering features of these processes are discussed to help clarify process details to be considered later.

Chemical and Engineering Aspects of $AlCl_3$ *Alkylation:* Alkylation of benzene with ethylene using aluminum chloride is a three-phase system consisting of two liquid phases and one gas phase. The liquid catalyst (or acid) phase, containing dissolved $AlCl_3$, and the liquid hydrocarbon phase, containing benzene and various ethylated benzenes, are mixed and emulsified. Sufficient pressure is employed for the gaseous ethylene to dissolve and react almost as soon as it contacts the emulsion.

The alkylation and dealkylation reactions occur primarily in the acid phase or at least at the acid interface. These reactions are reversible, and Fig. 7-2 shows how the equilibrium composition of a mixture of benzene, ethylbenzene, and polyethylbenzenes is obtained as the ratio increases to 1.0. In commercial reactors, the equilibrium composition is only approached; the polyethylbenzene content is generally higher than the equilibrium value since polyethylbenzenes are recirculated.

Most of the aluminum chloride in the catalyst mixture forms complexes with various aromatic hydrocarbons, and the composition of the mixture on a weight percentage basis is often approximately as follows (46): $AlCl_3$ (combined), 26 percent; $AlCl_3$ (free), 1 percent; benzene and ethylbenzene, 48 percent; and high-molecular-weight hydrocarbons that

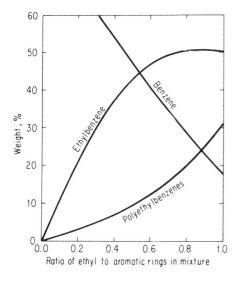

Fig. 7-2 Effect of reactant ratio on equilibrium composition.

are mainly polyethylbenzenes, 25 percent. A small amount of HCl, which complexes with both $AlCl_3$ and aromatics, is also needed as a promoter. Rather wide variations in catalyst composition occur, depending on the method of preparing the catalyst mixture and on the operation of the reactor system. Such variations have a significant effect on the activity of the catalyst, a high activity catalyst being defined as one resulting in a high ratio of ethylbenzene to polyethylbenzenes for both alkylation and dealkylation reactions.

Two methods are commonly employed for the initial preparation of the acid complexes, which have viscosities similar to heavy oils, are usually brown to black in color, and have densities greater than 1.0 g/cm³. By one method, solid aluminum chloride is suspended in an aromatic hydrocarbon such as benzene or ethylbenzene. When ethylene is bubbled through the mixture, an acid complex forms. In the second method used industrially (2), powdered aluminum is suspended in liquid benzene. Anhydrous HCl is bubbled through the mixture at about 55° C, and most of the aluminum reacts within 2.5 h, forming aluminum chloride. The benzene is then drained from the mixture, and fresh benzene is added. This mixture is then heated to 55 to 70°C, and ethylene is added, resulting in the formation of an acid complex. Use of a powdered aluminum alloy containing several percent of copper minimizes the formation of naphthenic and other nonaromatic hydrocarbons in the catalyst complex (compared to catalyst mixtures prepared using plain aluminum powder) (7).

As the catalyst mixture is used for alkylation, undesired impurities slowly build up in it, causing the catalyst activity to decrease. Water, sulfur compounds, and certain trace hydrocarbons in the feedstock act to reduce the catalyst activity. Several highly detrimental by-products of the reaction include polynuclear hydrocarbons (alkylated anthracenes, fluorenes, and phenanthrenes), naphthenic, and other nonaromatic hydrocarbons. The last may be formed by dimerization and polymerization of ethylene. Electroconductivity (43), fluorescence (6), radiation-energy absorbance (52), and viscosity have all been shown to be of value in estimating the activity of the catalyst complex. Commercially, extra AlCl$_3$ and HCl are added to deactivated catalyst mixtures to revive their activities.

Relatively pure benzene and ethylene are normally used as reactants. *Styrene-grade* benzene has a purity slightly greater than 99 percent, a boiling range of about 1°C, and a minimum freezing point of 4.85°C. Ethylene purity is not as a rule critical but usually is above 90 percent. Paraffinic impurities are unreactive but may complicate the recovery steps of the process. The reactants should be free of acetylene, sulfur compounds such as thiophene, other unsaturates, and in most cases moisture, since all are detrimental to catalyst life. The moisture reacts with AlCl$_3$ to form HCl and alumina. The HCl required as a cocatalyst was produced in a German plant by adding controlled amounts of moisture (8), but in most plants either free HCl or ethyl chloride is used. The latter decomposes in the reactor, forming ethylene and HCl.

Relatively little has been reported about the character of the emulsion present in the commercial reactors. Some agitation is provided in at least some reactors by bubbling ethylene up through the emulsion, but most of the ethylene is absorbed and reacted. Since the emulsion is maintained at the boiling point of the hydrocarbon phase, benzene and some ethylbenzene are vaporized because the alkylation reaction is highly exothermic. Agitation caused by the ethylene gas and the boiling produces rather vigorous agitation in the emulsion.

The overall weight ratio of the acid phase to benzene feed often is maintained between about 1:1 to 2:1 (2, 7), but in many reactors this ratio varies significantly with height. The acid phase, which has a higher density than the hydrocarbon phase, is generally present in greater proportions in the lower parts of the reactor and is the continuous phase there. The overflow from the reactor is often mainly the hydrocarbon phase, containing a small amount of catalyst. The hydrocarbon phase is hence the continuous one in the top portion of the reactor.

The ethylene generally enters the reactor through a sparger located near the bottom of the reactor and there contacts primarily the acid phase. When mixing in the emulsion is poor, high local concentrations of ethylene are present near the sparger and considerable amounts of by-products may form. In summary, the concentrations of the dissolved reactants are critical relative to the kinetics of the various reactions, the relative ratio of monoethylbenzene and of diethylbenzenes (and higher polyethylbenzenes) formed, and the eventual composition of the acid phase.

Kinetic expressions for the alkylation reactions have not yet been reported. Some rather qualitative data have been presented for a semi-batch reactor operated at temperatures lower than those used commercially and with such low amounts of AlCl₃ that only one liquid phase was present. Increased reaction temperatures and concentrations of AlCl₃ resulted in faster rates of reaction (27). In commercial reactors, the ethylation of benzene is said to be virtually instantaneous, resulting in almost complete disappearance of ethylene (62). Transalkylation reactions between benzene and overalkylated benzenes are relatively slow, however, and higher temperatures are sometimes employed to obtain reasonable rates of reaction. The controlling step for the primary ethylation reaction is probably the transfer of the reactants to the acid phase. Since benzene is dissolved in appreciable quantities in the catalyst phase, the transfer of ethylene from the gas phase to the acid phase is likely to be at least partially controlling.

The multiple roles of the important operating variables for benzene ethylation presumably are similar to those for isobutane alkylation (40). In that case agitation, residence time of reactants in the emulsion, pressure, and the ratio of benzene to ethylene in the feed all affect primarily the concentrations of reactants dissolved in the catalyst phase. Temperature and composition of the acid phase affect both the transfer steps in the process and the kinetics and relative importance of the various reactions. Temperature, for example, has a complex effect on the character of the emulsion because of changes in the viscosities and interfacial tensions. It then indirectly affects the interfacial areas between phases. Temperature in addition then controls to some extent the compositions and amounts of hydrocarbons dissolved and reacted in the acid phase. Analyses of how the other variables affect alkylation reactions are also presented (40).

Alkylation Processes Using AlCl₃-Type Catalysts: The flowsheet used by most companies is relatively similar to the ones reported by Mitchell (46) and Boundy and Boyer (8) for World War II plants. Major improvements have been made since then, however, in designing and operating portions of the process and in the sequence of separating the product mixture. Figure 7-3 shows the typical flowsheet for a modern plant.

The benzene feed, which includes recycle benzene, must first be dried to at least 2 to 3 ppm. Solid desiccants have been used for this purpose, but an azeotropic distillation or stripping technique is generally preferred. In this latter method, the overhead gas stream from the column is condensed to form a heterogeneous mixture of wet benzene and water. The water layer is separated, and the wet benzene is recycled to the wet-benzene feed tank. The bottom product from the column is "dry" benzene, which after cooling and possible storage is continuously

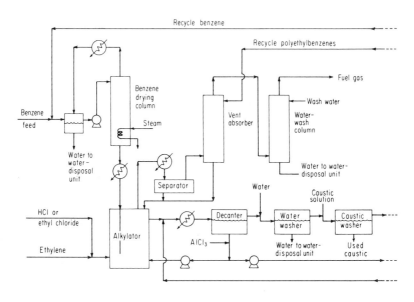

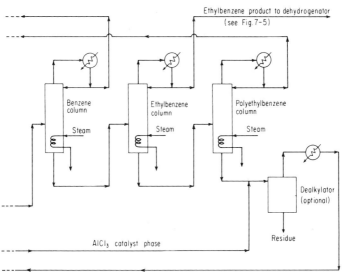

Fig. 7-3 Ethylbenzene production of benzene and ethylene using AlCl$_3$ catalyst.

metered to the alkylation reactor. A dip tube whose opening is located slightly above the sparger for the ethylene feed is frequently employed for introducing the benzene.

The ethylene feed is generally dried using either molecular sieves or alumina. HCl (or ethyl chloride) is added to the ethylene stream on about a 1:100 mole ratio (2, 7), and the mixture is then sparged into the bottom portion of the emulsion. When ethyl chloride is employed, ethylene feed is often employed to vaporize the required amounts of ethyl chloride.

A reactor used by the Dow Chemical Company is essentially a large cylindrical vessel or tower filled with the reaction mixture to a height of about 35 ft (18, 46). Since the emulsion is highly corrosive, especially if any moisture gets into it, the reactor is either glass- or brick-lined. Dow has for many years used brick lining in at least some of their reactors. Other feed streams entering the reactor near the bottom include recycle catalyst complex and a recycle stream of overalkylated benzenes. AlCl$_3$ can be fed by a screw feeder into the top of the reactor. Sometimes it is added to the overalkylated benzene stream. The Dow reactors are operated at a temperature of about 95°C and a pressure just slightly above atmospheric, about 5 lb/in^2 gage.

German ethylbenzene plants operated during the early 1940s were similar to those developed at the same time in this country.[*] A German reactor used to produce ethylbenzene at the rate of 18 million lb/year was 40 ft high, was filled with emulsion to a level of approximately 32 ft, and was 3.75 ft in diameter. About 6 lb of ethylbenzene per cubic foot of emulsion per hour was then produced in this reactor. Both the ethylene and benzene were bubbled into the bottom of this reactor.

The reactor of the Badger-Cosden-Carbide (BCC) process is thought to operate at pressures up to 140 lb/in^2 gage (9.5 atm) and at temperatures of about 130°C. Two advantages are claimed for such operation:

1. The chemical kinetics is faster, especially for the disproportionation reactions between benzene and diethylbenzene to form ethylbenzene.

2. Low-pressure steam can be generated, and hence most of the exothermic heat of reaction can be recovered.

The reactor design is complicated because of the higher pressures and because the emulsion mixture becomes more corrosive at higher temperatures.

Alkylation vessels for the production of at least 130 million lb/year of ethylbenzene have been built. Modern reactors are designed for rapid and effective dispersion of the reactants into the emulsion, and they are provided with better mixing, or agitation, of the phases than

[*] See Ref. 8, pp. 893–900.

the earlier reactors. Ethylbenzene production levels of at least 10 lb per cubic foot of emulsion per hour are probably obtainable now.

The two exit streams from a Dow reactor consist of an overflow stream of the emulsion and also a gas stream. The gas stream consists mainly of the inert hydrocarbons that enter in the ethylene feed, benzene, some ethylbenzene, and HCl. Most of the benzene and ethylbenzene are condensed in a particle condenser, and the condensate is allowed to flow back (or reflux) to the reactor. The uncondensed gases are scrubbed in a relatively small absorbing column at about 40°C, using the recycle light polyethylbenzenes as the absorbent. By this technique, most residual benzene, ethylbenzene, and ethylene are recovered and recirculated to the alkylation unit. The gas stream from the absorber is primarily a mixture of light paraffin hydrocarbons and HCl. This gas stream is washed with water to remove HCl, and the remaining gases are frequently used as fuel gases.

The liquid emulsion from the reactor is first cooled to about 40°C, and the hydrocarbon and acid phases are allowed to settle by decanting. Frequently the hydrocarbon phase will retain about 1 percent of finely dispersed droplets of acid complex that do not readily separate. Passing the hydrocarbon phase through a packed bed of glass wool, silicon dioxide (sand), or polyethylene granules helps coalesce the acid complex (44). Two cylindrical vessels positioned horizontally were provided as decanters in the plants described by Mitchell (46). The recovered acid complex is recycled to the reactor. In older units, a small portion of the complex is pumped to a dealkylator (described later).

Granulated $AlCl_3$ is sometimes added to the recycle acid stream in order to maintain a high catalyst activity. In other plants, $AlCl_3$ is fed continuously to the reactor using a combination of screw conveyor and hopper. Precautions must always be taken to prevent moisture contamination of $AlCl_3$, which must have a purity of at least 97.5 percent (46); impurities have a detrimental effect on the reaction.

The hydrocarbon product from the decanter contains ethylbenzene, benzene, diethylbenzenes, polyethylbenzenes and tars, and small amounts of $AlCl_3$ and HCl; these compounds are arranged in the usual order of importance. Water washing followed by 20 percent caustic washing quickly destroys and neutralizes residual catalyst (18). Small particles of alumina formed probably need to be filtered from the dilute acid solution before it is neutralized and prepared for discard. Caustic neutralizes residual acids. In the old process described by Mitchell, only one wash with 30 to 50 percent caustic was provided, but such a procedure would require high caustic consumption. Furthermore at such high concentrations after some salt (sodium chloride) is formed, the salt begins to drop out of solution, plugging lines, pumps, etc.

The washing techniques used are relatively standard. The hydrocarbon and desired wash liquid can be combined and passed through a centrifugal pump to provide intimate contact between the two phases. A large horizontal cylindrical vessel is then effective for separating the two phases. More elaborate contacting devices and separating equipment such as centrifuges may now be used in some cases.

The washed hydrocarbon mixture is separated in a series of distillation columns. In the early processes described by Mitchell (46), the lighter cuts were separated from the polyethylbenzenes and tars in the first column of the separation sequence. Two reboilers were needed for this column since fouling because of unseparated acid complex was then a serious problem. Effective separation of the acid complex permits an arrangement (shown in Fig. 7-3) in which the first column removes the wet benzene, the second removes the ethylbenzene, and the third separates the light polyethylbenzenes from the residual hydrocarbons and tars.

Twenty actual plates, probably bubble-cap plates, were used in the early columns (46) for the separation of benzene from ethylbenzene and polyethylbenzenes. A reflux ratio of 3:1 and essentially atmospheric pressure were provided. A column with similar design and similar operating conditions should be quite satisfactory for the first column in Fig. 7-3; sieve plates, or other improved plates, are generally used now.

The ethylbenzene column described by Mitchell had 58 plates and was operated at slightly above atmospheric pressure with a 3:1 reflux ratio. The ethylbenzene product from the top of this column must be essentially free of diethylbenzenes, since they form divinylbenzenes during subsequent dehydrogenation. These latter compounds polymerize rapidly and cause serious maintenance and operating problems in the equipment used to produce styrene. The ethylbenzene stream is generally washed with caustic and then dried (not shown in Fig. 7-3). These operations remove trace amounts of HCl and reduce the chloride content of the ethylbenzene to 10 ppm. The ethylbenzene product normally is at least 99.9 percent pure.

The third column in Fig. 7-3 is for the separation of the lighter polyethylbenzenes from the heavier polyethylbenzenes and the tars. Mitchell describes such a column that contains 10 baffle plates and was operated at 50 mm Hg absolute pressure, with a 0.5:1 reflux ratio, and with a top temperature of 135°C. Corrosion is generally not a problem in the distillation train, and low-carbon steel is a common material of construction.

If any propylene should enter the reactor system, cumene would be produced. In such a case, the cumene should be separated from the ethylbenzene since α-methylstyrene, an undesired contaminant for styrene, would be formed during styrene manufacture.

A dealkylator, as shown in Fig. 7-3, is sometimes used in some plants, especially older ones. If such a unit is provided, the bottom product from the third distillation column is pumped to it. Here the higher polyethylbenzenes are contacted with the liquid acid phase (containing dissolved $AlCl_3$ and benzene). At 200°C and higher, the polyethylbenzenes react rapidly with the benzene to form ethylbenzene, diethylbenzenes, and lighter polyethylbenzenes. The overhead stream from the dealkylator is recirculated to the alkylation system. A tar-$AlCl_3$ residue is removed from the bottom of the dealkylator. Mitchell (46) indicates that rotating scraper blades have been used to keep the walls of the dealkylator clean. In some modern units, most of the bottom product of the third column is recycled directly to the alkylator.

Process details, including the dimensions for major pieces of equipment in the ethylbenzene-styrene plant operated at Schkopau, Germany, during the early 1940s, have been reported by Boundy and Boyer.[*]

Consumption of $AlCl_3$ and ethyl chloride (when used as the cocatalyst) is approximately 0.01 to 0.03 and 0.005 lb, respectively, per pound of ethylbenzene produced (26). These values of course depend on the operating conditions and on the level of catalyst activity maintained in the reactor.

Optimization of Processes Using $AlCl_3$ *Catalyst:* When the ratio of ethylene to benzene fed to the alkylation reactor is increased, more ethylbenzene is produced (see Fig. 7-2); however, more overalkylated products are also formed. At low feed ratios, increased benzene must be recycled; at high ratios, more diethylbenzenes and higher polyethylbenzenes must be recovered and recycled to be dealkylated. Some undesirable products including tars are formed during dealkylation, resulting in decreased yields of ethylbenzene. Hence there is an optimum feed ratio to be used.

The ratio of ethylene to benzene fed to the reactor, the techniques used for recycling polyethylbenzenes, and the operation of the dealkylator all affect the composition of the reaction mixture in the alkylation vessel at steady-state conditions. This composition is frequently reported as the ratio of ethyl group to aromatic rings. In the ethylbenzene units employed during World War II, The Dow Chemical Co. operated so that this composition ratio was 0.58 (46). With more effective utilization of energy (because of more effective energy-recovery techniques) and with the improved distillation columns now available, the optimum ratios have probably tended to decrease and may be as low as 0.5 to 0.55 in at least some cases. Lower ratios also minimize the need for a separate dealkylation unit.

Ethylbenzene yields for World War II plants were reported to be 95.5 and 96.8 percent, based on the benzene and ethylene feeds, respectively. Improved operation techniques have certainly raised these values for modern plants. Although current values have not been reported in the literature, yields of 98 percent or greater are probably obtained for both feeds.

Alkar Process   In the late 1950s, Universal Oil Products Co. (UOP) announced their Alkar process, designed primarily for the production of ethylbenzene. Ethylene feedstocks containing as little as 8 to 10 percent ethylene can be utilized. The Alkar process is currently employed in over 10 units scattered throughout the world, including the United States (35), and more units are reportedly being built or designed.

The patent by Fenske (22) probably describes the basic process, which uses a fixed bed of substantially anhydrous inorganic oxides (including silica, alumina, silica-alumina, and many others) that is activated with boron trifluoride, BF_3. Bloch (5) describes a dealkylation procedure with the same catalyst.

BF_3-modified anhydrous α-alumina is a most satisfactory alkylation catalyst when small amounts of BF_3 are continuously introduced into the reactor (22). The BF_3 can be mixed with the ethylene-containing feed gas to the reactor. Most BF_3 is recovered from the product stream and is recirculated; less than 1 lb of BF_3 is "used" per 2,000 lb of ethylbenzene product.

Many ethylene-containing streams can be satisfactorily used in the process. Light paraffins, hydrogen, carbon oxides, and nitrogen do not interfere with the reaction or the catalyst's activity (28, 60); However, heavier olefins, including propylene and butenes, are also alkylated, producing the expected higher alkylbenzenes. In some refineries, the off-gas from a catalytic cracking unit contains 8 to 10 percent ethylene and perhaps 1 to 2 percent propylene. Such a gas stream reacts to produce both ethylbenzene and cumene (isopropylbenzene). Both water and sulfur compounds such as hydrogen sulfide are detrimental to catalyst life, and careful drying of the feed streams is necessary.

Three phases are generally present in the reactor used for the Alkar process: a liquid phase, consisting mainly of benzene, ethylbenzene, polyethylbenzenes, and dissolved ethylene and BF_3; a gas phase, containing ethylene and light inert gases; and the solid catalyst. The operating conditions used commercially have not been reported, but presumably fairly moderate temperatures and pressures are employed. It is assumed that the main alkylation reaction occurs after the reactants transfer to the catalyst surface. BF_3 from the feed stream maintains a high catalyst activity.

A flowsheet for an ethylbenzene plant using BF_3-modified alumina catalyst has been discussed by Fenske (22). Excess benzene (on a mole bases) is provided, and all ethylene and other olefins normally react in a single pass through the catalyst bed. The main products are monoalkylated benzenes, but some polyalkylated aromatics are also formed. The product stream is separated by distillation to yield BF_3, benzene, ethylbenzene, and polyethylbenzenes. BF_3 and benzene are recycled to the reactor, and polyethylbenzenes are sent to a dealkylation reactor to produce more ethylbenzene.

UOP has made estimates for ethylbenzene production of 56,000 tons/year in a central European country; the ethylbenzene will then be used to produce styrene at 50,000 tons/year (60). When a relatively low-purity gas containing 18.5 percent ethylene is used, the total erected cost of an Alkar plant is about $2.05 million. When high-purity (97.5 mole percent) ethylene is used, however, the cost is $1.2 million. The product stream is of course separated more easily when relatively little inert gas is present.

UOP claims that the operating costs for the process are low; Grote and Gerald (28) give an example in which these costs, excluding hydrocarbon feedstocks, total 1.07 cents per pound of ethylbenzene produced in a small plant. Even lower costs would be expected in a larger plant. Yields of ethylbenzene based on both benzene and ethylene are high, approaching 100 percent, and are higher than those obtained in $AlCl_3$-type processes. Corrosion in the Alkar process is minimal, and plain carbon steel is used throughout the entire plant.

Ethylbenzene Production Using Vapor-Phase Processes Vapor-phase processes for production of ethylbenzene are still of some commercial importance, especially for smaller plants using rather impure feedstocks. The vapor-phase processes developed by UOP and previously used at the Sinclair-Koppers Co. (earlier The Koppers Co.) plant at Kobuta, Pennsylvania, have been described in considerable detail (23, 26). First a solid phosphoric acid on kieselguhr and later silica-alumina were employed there as catalysts.

In these processes, liquid benzene is pumped using a multistage vertical centrifugal pump to high pressures, and it is then mixed with gaseous ethylene. The mixture is then heated and vaporized. Heat exchange is provided first with hot benzene vapors (the overhead product in the benzene column), second with the hot product stream from the reactor, and third with flue gases in a gas-fired furnace. The heated mixture then enters the reactor.

The product mixture from the reactor is first cooled, and most if not all of it condenses because of the high pressures involved and because the amounts of unreacted ethylene and light ends in the product stream are quite low. The condensate contains unreacted benzene (generally used in large excesses in vapor-phase processes), ethylbenzene, polyethylbenzenes, and tars or heavy residues. The product mixture contains some acid gases, particularly if a phosphoric acid catalyst is used, and in such a case the acids are removed by washing with caustic or by using packed caustic beds.

The liquid hydrocarbon product at about 58 atm pressure is fed to the benzene column maintained at 4 atm pressure. When the pressure is suddenly reduced, most of the light ends (including ethylene) and some benzene are flashed. The two product streams from the top of the benzene column are liquid benzene, which is recycled, and a gas stream containing ethylene and light inert gases (introduced as impurities in the feedstock). This gas stream has a low ethylene content, and most of it is used as fuel. Part of the gas, however, is recycled to maintain a constant level of inerts in the ethylene feed stream to the reactor.

The heavy-hydrocarbon liquids are separated by conventional multi-column distillation to produce ethylbenzene, polyalkylbenzenes, and tars or residues. The ethylbenzene product has a purity of 99.9 percent or greater. A dealkylation system is always provided to convert polyethylbenzenes to ethylbenzene.

Packed-bed reactors have been employed in which apparently no provisions were made for removing the heat of reaction. In such a case, a significant temperature rise occurs as the gases pass through the bed. Techniques that could be used for removing the heat include a series of packed beds with heat removal between beds and reactors built similar to tube-and-shell heat exchangers. Steam generated by such techniques would be available for preheating the feed streams or for furnishing heat in the distillation section of the process.

Catalytic Condensation Process: The catalytic condensation process of UOP uses a solid phosphoric acid catalyst like that widely employed for polymerization of propylene to form C_9 to C_{12} olefins (62). Operating conditions required for this catalyst are discussed by Ipatieff and Schmerling (36), and high levels of ethylation (with ethylene) are obtained at conditions varying from 325°C and 42 atm (600 lb/in² gage) pressure to 280°C and 62 atm (900 lb/in² gage). Lower temperatures (about 280°C) and feed ratios of benzene to ethylene of at least 4:1 are needed for long catalyst life and for obtaining a high ratio of ethylbenzene to polyethylbenzenes in the product stream.

The life of the phosphoric acid catalyst is increased by adding small amounts of ethanol or water to the feed stream. The ethanol decomposes to form ethylene and water, and the water promotes a high and relatively constant level of catalyst activity.

Loss of activity eventually results as carbonaceous deposits form on the catalyst. Controlled burning followed by steaming is successful for regenerating the catalyst. A small loss of phosphoric acid occurs as the catalyst is used or regenerated. Such a loss does not seriously deactivate the catalyst, but the freed acid does cause some corrosion problems.

An improved phosphoric acid catalyst developed by UOP was used in the ethylbenzene plant of El Paso National Gas Co., Odessa, Texas (37). This catalyst is claimed to be highly effective as an alkylation catalyst, and polmerization reactions with ethylene are slight. Some dimerization of ethylene occurs, however, since butylbenzene is a by-product. The catalyst is still reasonably active after 7 months of operation.

The improved phosphoric acid catalyst is said to be less corrosive than the previous one. Lean ethylene streams can be processed successfully, but the operating conditions employed have not been reported. Presumably they would be somewhat milder than those required for the original catalyst.

Phosphoric acid catalysts are not effective for the disproportionation reactions. Dealkylators in these processes have generally used $AlCl_3$-type catalysts.

Processes Using Silica-Alumina Catalysts: The Koppers plant at Kobuta, Pennsylvania, has more recently used a packed bed of acidic alumina on a silica-gel carrier (8, 23, 26, 62). The alkylation reaction is controlled at about 60 atm (900 lb/in^2 gage) and at 310 to 315°C (590 to 600°F). A fairly high ratio of benzene to ethylene is used in the feed stock to minimize formation of overalkylated products. With a 6:1 ratio, about 17 percent of the alkylated product consists of polyethylbenzenes, whereas with an 11:1 ratio less than 11 percent polyethylbenzenes is present (26). Higher ratios also minimize the amount of ethylene that dimerizes or polymerizes. Probably ratios of at least 8:1 are normally used with space velocities of about 2 vol of liquid benzene per volume of catalyst per hour. Three packed-bed reactors operated in parallel were used to produce ethylbenzene at over 120 million lb/year.

The dealkylator in this process uses the same silica-alumina catalyst as employed in the primary alkylation unit. The hydrocarbon feed stream to be dealkylated is diluted with steam, and this mixture is passed over the catalyst at essentially atmospheric pressure and temperatures of about 530°C. The diethylbenzenes are converted primarily into ethylbenzene and ethylene.

The life of the catalyst in the alkylator is long, often 70 days before reactivation is required. Carbonaceous deposits form on the catalyst, especially in the upper quarter of the bed. Feedstocks of improved purity increase the life of the bed (26), and acetylenic compounds in particular are undesirable. As carbonaceous deposits are formed, pressure drop through the bed increases. With a freshly reactivated catalyst, a pressure drop of almost 1.5 atm occurs, which rises to 12 atm after about 70 days. The inlet pressure of the reactor is maintained during the entire cycle at the same value, and the exit pressure is slowly decreased with time. The bed temperature is slowly raised as the catalyst is deactivated in order to maintain a constant production rate of ethylbenzene. Reversing the flow in a bed that is almost deactivated prolongs the life of the bed for about 2 weeks. The catalyst is easily and essentially completely regenerated using preheated air to burn off the deposits.

This process using an alumina-silica catalyst has the following advantages over the AlCl$_3$-type processes (26):

1. Fewer corrosion problems arise, permitting use of cheaper construction materials.

2. Catalyst consumption and costs are lower.

3. Neutralization problems, such as formation of rather stable emulsions, are avoided.

4. Drying the feedstocks is not necessary.

Disadvantages of the vapor-phase process are as follows:

1. Higher pressures require more complicated pumps and compressors.

2. Higher benzene-to-ethylene ratios mean that more benzene has to be recovered and recycled.

3. Dealkylation reactions are more difficult, and since essentially none occur in the main alkylation reactor, a dealkylation unit is required.

4. More ethylene dimerizes, and rather significant amounts of butylbenzenes are formed.

Ethylbenzene Separation from Refinery Cuts

Eight American companies use superfractionation to recover about 9 to 10 percent of the ethylbenzene produced in the United States from various refinery cuts. The term superfractionation is appropriate since ethylbenzene is separated from mixed xylene streams. Ethylbenzene has a normal boiling point only 2.3°C lower than that of p-xylene.

The ethylbenzene unit used by Cosden Petroleum Co. in Big Springs, Texas, has been widely publicized (1, 16, 17). The West Texas crude oil is first fractionated to obtain a straight-run naphtha fraction that is reformed using a UOP platformer. The mixed xylene fraction eventually obtained contains 28 percent ethylbenzene.

To make the desired recovery of ethylbenzene, a total height of 600 ft of fractionating columns is required. Cosden, who teamed up with Badger Manufacturing Co. in designing the unit, realized that a single 600-ft column was not practical. Even two 300-ft columns were not practical since they would sway so much during high winds that they would lose their liquid seals, especially on the top plates. The final design was three 200-ft columns; these three columns plus a 185-ft column for styrene purification were built in a quadrant. The four columns are joined laterally at six platform levels. The resulting group of columns can withstand winds up to 100 mi/h.

Since pressure drops in the columns, which contain about 350 plates (17), are substantial it was necessary to obtain reliable vapor-liquid equilibrium data over a relatively wide range of pressures. The amount of material in the columns at steady-state operation is approximately 1.5 times the daily charge, and control for such a column is very sluggish (45); thus 1 or even 2 days may pass before an operating adjustment is noticed. A choice of over 100 feed plates is provided in the Cosden unit to allow for changes of the feed composition.

The three columns are connected in series. The vapor line from the top of the first column is connected to the bottom of the second column. The liquid feed from the second column is pumped to the top of the first column. A similar arrangement is made between the second and third columns. A high reflux ratio is used in these columns, and 99.7 percent pure ethylbenzene has been obtained. Paraxylene is an impurity but does not cause any real problems in styrene production.

The ethylbenzene separated in the Cosden fractionators totaled about 20 million lb/year. Cosden claimed in 1960 that their total ethylbenzene costs were about 1.5 cents less per pound by fractionation than by an alkylation route (1). Although several companies continue to fractionate for xylenes and ethylbenzene, others would rather use these components as blending components for the production of high-quality gasolines. Ethylbenzene has an unleaded research octane number of about 107.4 and the xylenes from 107.4 to 117.5.

Catalytic Dehydrogenation of Ethylbenzene

Styrene is produced predominantly, if not exclusively, by the catalytic dehydrogenation of ethylbenzene. Noncatalytic dehydrogenation is of no commercial interest because of the high temperatures required, large amounts of undesired by-products, and low overall yields of styrene (63). Processes involving dehydrogenation by oxidation techniques have been proposed, but apparently no units are currently being employed commercially.

The basic catalytic process is simple in principle. A gaseous mixture of ethylbenzene and steam is passed at high temperatures through a packed bed of catalyst pellets. Much of the ethylbenzene reacts to form styrene and hydrogen. By-products formed in relatively small amounts include toluene, benzene, and methane.

The catalysts employed commercially are mainly ferric oxide, Fe_2O_3, promoted with potassium oxide (or potassium carbonate), chromic oxide, and sometimes other materials (38, 49). Shell Chemical Co. markets two popular catalysts, Shell 105 and Shell 205, and reports their compositions as follows:

Shell 105: Fe_2O_3, 87%; KOH, 10%; Cr_2O_3, 3%

Shell 205: Fe_2O_3, 90%; K_2CO_3, 6%; Cr_2O_3, 4%

Catalysts that have been claimed to be superior in at least certain aspects will be discussed later.

The approximate range of operating conditions or variables for commercial reactors are as follows:

Temperature, °C.. 580–660
Steam-to-ethylbenzene weight ratio in feed stream................ 1:1–2.5:1
Pressure, atm absolute....................................... 1–2
Liquid hourly space velocity, volumes of liquid ethylbenzene per
   volume of catalyst bed per hour............................. 0.4–2.0
Conversion of ethylbenzene per pass, %........................ 40–70

Conversions per pass of about 40 to 45 percent were widely used in the past, but some newer units now have conversions in the 60 to 70 percent range. Major reduction in operating costs per pound of styrene have as a result been realized. Residence times of the reactants in the catalyst bed often range from about 0.4 to 1.0 s.

Chemical Factors Affecting Reactor Design Side reactions result in by-product formation that often accounts for 10 to 15 percent of the ethylbenzene that reacts. An understanding of the factors affecting these side reactions is helpful for designing and operating dehydrogenators so that production of by-products can be minimized. At temperatures below about 630°C, most reactions occur on the catalyst surface, but gas-phase reactions become relatively more important at higher temperatures.

The following reactions have been suggested (47, 55) as the most important.

$$(1) \quad C_6H_5\text{---}C_2H_5 \rightarrow C_6H_5\text{---}C_2H_3 + H_2$$
Ethylbenzene   Styrene

$$(2) \quad C_6H_5\text{---}C_2H_5 \rightarrow C_6H_6 + C_2H_4$$
Benzene   Ethylene

$$(3) \quad C_6H_5\text{---}C_2H_5 + H_2 \rightarrow C_6H_5\text{---}CH_3 + CH_4$$
Toluene   Methane

$$(4) \quad \tfrac{1}{2}C_2H_4 + H_2O \rightarrow CO + 2H_2$$

$$(5) \quad CH_4 + H_2O \rightarrow CO + 3H_2$$

$$(6) \quad CO + H_2O \rightarrow CO_2 + H_2$$

$$(7) \quad C_6H_5\text{---}C_2H_5 \rightarrow 8C + 5H_2$$

$$(8) \quad 8C + 16H_2O \rightarrow 8CO_2 + 16H_2$$

All catalytic reactions involve several steps, e.g., adsorption, desorption, and surface reactions. Most of the above equations are the sum of several intermediate steps. Formation of carbon or carbonaceous deposits on the catalyst surface are undoubtedly more complex than shown by reaction 7 above.

The choice of reactions above, and perhaps especially reaction 5, is open to question. Methane is a relatively stable hydrocarbon, and other hydrocarbons, including ethylbenzene, are probably more reactive with steam by the water-gas reaction.

Commercial catalysts are often called *self-regenerative* since the potassium salts promote the destruction of carbon deposits. Some of these deposits are undoubtedly caused by polymerization or adsorption of styrene, and some of the active catalyst sites are covered. Steam reacts with these surface deposits such as shown by reaction 8 to form carbon oxides and hydrogen. When a new catalyst is used, the rate of deposition of these materials is at first greater than the rate of removal. The concentration of surface deposits initially increases, causing partial deactivation of the catalyst (10). After the catalyst has been operating for an extended time, a steady state is reached and the rate of deposition is essentially balanced by the rate of removal.

Commercial catalysts have been used up to 2 years in some cases, before they need to be replaced (55). With use some promoters in the pellets are lost, and the character and area of the catalytic surface gradually change. The aging and subsequent activity of the catalyst depend on its position in the bed since the temperature, rates of reaction, and composition of the reaction gases all vary with position. As a catalyst ages, some changes are normally made in the operating variables to help compensate for loss of activity. Temperature, for example, is raised (8, 46, 49).

Both $\tfrac{1}{8}$- and $\tfrac{3}{16}$-in catalyst pellets are used commercially. The smaller pellets exhibit greater catalyst activity but cause greater pressure drops of the gases passing through the bed. Pressure drops in commercial units are about 25 percent higher than calculated values, probably because of fines in the bed (24). Resistance to flow often changes as the bed is used, and pressure drops frequently vary from 0.3 to 1.0 atm. Since the exit pressure should never be less than atmospheric because of the potential danger of atmospheric leaks, the inlet pressure to the bed normally has to be changed somewhat as the bed is used.

Both laboratory (10) and plant (55) data for dehydrogenation have been modeled using part of reactions 1 to 8 (47, 55). These models implicitly include major simplifying assumptions, e.g., that mass-transfer steps of reactants and products to and from the catalyst surface are not controlling and that the activity of the catalyst is constant throughout the bed. Although these models are quite empirical in nature and of limited value for making conclusions about the chemical mechanism, the models do a good job in representing the data and in indicating the relative yield of styrene and by-products. Summarizing, increased styrene yields are obtained at lower temperatures, lower ethylbenzene conversions, and lower partial pressures of hydrocarbons (as obtained by lower total pressures in system and by use of higher steam-to-ethylbenzene ratios in the feed). Figure 7-4 shows one example of how conversion and total pressure affect styrene yields.

Catalysts or methods of pretreating catalysts have been reported (39, 50) that result in improved operating characteristics and often in higher styrene yields than shown in Fig. 7-4. One catalyst gave a yield of 92.8 percent for laboratory runs with an ethylbenzene conversion of 61.1 percent. Whether these catalysts are of commercial interest depends to a considerable extent on their life or aging characteristics, which have not been adequately publicized.

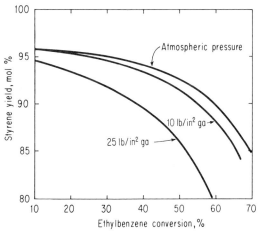

Fig. 7-4   Effect of conversion and pressure on styrene yields.

Temperature has a complex effect on the reaction sequence. It directly affects the kinetic rate constants for the various reaction steps. Furthermore, it affects the surface concentrations of "carbon" deposits, ethylbenzene, and styrene on the catalyst surface, thus controlling the catalyst activity. Styrene presumably adsorbs more strongly than ethylbenzene on the catalyst, but reliable quantitative information is lacking. Temperature control is complicated since the dehydrogenation reactions are highly endothermic. The overall reactions, including side reactions, have an endothermicity of about 31.8 to 33.5 kcal/g mol of ethylbenzene reacted. Consequently the reactants cool rapidly as the reaction progresses unless heat is transferred to the system. Such transfer to the catalyst bed complicates the reactor system, and significant temperature gradients normally occur in the catalyst bed of a reactor.

Steam that is premixed with ethylbenzene plays several roles in the reaction sequence:

1. It often provides much of the heat of reaction, and it promotes more isothermal operation.

2. Steam reacts in water-gas reactions, such as reactions 4, 5, 6, and 8, and it helps maintain relatively clean catalyst surfaces.

3. Steam lowers the partial pressure of the hydrocarbons, resulting in improved styrene yields.

4. Steam reduces the adsorption of hydrocarbons and subsequent formation of carbon deposits on the surface.

5. Steam adsorption on the catalyst may possibly affect catalyst activity.

Increased ratios of steam to ethylbenzene in the feedstock promote high styrene yields. Steam, however, is relatively expensive and is a major operating cost in the process, so that an optimum ratio based on economic considerations must be chosen.

Commercial Catalytic Dehydrogenators   Major modifications of the ethylbenzene dehydrogenation step have been realized within the last few years to permit significantly lower cost for styrene production. In particular, new catalysts and new reactor designs have been used to obtain much higher ethylbenzene conversions per pass.

Two relativly different types of dehydrogenator reactors have been used commercially since the 1930s, when they were developed, the adiabatic reactors developed by Dow Chemical and the so-called isothermal ones of BASF. The ethylbenzene feed is always a mixture of fresh and recycle ethylbenzene. The recycle stream is generally not completely purified because of economic considerations, and it contains styrene, benzene, and toluene. The ethylbenzene stream to the dehydrogenator often has the following approximate composition on a weight percent basis: ethylbenzene, 98.5 percent; styrene, 1.25 percent; benzene, 0.05 percent; and toluene, 0.20 percent.

Adiabatic Reactors (Single Unit per Line): In older plants in particular, an ethylbenzene-steam mixture is passed through a single adiabatic reactor, whereas more modern plants often have two or three reactors in series. Plants using only one reactor in a line have been described (8, 46, 55, 62). Numerous variations are employed commercially, but a typical unit is described here (see Fig. 7-5).

The ethylbenzene feed and about 10 percent of the steam to be used are mixed and then heated to about 510 to 535°C in two heat exchangers using the hot product gases from the dehydrogenator.

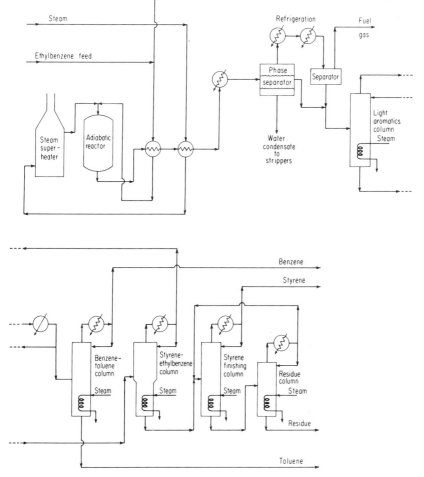

Fig. 7-5 Flowsheet for dehydrogenation of ethylbenzene (adiabatic reactor).

The remaining 90 percent of the steam is superheated in a furnace to 710 to 750°C; 750° is approximately the maximum temperature readily obtainable in such a unit. The superheating coils for the steam consist of several vertical tubes connected in series by U bends (46). Gas burners are located throughout the combustion zone of the furnace (or superheater), and heat transfer to the coil is by both radiation and convection. Skin temperatures of the coil are relatively high, and the tubes are generally constructed with high-nickel steels. The air used for combustion of fuel in the burners is preheated in the convection zone of the furnace. The resulting superheated steam is then mixed with the hot-steam–ethylbenzene mixture using a concentric mixing tube (46). Mixing should be rapid and thorough to minimize local hot spots that promote vapor-phase reactions.

The resulting mixture, containing about 2.5:1 weight ratio of steam to ethylbenzene at a temperature of approximately 620°C, enters the top of the dehydrogenation reactor. This mixture is contacted with the catalyst bed at liquid hourly space velocities of approximately 0.45 to 0.65 vol of ethylbenzene per volume of catalyst bed. No attempt is made to transfer heat in the bed. As the gases pass downward through the bed, about 40 to 45 percent of the ethylbenzene reacts as the mixture cools to about 575 to 585°C, or perhaps even lower when the catalyst is new. Styrene yields based on ethylbenzene reacted vary from about 88 to 91 percent.

The reactor used by the Polymer Corp. of Sarnia, Ontario, is a typical adiabatic reactor. It has a catalyst bed that is 5.3 ft deep and an ID of 6.4 ft (55). When it was operated with an aged catalyst at less than ideal conditions, it produced annually about 30 million lb of styrene. With more ideal operating conditions, it would be likely to produce 35 to 40 million lb/year.

The reactor shell and the inlet and outlet lines of the reactor are constructed with a high-chrome steel. Nickel alloys are not satisfactory since they promote side reactions resulting in carbon deposits. Several design features of these adiabatic reactors are of importance:

1. The reactors should be designed to minimize undesired gas-phase reactions. Gas spaces in the reactor above and below the catalyst bed are small.

2. Uniform flow of the gases through the packed bed is desired. Distribution rings or similar devices may be helpful, especially for larger diameter reactors.

Mitchell (46) has described some of the constructional details for early adiabatic reactors, including refractory-lined catalyst case, distribution rings for feed gases, alloy-steel screens for removing exit gases, thermocouple wells, catalyst-loading device, and insulation. Since heat losses are most undesirable, effective insulation must be provided around the 0.25-in steel shell used to provide a gastight seal.

Adiabatic Reactors (Units Operated in Series): Two or more adiabatic reactors are employed commercially in several newer plants. Additional steam and sometimes additional ethylbenzene is mixed with the gas stream between reactors. As a result, higher ethylbenzene conversions are realized without employing excessive amounts of steam or preheating the ethylbenzene feed to undesired high temperatures.

Lovett and Jones (41) of Monsanto Co. have described the operation of two adiabatic reactors in series. Between 40 and 60 percent of the total steam used is added to the ethylbenzene feed before the first unit. The remainder of the steam is mixed with the product stream from the first unit, and the combined mixture is then fed to the second unit. The overall ratio of steam to ethylbenzene at optimal conditions may be as low at 2:1 with this two-bed adiabatic system. Somewhat lower maximum temperatures occur in this arrangement, compared to a single-unit system, resulting in higher styrene yields for a given ethylbenzene conversion.

Use of different catalyst or a different size of catalyst pellets in each bed of the system is sometimes beneficial. MacFarlane (42) has proposed that the catalyst for the first unit should provide a high styrene selectivity. He reports a catalyst that gives a 97.5 percent yield at a conversion of 34 percent. For the second unit, he recommends a high-activity catalyst such as Shell 105 catalyst. In an example for a two-bed system, an overall yield of 91.9 percent was obtained with a conversion of 64.3 percent. His results were obtained under ideal conditions of about 600°C and a total pressure just slightly greater than atmospheric. Lower yields would be expected in commercial reactors with maximum temperatures higher than 600°C operated at higher pressures. Yields at least as high as 88 to 90 percent, however, seem possible for plant reactors with conversion of 60 percent.

A modified three-bed adiabatic reactor has been designed by UOP. The feed mixture of ethylbenzene and steam for each bed is introduced through a central supply pipe, and the product gases flow radially outward to the wall of the reactor vessel, where they are then directed to the central supply pipe for the next bed (59). Additional steam is probably added between beds. Such an arrangement of flow minimizes pressure drop since the cross-sectional area for the flowing gases increases as the volume of the gas increases (because of the dehydrogenation reaction).

Modified adiabatic reactors containing two or more catalytic beds in series have been designed in which the effluent from the first bed is heated in a heat exchanger before entering the next bed (33). Methods of mixing and preheating ethylbenzene and steam before a reactor are important. Gas-phase reactions are reported to be relatively insignificant when the residence time of a mixture in a heater (or furnace) is at most about 0.5 s (58). These heating methods are also of interest for isothermal reactors, discussed next.

Isothermal Reactors: What are essentially shell-and-tube heat exchangers are used for construction of isothermal reactors (8, 48, 49). The tubes generally are 8 to 12 ft long, with a diameter of 4 to 8 in. The catalyst is packed inside the tubes, and hot flue gases on the shell side of the reactor provide heat to maintain rather isothermal conditions in the catalyst bed. Some of these reactors have for many years been operated at ethylbenzene conversion levels of about 40 percent. Conversion levels in more modern isothermal reactors have not been published, but apparently technology employed in the more modern adiabatic reactors could be adapted for isothermal units to get conversions above 40 percent, as will be considered in more detail later.

Isothermal reactors have traditionally had two important advantages over adiabatic reactors. Much less steam is needed for the isothermal reactors since considerably lower steam-to-ethylbenzene weight ratios are used—often about 1.2:1. Second, the maximum temperatures in isothermal units are relatively low, generally being in the range from about 550 to 620°C. As a result, styrene yields are often higher by about 2 to 3 percent than those for adiabatic reactors. The big disadvantage of the isothermal units is their higher capital costs.

As in the adiabatic reactor, the gas spaces that are unfilled with catalyst pellets are minimized. Catalyst pellets are sometimes added to the header volumes in the reactors both above the top tube sheet and below the bottom tube sheet. A perforated plate is used in the bottom header as a support for the catalyst pellets. Heat transfer to the catalyst pellets in the top and bottom headers is generally relatively poor, and adiabatic operation is approximated there.

Ohlinger and Stadelmann (49) described the BASF isothermal reactor operated with ethylbenzene conversions of about 40 percent. During the first 3 months of the operation after a new catalyst charge was added to a reactor, the average operating temperatures in the reactor were 560 to 590°C, and the styrene yield was 94 percent. As the catalyst aged, the temperatures were slowly increased to maintain the desired conversion of ethylbenzene, and styrene yields then decreased. For a catalyst 13 to 18 months old, the temperatures varied between 600 and 620°C, and yields were in the range of 89 to 91 percent.

For an isothermal reactor unit, first all the steam and ethylbenzene are mixed, and the mixture is then heated in several heat exchangers to approximately 585°C. In the first exchanger, heat is transferred from the hot product gases leaving the reactor. In the last exchanger, a flue-gas stream at about 620°C is employed for heat transfer; this gas stream is the one that had been used for heating the reactor. Operation of the last exchanger is critical. If the temperature of the flue gas is too high, skin temperatures of the tubes in the reactor are high enough to cause undesired side reactions.

Important design and operating considerations for isothermal reactors include the inlet temperatures for both the feed stream and the flue gases used for heat-transfer purposes, the dimensions of the tubes, the liquid hourly space velocity of the ethylbenzene, the ratio of steam to ethylbenzene in the feed, the inlet pressure for the feedstock, the type and size of the catalyst pellets, and the method of contacting the flue gases with the tubes in the reactor. The flue gases can be contacted in essentially either a cocurrent or countercurrent direction relative to the gas flow inside the tubes. Cocurrent flow is widely used since the temperatures of the inside walls of the tubes are lower, resulting in fewer side reactions.

Significant temperature variations are present inside the so-called isothermal reactor. Variations of about 30 to 50°C in both the radial and axial direction occur in the tubes of commercial reactors. The temperature gradients of course depend on both design and operation of the reactor. Increasing the temperature of the gases as they flow through the bed is reported to give higher styrene yields and/or to allow higher ethylbenzene conversions, up to 60 percent (31).

Some factors that must be considered in designing a reactor are as yet hard to quantize, yet they are important in the overall operation. The aging of the catalyst, relative importance of gas-phase reactions, and the role of the tube wall are examples. If the tube diameters are too large, excessively hot walls occur because of the high heat fluxes. Reactants near the wall then react at undesired high temperatures. If, on the other hand, too small a tube is used, the void volumes near the tube wall increase in relative magnitude to allow relatively more gas-phase reactions. It can also be asked whether the tube walls are not at least slightly catalytic in nature.

Using improved catalysts or combination of catalysts probably will permit optimum ethylbenzene conversions above 40 percent. The top and bottom sections of the tubes in the reactor might be packed with different catalysts, for example (42).

Low steam-to-ethylbenzene ratios in the feed as used in the past with isothermal reactors may not permit ethylbenzene conversions as high as obtained in the newer adiabatic reactors. For a 1.2:1 feed ratio, the partial pressure of ethylbenzene in the feed stream is over 50 percent greater than it is at a 2:1 feed ratio, assuming a constant total pressure for both reactors. Styrene yields are apparently reduced significantly with increased partial pressures of hydrocarbons for especially high ethylbenzene conversions, as indicated by the kinetic models (47, 55) and also by Fig. 7-4.

Recovery of High-Purity Styrene   Recovery of styrene from the gaseous product stream leaving the catalytic dehydrogenator is accomplished by first cooling and condensing the styrene, ethylbenzene, and other aromatics. The hydrocarbon condensate is then separated into the pure compounds by distillation. These recovery steps are relatively complicated, accounting for a large share of the total operating and capital expenses of a styrene plant.

One problem that must always be considered during the recovery operation is partial polymerization of styrene that occurs, especially at higher temperatures. Polymerization is minimized by lowering the time that styrene is at higher temperatures, reducing temperatures in the equipment, and adding inhibitors to minimize polymerization.

Condensation of Product Stream: Cooling and condensation of the product stream from the dehydrogenators are in several heat exchangers arranged in series. Recovery of most of the sensible and possibly even some latent heat from the product stream is essential for reducing the energy requirements of the process. Several methods for arranging the heat exchangers and the specific coolants to be used have been reported (8, 48, 49, 62). The variations noted in these methods depend to some extent on the amount of steam in the product stream (and indirectly on the type of dehydrogenator used).

Four heat exchangers arranged in series have been used in some plants, as shown in Fig. 7-5. Coolants used in the first two exchangers are the ethylbenzene and the steam feedstocks, respectively. In the last exchangers or condensers, most of the hydrocarbons and water are condensed as the mixture is cooled to about 50°C. Water can be used as the coolant in the last heat exchanger, but air is frequently cheaper. The product from the condenser consists of liquid hydrocarbon, water, and gas phases, which are separated.

The overall heat-transfer coefficients in these exchangers are relatively small since the streams are gaseous during most of the heat-transfer sequence. Fouling of the transfer surfaces results particularly from styrene polymerization and because of adsorption and subsequent surface reactions. High-chrome steel is used in the exchangers to minimize fouling. Tar deposition and surface polymerization are quite pronounced when condensation begins. The exchangers must be cleaned occasionally.

The amount of heat transferred in the first two exchangers is available for further use, and techniques for generating low-pressure steam and providing reboiler heats have been patented (3, 11, 30, 53). Useful heat cannot be recovered, however, from the last two exchangers.

The gas leaving the condenser is primarily hydrogen and carbon dioxide (from the water-gas reaction) plus lesser amounts of methane, ethylene, and uncondensed styrene and ethylbenzene. It is generally refrigerated and sometimes compressed to obtain further condensation and recovery of the aromatics. The remaining gas may be processed for recovery of hydrogen, or it may be used as a fuel.

Distillation of Crude Hydrocarbons: The crude hydrocarbon liquid is separated by distillation. It contains from about 35 to 70 percent styrene (depending on the ethylbenzene conversion level of the dehydrogenator); small amounts of benzene, toluene, and tars; and the remainder ethylbenzene. Three special precautions are taken to minimize styrene polymerization in the distillation equipment:

1. Vacuum distillation provides rather low temperatures. With a vacuum, air leaks are always possibilities. Such leaks cause distillation problems and promote polymerization and oxidation of hot styrene (8).
2. Equipment, such as reboilers, is designed to minimize the residence time of the styrene.
3. Inhibitors are added in small quantities in various streams of the columns. Commercial inhibitors include sulfur, nitrophenols, hydroquinone, monoxime of paraquinone, and para-*t*-butylcatechol (TBC).

Considerable details have been published concerning the distillation equipment and procedures used in several commercial plants (8, 25, 48, 49, 62). The order of separating the hydrocarbons is different in various plants, but that shown in Fig. 7-5 is popular with some companies. In this case, the benzene and toluene are first separated from the heavier hydrocarbons, which include styrene and ethylbenzene. A column using 30 bubble-cap trays was used in early designs (8) and was operated with a pressure at the top of the column of 175 mm Hg absolute and with a reflux ratio of 12:1. The reboiler temperature in the column was 96°C, and little polymerization occurred there since styrene was still diluted with ethylbenzene and heavier cuts and since sulfur was used as an inhibitor.

Sulfur is added in small quantities to the reflux stream or the feed of this column. When it is added to the reflux, sulfur is present throughout the column, leaving with the bottom product. In more modern designs of this column, trays that have better performance characteristics than bubble-cap trays are used. Sieve trays, for example, result in lower pressure drops per tray; low pressures and hence low temperatures are therefore easier to obtain in the reboiler.

Styrene-Ethylbenzene Separation: Separation of styrene and ethylbenzene is by far the most difficult step of the separation, and it accounts for a major share of the expenses of the entire process. Significant reductions in the cost of this step have been realized, however, because of improved sieve trays and because of the much higher ethylbenzene conversions now being obtained in newer dehydrogenators.

When bubble-cap trays were used in the early installations of the 1930s and 1940s (46), two columns were required. At that time one column was insufficient since the pressure drop through the trays would have caused too high a pressure and hence too high a temperature in the reboiler. Excessive polymerization of styrene in the reboiler would have resulted. Steam requirements for the operation of two columns are high, above 2.0 lb per pound of styrene.

Hence there was a strong incentive for finding a tray for this separation that would result in small enough pressure drop per tray to permit a single column to be designed. Frank et al. (25) have published details on a single column with Linde sieve trays used by The Dow Chemical Co. beginning in 1965, to produce about 160 million lb/year of styrene. Two such columns are used in their plant at Midland, Michigan. Each column contains 70 Linde sieve trays and operates at a reflux ratio of about 6.15:1. Steam requirements are approximately 1.31 lb per pound of styrene. The Dow columns have a larger diameter (17 ft) in their top sections than the smaller (14.5 ft) bottom sections. Such an arrangement is necessary because of the severalfold differences in the gas densities caused by differences in pressures. Sulfur is added in small quantities to the reflux line of the column.

The maximum allowable temperature in the reboiler for this column is approximately 105°C (25), obtainable with a reboiler pressure of 220 mm Hg. The minimum overhead pressure is determined by the temperature of the coolant used for the reflux condenser of the column. When water is used, this pressure is estimated to be about 35 mm Hg; whereas about 50 mm Hg, corresponding to a reflux temperature of 53°C, seems reasonable when air is used. Air-cooled condensers are used with the Dow column, and the maximum allowable pressure drop in the column is approximately 170 mm Hg.

Theoretical plates required for the desired separation of ethylbenzene and styrene of course vary with the reflux ratio. About 66 theoretical plates are required at a reflux ratio of 6:1, whereas 48 plates are required at 8:1 (25). The overall plate efficiencies of Linde sieve trays are high, averaging over 90 percent, and pressure drops are less in most cases than 2.8 mm Hg per theoretical tray. In general, relatively low reflux ratios are used since the steam requirements of the column are almost directly proportional to the reflux ratio.

Precautions are taken in the reboiler to minimize the temperatures of the heat-transfer surfaces and hence minimize polymerization and fouling. Relatively low temperature differences are provided for heat transfer. The suggestion has been made that Linde high-flux tubing would be especially good for heat-transfer surfaces there (19). The volume of the styrene-rich liquid in the reboiler is small, to minimize the residence time of the liquid there. The liquid bottoms are rapidly recirculated through vertical tubes in the reboiler (8, 25). The reboiler must be cleaned periodically, and more than one reboiler is provided per column in at least some cases.

The Dow column is operated so that almost no ethylbenzene (0.1 percent or less) is present in the styrene-rich bottom product. The distillate product, however, contains about 97 percent ethylbenzene and the remainder essentially styrene. This last stream is recirculated to the dehydrogenator. Most of the styrene recycled is eventually recovered in the styrene product stream.

Frank et al. (25) have designed a styrene-ethylbenzene column using Linde sieve trays with a styrene capacity of 0.5 billion lb/year. The optimum column apparently to be used in conjunction with a dehydrogenator to convert about 48 percent of the ethylbenzene would have a top diameter of 30 ft and a bottom diameter of 25 ft and would operate at a reflux ratio of about 6.5:1. Maintaining a uniform height of liquid across the entire tray of such a large vacuum column during high winds requires careful design. Failure to obtain such a uniform height results in nonuniform gas flow through the holes in the plates, seriously reducing plate efficiencies.

Steam requirements and the column cross-sectional areas are essentially directly proportional to the amount of ethylbenzene in the feedstock to the column. Estimates were made as shown in Table 7-1, based on the plant data, reported by Frank et al. (25), of the steam requirements and of the relative cross-sectional areas needed for the column as ethylbenzene conversions in the dehydrogenator were varied from 40 to 70 percent. In these estimates the same reflux ratio and 90 percent styrene yields are assumed in all cases. From the composition of the feedstock in the example reported by Frank et al. the ethylbenzene conversion in the Dow dehydrogenator is estimated as 48 percent. Table 7-1 clearly indicates that major distillation savings are possible at high ethylbenzene conversions.

Monsanto Co. separates the crude-hydrocarbon mixtures in a different order (14) than that shown in Fig. 7-5. They use the first column as the styrene-ethylbenzene splitter. The top and bottom products from

TABLE 7-1   Effect of Ethylbenzene Conversion in Dehydrogenator on Operation of Styrene-Ethylbenzene Column

Ethylbenzene conversion, %	Steam required per pound of styrene, lb	Ratio of cross-sectional area to area at 40% conversation
40	1.83	1.00
48	1.31*	0.72
50	1.21	0.67
60	0.81	0.45
70	0.52	0.39

* Data of Frank et al. (25).

this column are then fractionated further.   An advantage of this method is that most of the styrene enters only two columns instead of the three shown in Fig. 7-5.

The styrene-ethylbenzene splitter used in the 1.3 billion lb/year styrene plant (35) of Monsanto Co. at Texas City, Texas, takes advantage of both improved sieve plates and high ethylbenzene conversions.   The column (see Fig. 7-6), put in operation in 1971, appears to be about 30 ft in diameter at the top and 25 ft at the bottom.   Such a huge column is expensive, and the assembled (but unerected) column is estimated to cost $1.0 million.   As a first approximation, the steam requirement is about 0.5 lb per pound of styrene.   This value was determined by comparing announced styrene capacities of the Monsanto and Dow columns and assuming identical vapor rates (on a cross-sectional-area basis) for the two columns.

Finding even more improved plates for styrene-ethylbenzene columns (to obtain even lower pressure drops per theoretical plate) would allow some or all of the following advantages.

1. Lower reflux ratios, causing reduced steam requirements

2. Lower pressures (and temperatures) in the reboiler, and hence still less styrene polymerization there

3. Use of more plates in the column to further reduce the styrene content of the ethylbenzene

Other types of trays or packed columns have been seriously considered (4, 20, 21, 57).   Pressure drops per height equivalent of a theoretical tray are relatively small for columns packed with Pall rings, but several factors have caused concern:

1. The cost of a packed column would be higher than that for present columns.

2. Maintaining uniform distribution of the liquid and gas streams across the cross-sectional area of large columns would require a special distribution device for the reflux at the top of the packing.

3. Foaming is sometimes a problem in packed columns.

4. Performance of a packed column may change with time because of polymer formation, settling of packing, or corrosion.

Styrene Finishing and Storage: High-purity styrene is rather easily recovered from the bottom product of the styrene-ethylbenzene column using the styrene-finishing column. The overhead distillate of this column is commercial polymerization-grade styrene that is at least 99.6 percent pure. The main impurity is ethylbenzene, and it contains only 5 to 10 ppm of aldehydes, peroxides, or chlorides and no polymers. Small amounts of TBC are added to the reflux line as an inhibitor for polymerization.

Storage of liquid styrene has been discussed in considerable detail by Shelley and Sills (56). TBC is used as an inhibitor in concentrations

Fig. 7-6 Plant to produce 1.3 billion lb/year of styrene.   The huge column separates ethylbenzene and styrene. (*Monsanto Company*)

of about 10 ppm, and the styrene is normally stored under air which makes TBC effective. Dow Chemical considers it good practice to insulate and refrigerate styrene storage tanks where ambient temperatures exceed 80°F (27°C) for extended periods. Design features of a storage tank 70 ft in diameter have been described.

Styrene storage tanks are generally internally coated. The coating on the upper portion of the tank may be an epoxy-type resin, and a rust-resisting inorganic zinc silicate material is suitable for the lower sections of the tank. Rusty or complicated surfaces promote polymer buildup. In uncoated tanks, blanketing the tank with an inert gas is recommended.

Miscellaneous Separations and Recovery: The bottom product of the styrene finishing column is mainly tars, polymers, and sulfur. These bottoms are usually redistilled in a small column to recover some styrene, which is recycled to the styrene-finishing column. The tars are of limited value as a fuel because of their high sulfur content, and disposal of the substantial amounts produced in a large styrene plant is a problem. Walker (61) recommends a high-temperature, high-pressure, catalytic-hydrogenation process for production of benzene, toluene, and numerous other light aromatic hydrocarbons.

A benzene-toluene column is used to separate the benzene and toluene mixture obtained as the distillate from the first column. Such a separation is generally at about atmospheric pressure, and a column containing 40 plates and with a reflux ratio of 3:1 makes a rather clean separation (8). Benzene with less than 0.1 percent toluene is recycled to the ethylbenzene plant for alkylation. The bottom product is toluene with perhaps a trace of ethylbenzene. High-purity toluene can be obtained if desired by further distillation of the bottom product.

Optimization of Catalytic Dehydrogenation Plants   Careful analysis and economic evaluation of the many operating variables must be made in selecting the optimum design for the commercial dehydrogenation plant. Several companies are very active in improving their design techniques, and some of their results have been published.

Several models developed for catalytic reactors are useful for design purposes. These models include terms for the main and secondary reactions, heat transfer, mass transfer, and pressure throughout the catalyst bed. Models for an adiabatic reactor are simpler than those for isothermal reactors since heat- and mass-transfer terms usually can be ignored. There is of course some heat (and mass) transfer, especially in an axial direction in an adiabatic reactor. Models developed for both types of reactors need a large computational facility for solving.

Sheel and Crowe (55) have modeled the adiabatic reactor of the Polymer Corp. For a given ethylbenzene feed rate, they calculated the amounts of styrene and by-products obtained and also the profit for a wide variety of steam rates, steam temperatures, and temperatures of the inlet feed mixtures. They were thus able to determine "optimum" conditions for the most profitable operation. Their model did not of course take into account all variables (including catalyst aging), but their results clearly indicate how to proceed for improved operation. They found that the preferred temperature of the inlet mixture of ethylbenzene and steam should be in the range of 645 to 650° (depending somewhat on the other variables); these temperatures are essentially identical to those reported by Mitchell (46) many years ago.

Foy (24), using the kinetic equations of Sheel and Crowe, has developed a complicated model for isothermal reactors. Foy assumed the following constraints:

1. A 40 percent ethylbenzene conversion per pass
2. Outlet reactor pressure of 5 lb/in² gage
3. A 2-ft adiabatic catalyst zone below the bottom tube sheet
4. Tubes arranged on a triangular pitch of 1.25 times the outside diameter of the tube used

He then made calculations using the following parameters:

```
Outside tube diameter, in................  4, 6, or 8
Tube length, ft.........................  8, 10, or 15
Liquid hourly space velocity, volume
    of liquid ethylbenzene per
    volume of catalyst per hour ..........  0.5, 1.0, or 2.0
Steam-to-ethylbenzene feed ratio.........  1.0:1, 1.2:1, or 1.4:1
```

Temperatures were calculated for the inlet flue gas and also as a function of tube length for the flue gas, tube wall, boundary layer inside the tube, and the center (or axis) of the catalyst bed. The pressure required for the inlet feed and styrene yields were also determined. The highest yield values found were slightly greater than 90 percent. Such values are 2 to 3 percent less than those reported by Ohlinger and Stadelmann (49) for a commercial isothermal reactor. Foy postulated some inaccuracies, which resulted in extrapolating the kinetic equations of Sheel and Crowe (55) to a rather different set of operating conditions. In addition, these kinetic equations were perhaps developed with data from a relatively old and inactive catalyst.

The results of Foy (24), however, should give a good indication of the conditions and design variables for obtaining the highest styrene yields in an isothermal reactor. These are a space velocity of 0.5 vol of ethylbenzene per volume of catalyst per hour, 4-, or 6-in-OD tubes 8 to 10 ft long, and a feed ratio of 1.2:1 or 1.4:1. Published results

TABLE 7-2   Costs for Ethylbenzene-Dehydrogenation–Styrene-Finishing Plant (9)

Capacity: 0.5 billion lb/yr styrene (60,000 lb/h)

Capital Costs	
	Millions of dollars
Superheater for steam............................	$1.1
Adiabatic reactor................................	0.2
Heat exchangers................................	1.1
Air condensers..................................	1.7
Distillation equipment...........................	3.8
Total..	$7.9

Hourly Operating Expenses	
	Dollars
Ethylbenzene (at 4.1 cents/lb)...................	$2,933*
Steam...	199
Fuel gas.......................................	78
Dehydrogenation catalyst.......................	16
Distillation equipment..........................	98
Other expenses†...............................	180
Subtotal..................................	$3,504
Credits for benzene, toluene, and hydrogen........	284
Net total..................................	$3,220

* Includes $2,500 for ethylbenzene used stoichiometrically to form styrene and $430 for ethylbenzene that forms by-products.
† Includes labor, depreciation, and miscellaneous expenses.

for industrial isothermal reactors confirm this general range of variables (48, 49). Somewhat higher steam-to-ethylbenzene ratios would probably be required, however, if ethylbenzene conversions of 50 percent or higher were employed.

Buzzelli (9) has optimized an integrated ethylbenzene-dehydrogenation styrene-finishing plant with a styrene production capacity of 0.5 billion lb/year. Costs for this plant are reported in Table 7-2.

Several interesting features of the economics of the dehydrogenation process are as follows:

1. Ethylbenzene costs are over 90 percent of the total operating costs. About 13 percent of the total costs are caused by ethylbenzene lost in side reactions, and improved yields would result in significant cost reductions.

2. The distillation and steam-superheating sections of the plant account for a large fraction of the capital costs. Excluding ethylbenzene costs, these sections also account for much of the operating expenses.

3. Dehydrogenator and catalyst costs are small.

4. The total operating costs are 5.4 cents per pound of styrene. The cost for the theoretical amount of ethylbenzene required (assuming 100 percent yields) is 4.2 cents per pound of styrene. All other costs, including ethylbenzene yielding by-products, total 1.2 cents/lb.

Information has been reported on a 0.5 billion lb/year styrene unit that includes both production and dehydrogenation of ethylbenzene (14). Total capital investment for this plant, which uses a Monsanto design, is approximately $13 million. The benzene and ethylene costs (at about 3 cents/lb each) total $17.8 million/year for the plant. Other operating expenses, including utilities, catalyst, labor, and chemicals, are $3.2 million. All operating expenses excluding plant payout and profit total $21 million/year, which is equivalent to 4.2 cents per pound of styrene.

This Monsanto-designed plant is said to use only 4.6 lb of steam per pound of styrene whereas previous styrene plants use as much as 6.6 lb. A reduction in steam consumption by this amount results in a saving of about $0.5 million/year. These process modifications are said to include improved catalytic dehydrogenators (rather than conventional adiabatic reactors) and an improved styrene-ethylbenzene column.

Dehydrogenation of Ethylbenzene by Oxidation Techniques

Two processes have been developed in which ethylbenzene is converted to phenylmethylcarbinol C_6H_5—$CHOH$—CH_3, sometimes called methylbenzyl alcohol. An oxidation step is one of the reactions required to obtain this alcohol, which is then dehydrated to produce styrene and water.

Union Carbide Corp. operated a process for several years that produced both styrene and acetophenone, C_6H_5—CO—CH_3, from ethylbenzene (54). An advantage of this process is the relatively easy separation of styrene from the product stream of the dehydrator. Unfortunately the overall yields of styrene based on ethylbenzene used are far too low to make this process competitive with current processes using a catalytic-dehydrogenation step.

The Halcon process produces both styrene and propylene oxide using ethylbenzene and propylene as the feedstocks. It is very similar to related propylene oxide processes developed by Scientific Design. The Halcon process consists of three major chemical steps, but relatively few process details are available. From patents (29) and other literature (34), the process can be summarized as follows:

1. Ethylbenzene is oxidized using oxygen or air by a liquid-phase process to produce primarily ethylbenzene hydroperoxide, C_6H_5—CHOOH—CH_3. Some acetophenone, phenylmethylcarbinol, and benzoic acid are also produced. Improved yields of hydroperoxide are formed with relatively low partial pressures of oxygen in the entering gas stream and with high agitation. Good transfer of the oxygen to and throughout the liquid phase promotes high yields of the desired products.

2. Ethylbenzene hydroperoxide is reacted with propylene in the second step of the process. Soluble molybdenum catalysts, such as molybdenum naphthenate, are preferred for the liquid-phase reactions. Propylene dissolves in the liquid phase, and major reaction products are phenylmethylcarbinol and propylene oxide. Temperatures used are in the range of approximately 100 to 130°C. Almost all the hydroperoxide reacts in 1 to 3 h residence time.

3. After separation, mainly by distillation, of propylene oxide, unreacted ethylbenzene, and undesired by-products, the phenylmethylcarbinol is dehydrated over a supported titania catalyst at 200 to 250°C.

4. Distillation of the product mixture from the dehydrator yields high-purity styrene. The unreacted carbinol and acetophenone are recovered; acetophenone is probably hydrogenated to produce more phenylmethylcarbinol. The entire carbinol stream is then recycled to the dehydrator.

The overall styrene yields based on the ethylbenzene feed have not been reported for the Halcon process. The oxidation of ethylbenzene in the first step of the process may be the critical one. More information is required before the Halcon process can be compared to current styrene processes. Propylene oxide yields and selling price will also determine to a considerable extent the importance of the process.

Conclusions

The major operating expenses for production of first ethylbenzene and then styrene are those of the feedstocks, generally benzene and ethylene. Other operating expenses account for only a relatively small fraction of the total.

The most critical step in the overall styrene process is the dehydrogenation of ethylbenzene. Although major improvements have been realized recently for this step, perhaps even higher yields of styrene at still higher conversions of ethylbenzene can be obtained in the future. If so, the styrene costs can perhaps be reduced 0.1 to 0.2 cent/lb. Since modern styrene plants must usually be huge to be competitive, reliability of all portions of the process equipment is essential.

Literature Cited

1. Anderson, E. V., R. Brown, and C. E. Belton: Styrene: Crude Oil to Polymer, *Ind. Eng. Chem.*, **52**:550 (1960).
2. Ashmore, R. D. (to Monsanto Co.): Alkylation of Aromatic Compounds, U.S. Pat. 2,948,763 (Aug. 9, 1960).
3. Berger, C. V. (to Universal Oil Products Co.): Catalytic Conversion Process, U.S. Pat. 3,515,765 (June 2, 1970).
4. Billet, R.: Recent Investigations of Metal Pall Rings, *Chem. Eng. Prog.*, **63**(9):53 (1967).
5. Bloch, H. S. (to Universal Oil Products Co.): Alkylation-Transalkylation Process, U.S. Pat. 3,183,233 (May 11, 1967).
6. Bodre, R. J. (to Monsanto Co.): Control of Alkylation Catalyst Activity, U.S. Pat. 3,277,195 (Oct. 4, 1966).
7. Bodre, R. J. (to Monsanto Co.): Hydrocarbon Conversion, U.S. Pat. 3,381,050 (Aug. 30, 1968).
8. Boundy, R. H., and R. F. Boyer: "Styrene: Its Polymers, Copolymers, and Derivatives," Reinhold, New York, 1952.
9. Buzzelli, D. T.: A Hybrid Computer Optimization of an Integrated Ethylbenzene Dehydrogenation–Styrene Finishing Facility, *Professors' Workshop Ind. Monomer Polym. Eng.*, Midland, Mich., May 1969.
10. Carra, S., and L. Forni: Kinetics of Catalytic Dehydrogenation of Ethylbenzene to Styrene, *Ind. Eng. Chem. Process Des. Dev.*, **4**:281 (1965).
11. Carson, D. B., and K. D. Uitti (to Universal Oil Products Co.): Catalytic Conversion Process, U.S. Pat. 3,515,767 (June 2, 1970).
12. Solvent Uses Improve Ethylbenzene Outlook, *Chem. Eng. News*, Sept. 4, 1967, pp. 18–21.
13. Styrene Forecast: Rosy for Near Term, *Chem. Eng. News*, Sept. 22, 1969, pp. 22–25.
14. New Styrene Process Cuts Annual Costs 16 Per Cent, *Chem. Eng. News*, Sept. 29, 1969, p. 49.
15. Styrene: Toward 9 Billion Pounds, *Chem. Eng. News*, June 15, 1970, p. 11.
16. Fractionation Taps New Source of Styrene, *Chem. Eng.*, June 1957, pp. 160–162.
17. Polystyrene via Natural Ethyl Benzene, *Chem. Eng.*, Dec. 1, 1958, pp. 98–101.
18. Coulter, K. E., H. Kehde, and B. F. Hiscock: Styrene and Related Monomers, in A. Standen (ed.), "Kirk-Othmer Encyclopedia of Chemical Technology," 2d ed., vol. 19, pp. 55–85, Interscience, New York, 1969; in E. C. Leonard (ed.), "Vinyl and Diene Monomers," pt. 2, Wiley-Interscience, pp. 479–576, New York, 1971.
19. Creighton, R. L. (Linde Division, Union Carbide Corp., Tonawanda, N.Y.): personal communication, 1972.
20. Eckert, J. S.: Selecting the Proper Distillation Column Packing, *Chem. Eng. Prog.*, **66**(3):39 (1970).
21. Fair, J. R.: Comparing Trays and Packings, *Chem. Eng. Prog.*, **66**(3):45 (1970).
22. Fenske, E. R. (to Universal Oil Products Co.): Alkylation-Transalkylation Process, U.S. Pat. 3,200,163 (Aug. 10, 1964).
23. Foster, A. L.: Ethylbenzene: Mother of Petrochemicals, *Pet. Eng.*, **25**(10):C-3 (1953).

24. Foy, R. H.: Process Design of an Isothermal-Type Ethylbenzene Dehydrogenation Reactor, *Professors' Workshop Ind. Monomer Polym. Eng., Midland, Mich., May* 1969.
25. Frank, J. C., D. R. Geyer, and H. Kehde: Styrene-Ethylbenzene Separation with Sieve Trays, *Chem. Eng. Prog.*, 65(2):79 (1969).
26. Garner, F. R., and R. L. Iverson: How Ethylbenzene Is Made by the High-Pressure Process, *Oil Gas J.*, 53(25):66 (1955).
27. Govindarao, V. M. H., P. N. Deshpande, and N. R. Kuloor: Ethylation of Benzene: A Statistical Study, *Ind. Eng. Chem. Process Des. Dev.*, 7:573 (1968).
28. Grote, H. W., and C. F. Gerald: Alkylating Aromatic Hydrocarbons, *Chem. Eng. Prog.*, 56(1):60 (1960).
29. Halcon International, Inc.: Preparation of Propylene Oxide from Ethylbenzene Hydroperoxide, Netherlands Pat. Appl. 6,500,118 (July 8, 1965); Netherlands Pat. Appl. 6,602,321 (Sept. 5, 1966).
30. Hallman, N. M., and D. J. Ward (to Universal Oil Products Co.): Catalytic Conversion Process, U.S. Pat. 3,515,764 (June 2, 1970).
31. Hatfield, C. G., and G. H. Lovett (to Monsanto Co.): Dehydrogenation of Alkylated Aromatic Hydrocarbons, U.S. Pat. 3,100,807 (Aug. 13, 1963).
32. Heberle, W. J.: Styrene Polymers: The Major European Market for Styrene Monomer, *68th Nat. Meet., Amer. Inst. Chem. Eng. Houston, Tex., March* 1971.
33. Huckins, H. A., H. Gilman, and T. W. Stein (to Halcom International, Inc.): Process for Producing Styrene, U.S. Pat. 3,330,878 (July 11, 1967).
34. Making Propylene Oxide Direct, *Hydrocarbon Process.*, April 1967, pp. 141–142.
35. Styrene Flowsheets, *Hydrocarbon Process.*, November 1967, p. 144; November 1969, pp. 148, 234–235; November 1971, pp. 125, 206, 207.
36. Ipatieff, V. N., and L. Schmerling: Ethylation of Benzene in Presence of Solid Phosphoric Acid, *Ind. Eng. Chem.*, 38:400 (1946).
37. Jones, E. K.: New Catalyst Converts Higher Percentage of Ethylene to Make Ethylbenzene, *Oil Gas J.*, 58(9):80 (1960).
38. Kearby, K. K.: Catalytic Dehydrogenation, *Ind. Eng. Chem.*, 42: 295 (1950).
39. Lee, E. H. (to Monsanto Co.): Dehydrogenation Catalyst and Process, U.S. Pats. 3,100,234 (Aug. 6, 1963); 3,179,706 (Apr. 20, 1965); 3,179,707 (Apr. 20, 1965); 3,306,942 (Feb. 28, 1967); 3,387,053 (June 4, 1968).
40. Li, K. W., R. A. Eckert, and L. F. Albright: Alkylation of Isobutane with Light Olefins Using Sulfuric Acid, *Ind. Eng. Chem. Proc. Des. Dev.*, 9:434–454 (1970).
41. Lovett, G. H., and E. M. Jones (to Monsanto Co.): Dehydrogenation of Alkylated Aromatic Hydrocarbons, U.S. Pat. 3,118,006 (Jan. 14, 1964).
42. MacFarlane, A. C. (to Monsanto Co.): Dehydrogenation of Ethylbenzene, U.S. Pat. 3,223,743 (Dec. 14, 1965).
43. McDonald, D. W. (to Monsanto Co.): Control of Friedl-Crafts Alkylation Catalyst by Measurement of Electroconductivity, U.S. Pat. 2,846,489 (Aug. 5, 1958).
44. McMinn, T. D. (to Monsanto Co.): Separation of Friedl-Crafts Catalyst Complex from Hydrocarbons with Glass Wool, U.S. Pat. 3,131,229 (Apr. 28, 1964).
45. McNeil, G. A., and J. G. Sacks: High Performance Column Control, *Chem. Eng. Prog.*, 65(2): 33 (1969).
46. Mitchell, J. E.: The Dow Process for Styrene Production, *Trans. Am. Inst. Chem. Eng.*, 42: 293 (1946).
47. Modell, D. J.: Optimum Temperature Simulation of Styrene Monomer Reaction, *68th Natl. Meet., Am. Inst. Chem. Eng., Houston, Tex., March 1971.*

48. Ohlinger, H.: "Polystrol," pp. 4–7, Springer-Verlag, Berlin, 1955.
49. Ohlinger, H., and S. Stadelmann: Development of the Dehydrogenation of Ethylbenzene to Styrene by BASF, *Chem. Ing. Tech.*, **37**: 361 (1965).
50. Pan, B. Y. K. (to Monsanto Co.): Dehydrogenation of Hydrocarbons with an Argon-treated Iron Oxide–containing Catalyst, U.S. Pat. 3,409,688 (Nov. 5, 1968).
51. Prescott, J. H.: Styrene Keeps on Growing, *Chem. Eng.*, June 2, 1969, pp. 48–50.
52. Roberts, G. L. (to Monsanto Co.): Control of Alkylation Catalyst Activity, U.S. Pat. 3,470,261 (Sept. 30, 1969).
53. Root, W. N., and K. D. Uitti (to Universal Oil Products Co.): Catalytic Conversion Process, U.S. Pat. 3,515,766 (June 2, 1970).
54. Sanders, H. J., H. F. Keag, and H. S. McCullough: Acetophenone from Ethylbenzene, *Ind. Eng. Chem.*, **45**: 2 (1953).
55. Sheel, J. G. P., and C. M. Crowe: Simulation and Optimization of Existing Ethylbenzene Dehydrogenation Reactor, *Can. J. Chem. Eng.*, **47**: 183 (1969).
56. Shelley, P. G., and E. J. Sills: Monomer Storage and Protection, *Chem. Eng. Prog.*, **65**(4): 29 (1969).
57. Stage, H.: Neuzeitliche Rektifizieranlagen zur Reinigung des Synthese-Styrols, *Chem. Ztg. Sonderdr.* **65**(4): 271–285 (1970).
58. Twaddle, W. W., A. A. Harban, and V. W. Arnold [to Standard Oil Co. (Indiana)]: Direct Production of Polystyrene from Petroleum By-Product Ethylbenzene, U.S. Pat. 2,813,089 (Nov. 12, 1957).
59. Uitti, K. D. (to Universal Oil Product Co.): Production of Styrene, U.S. Pat. 3,515,763 (June 2, 1970).
60. Universal Oil Products Co.: personal Communication, 1970.
61. Walker, H. M. (to Monsanto Co.): Recovery of Aromatic Hydrocarbons, U.S. Pat. 3,090,820 (May 21, 1963).
62. Ward, A. L., and W. J. Robert: Styrene, in R. E. Kirk and D. F. Othmer (eds.), "Encyclopedia of Chemical Technology," 1st ed., vol. 13, pp. 119–146, Interscience, New York, 1954.
63. Webb, G. A., and B. B. Corson: Pyrolytic Dehydrogenation of Ethylbenzene to Styrene, *Ind. Eng. Chem.*, **39**: 1153 (1947).

CHAPTER EIGHT

Production of Polystyrene and Styrene Copolymers

I. INTRODUCTION AND PROCESS CONSIDERATIONS

Polystyrene and other styrene-type plastics, first produced commercially in the 1930s, are a major family of plastics throughout the world (6, 9, 22, 23, 26). In the United States, production capacity was approximately 3 billion lb/year by 1970, and it is expected to grow at a substantial rate (12). The sales prices for general-purpose polystyrenes, also called *crystal* or *straight polystyrenes*, which are homopolymers of styrene, have dropped during this time; they currently range from about 11 to 13 cents/lb. Colored, impact, or high-temperature polystyrenes are several cents per pound more. ABS plastics [containing acrylonitrile (A), butadiene (B), and styrene (S)] range from 24 to 40 cents/lb.

Lower costs of polystyrene polymers since the late 1930s have resulted from reduced costs of styrene (currently about 7 cents/lb), improved polymerization techniques, and increased competition. Reduced operating expenses per ton of polystyrene product are generally realized in larger plants. At present, 16 companies produce and sell polystyrene in the United States, and rated capacities per company vary from about 12 to 700 million lb/year (20).

Styrene-type polymers have a wide range of desirable physical prop-

erties. Since they are also relatively cheap, they have found many applications. The major and also fastest-growing ones include packaging
of meat, poultry, eggs, cottage cheese, and oleomargarines; bottles and
jars for drugs and pharmaceuticals; appliances; radio and TV parts;
lighting fixtures; toys; and houseware items (20, 22, 23). Markets are
also rapidly expanding in furniture, automotive applications, and dispos-
able eating utensils (plates, bowls, serving trays, cups, and cutlery).

General-purpose (or crystal) polystyrenes have several very desirable
properties (4, 18). They are hard, rigid, and transparent, with considerable brillance because of their high refractive indexes. Furthermore,
they mold easily, have good electrical properties, are inert to many
chemicals, and are easy to color in a wide range of hues. Brittleness
and a relatively low softening temperature (about 85 to 95°C depending
to some extent on the molecular weight of the polymer) do limit their
use, however.

Polystyrenes are generally very resistant to weathering, and discarded
polystyrenes have sometimes resulted in litter problems. Bio-Degrad-
able Plastics, Inc. of Boise, Idaho is now manufacturing lids for cold-
drink cups that are made of a photodegradable and biodegradable polystyrene. A photosensitizer is added to the polystyrene, and the resulting
plastic begins to disintegrate within about 30 to 90 days when exposed
to the weather.

Copolymers that are primarily styrene have been developed to obtain
various improved properties. Copolymers prepared using either α-methyl-
styrene or maleic anhydride as a comonomer have higher softening
temperatures. Improved "chemical resistances," e.g., for foods, result
with styrene-acrylonitrile copolymers (SAN or PSAN). Rubber-modified polystyrenes, often designated as impact polystyrenes, provide improved impact resistance. Styrene-butadiene copolymers and ABS plastics are also important styrene-type polymers. Both ABS plastics and
rubber-modified polystyrenes are two-phase polymers in which an elastomeric phase is dispersed through a rigid plastic matrix, and they are
sometimes designated as alloy polystyrenes. Styrene is always used in
the preparation of the rigid polymer phases and also sometimes the
elastomeric phases.

Crystal polystyrenes are produced by polymerization of essentially
pure styrene. For the modified polymers, the styrene content varies
for rubber-modified polystyrenes from about 88 to 97 percent, for styrene-butadiene copolymers from 50 percent and above, and for SAN
copolymers from about 70 to 75 percent. SAN copolymers are frequently the solid-polymer matrix in ABS plastics.

Several types of polymerization processes are required to produce
the wide variety of styrene-type polymers.

Chemistry of Styrene Polymerization

Although many aspects of the polymerization of styrene are relatively straightforward, a review of the polymerization mechanism helps explain how variations in the structure or the polystyrene molecule and the physical properties of the final polymer can be realized. Important structural features of any polystyrene product include the average molecular weight, molecular-weight distribution, number of branches on chains, and stereospecificity of the repeating units in the chain.

Styrene is commercially polymerized by a relatively conventional free-radical chain mechanism, involving reaction steps for radical initiation, propagation, termination, and chain transfer. Both cationic and anionic techniques can also be used for styrene polymerization, but commercial applications are at most rather limited.

An excellent review of the free-radical polymerization mechanism has been given by Platt (28).

Initiation of Chain Reaction Chain-propagation reactions are initiated commercially either by initiators (or catalysts as they are frequently but mistakenly called) or by thermal initiation. Important commercial initiators for suspension polymerizations are benzoyl peroxide, di-*tert*-butyl peroxide, *tert*-butyl hydroperoxide, and 2,2'-azo-bisisobutyronitrile. Other peroxide initiators used commercially have been reported by Pennwalt (27). For emulsion polymerizations, potassium persulfate is often used as an initiator.

Initiators thermally decompose, thereby forming active free radicals that are effective in starting the main propagation reactions (discussed in more detail later). They are chosen to a considerable extent for their kinetics of decomposition, often expressed as their half-life. At higher temperatures, the rate of decomposition is of course higher. Side reactions of some initiators include chain-transfer steps that result in lower-molecular-weight polymers. An initiator is selected in part by a trial-and-error procedure since the complete chemistry of initiator decomposition and side reactions is not yet known.

Styrene is one of the very few vinyl monomers that reacts, especially for higher-temperature polymerizations, e.g., above 120°C, to form free radicals that initiate polymerization. The exact mechanism is still somewhat speculative, but styrene is known to form up to at least 15 oligomers (29) that often isomerize to produce several rather complicated structures. Major dimer products include 1,2-diphenylcyclobutane, 2,4-diphenyl-1-butene and a noncyclic triene compound (28). This triene may be the true initiator, reacting with styrene by means of a hydrogen-transfer step resulting in two free radicals. Hence, three styrene molecules may be involved in a single initiation step. A kinetic model based on this mechanism indicates that the observed reaction order for initiation should vary between 2.0 and 2.5, depending on reaction conditions. Such an order is consistent with styrene polymerizations in many solvents, but not in bulk polymerizations, which often have apparent orders of 1.0 to 1.5, possibly because of the gel or Trommsdorf effect (discussed later).

Since the overall kinetics of polymerization is highly dependent on the kinetics of initiation, better mechanistic models for the initiation sequence would be helpful. What role, if any, the surface of the polymerization vessel has on thermal initiation has not been reported. Yet solid surfaces are sometimes effective for the production as well as the destruction of free radicals.

A complicating feature in styrene polymerization may be the presence of small amounts of impurities that for reasons not yet completely understood sometimes inhibit the polymerization. Furthermore, the inhibitors added to styrene to minimize polymerization during storage may also result in a large induction period, especially at the start of lower-temperature polymerizations. During the induction period, the free radicals first formed apparently react with the inhibitor or impurities until these latter compounds have been converted to inert compounds. Inhibitors sometimes also act to reduce the molecular weight of the polystyrene. Fortunately, para-*tert*-butylcatechol, the most common inhibitor, often added at only the 10-ppm level, has little effect on the kinetics or molecular weight for polymerizations at higher temperatures above 120°C (6).

Purity of the styrene feedstock is obviously an important consideration for rapid polymerizations. The inhibitors and some undesired impurities have on occasion been removed by washing steps. The styrene liquid is contacted first with caustic solutions and next with water washes; finally it is dried. Such washings are not usually needed for current commercial processes.

Propagation Steps The propagation steps that occur during the growth of the polymer chain occur almost exclusively by a head-to-tail arrangement with the repeating units, $-C_8H_8-$, in predominantly atactic configurations. Nuclear-magnetic-resonance measurements indicate a few isotactic and syndiotactic groupings, with the latter more common (7). Possibly a more stereoregular polymer is produced as the temperature of free-radical polymerization is lowered. An anionic but as yet noncommercial polymerization with solid catalysts produces isotactic polystyrene (25).

Evidence has been summarized indicating that a few abnormal links occur in the polystyrene chain (28)—perhaps one in every 10,000 links. It has been suggested that such links may involve transfer of the free-radical site to the phenyl ring and that chain growth then proceeds from this site on the ring.

Polystyrene and styrene are miscible in all proportions, so that the viscosity of the reaction mixture increases markedly as polymerization progresses. The question has been raised whether the steps for transferring monomer to the growing chain are hindered seriously as the viscosity increases (28). Certainly heat transfer is difficult in such a case. Effective agitation by specially designed equipment is often necessary to obtain good temperature control, especially for bulk polymerizations.

The kinetics of propagation may also be affected by the methods for transferring the heat of polymerization from the growing chain to the surroundings, e.g., the solvent, unreacted styrene, or reactor surfaces. The rate of energy transfer from the growing chain may be relatively slow, especially in a viscous medium, so that a "hot radical" is on the chain. Such a radical may have quite different reaction rates with styrene. There is evidence that the rate constant for the propagation step does vary between bulk and solution polymerizations (28). The suggestion has also been made that a complex may form between a polymeric free radical and a styrene molecule, and hence the kinetics of the propagation reaction may be affected. More information is certainly needed to clarify the controlling features of the propagation steps.

Termination and Chain-Transfer Steps   Coupling (or combination) of two polymeric free radicals is the major terminating step for styrene polymerization at least at lower polymerization temperatures, such as $<100°C$ (4). Chain transfer to the styrene molecule is probably quite important at higher temperatures such as $>140°C$ (28). Disproportionation is presumably of minor importance in most cases. Chain transfer between a growing polymeric free radical and a polystyrene molecule occurs to a small extent when the concentration of polystyrene in the reaction vessel is high, i.e., when the extent of polymerization is high (3, 18). As a result, perhaps on the average up to two-tenths of a long branch occurs on each polystyrene molecule. Such long branches significantly increase the molecular weight of the resulting polystyrene molecules.

Segmental diffusion of the polymeric (high-molecular-weight) free radicals may be a rate-controlling step when the viscosity of the reaction mixture becomes large, which happens when polystyrene concentrations become high. In such a case, the apparent reaction rate constants for termination steps by coupling or disproportionation would decrease since two polymeric free radicals must diffuse and be brought together so that they can react. Yet diffusion of the relatively small styrene molecule is much less affected by changes in viscosity. Hence the kinetics of termination steps may be decreased to a considerably greater extent by increased viscosities, compared to the kinetics of propagation reactions. This effect often results in high rates of polymerization and in high-molecular-weight polymers; it is often referred to as the gel or Trommsdorf effect or as autoacceleration.

The viscosity of the reaction mixture becomes very high at high polymerization levels, especially for bulk polymerizations. The viscosity also depends on the molecular weight of the polystyrene and on the temperature of the reaction mixture.

Chain-transfer agents are sometimes employed commercially for reduction of the molecular weights of the polymers (17). Hydrocarbons, particularly the dimers of α-methylstyrene (24, 30), mercaptans, and disulfides, have all been used extensively. Although carbon tetrachloride is an effective chain-transfer agent, it often affects the quality of polystyrene adversely and so it is not employed commercially. Since the molecular weight and branching of the polymer are affected by the method of termination, a better job of "tailoring" the final polymer molecule might be possible if additional information on termination were available.

Structural Features of Polystyrene Molecules

Molecular weights for commercial polystyrenes frequently vary from number-average values of 40,000 to 180,000 and weight-average values of 100,000 to 400,000 (2, 6, 8, 22). As with other polyolefins, the molecular weight is important relative to the physical properties and the processibility of the polymer. Controlling the average molecular weight and the molecular-weight distribution is therefore important, and considerable information on how to do this is available. The rules vary, however, depending on the method of polymerization (28). For polymerizations involving thermal initiation the rules are summarized below (such polymerizations are at relatively high temperatures, for example, 120°C and higher, and are generally by bulk-polymerization processes):

1. The average molecular weight of the polymer decreases with increased polymerization temperatures. As in most free-radical addition polymerizations, increased temperatures cause the rates of the initiation and the termination steps to increase faster than the rates of the propagation steps.

2. The average molecular weight for isothermal polymerizations is relatively independent of the extent of polymerization, or styrene conversions, at levels up to at least 96.5 percent. Hence, the transfer of styrene molecules through the high-viscosity mixture or gel obtained at higher conversion levels does not seem to be a controlling factor. Long-chain branching, which would significantly increase the molecular weight of

the polymer, apparently is not very important up to at least 96.5 percent conversions.

3. The ratio of the weight-average to the number-average molecular weights of the polystyrene is essentially 2.0 for all isothermal polymerizations.

For polymerizations in which initiation is accomplished primarily by initiators, the type and amount of initiator employed and the method of adding the initiator are most important in affecting the concentration of free radicals during polymerization and hence in affecting the molecular weight of the polystyrene product. These latter polymerizations are normally at relatively low temperatures, less than 100°C where thermal initiation reactions are negligible; initiators are generally used in batch suspension processes.

When a fairly low-temperature initiator is all added at the start of a batch run, most of the initiator decomposes relatively quickly, forming free radicals. In such a case, the free-radical concentration near the beginning of the run is high but decreases as the run progresses since free radicals are destroyed by termination steps. As a result, both the average molecular weight and the ratio of molecular weights increase as the styrene conversion increases. The ratio often increases from 1.7 to greater than 2.0 as conversions increase to 90 percent or higher. More uniform concentrations of free radicals can often be obtained for batch runs by adding more initiator after partial polymerization or by using more than one initiator in the feedstock.

When an initiator-type polymerization process is operated with a steadily increasing temperature, two outcomes are possible, depending on the initiator employed and the starting polymerization temperature. When a fairly high-temperature initiator, i.e., one with a long half-life, is used, only relatively little initiator decomposes during the initial stages of the run and the free-radical concentration increases as the run progresses. Hence the average molecular weight of the polystyrene decreases with increased conversions. If, however, a low-temperature initiator, i.e., one with a short half-life, is used, the concentration of free radicals decreases as the reaction progresses, causing a higher-molecular-weight product to be produced as the temperature is raised.

Emulsion polymerization of styrene can be used commercially to produce polystyrene of very high molecular weight. Propagation occurs predominantly in the dispersed latex particles containing a shell that is monomer-rich (15). Since the initiators employed are soluble in the continuous water phase, chain termination by coupling is improbable and occurs primarily when a second initiator fragment (or free radical) enters the latex particle (28). The rate of termination is inversely proportional to the particle volume, and since the particle grows as polymerization progresses, the molecular weight of the polymer increases with conversion. Therefore, factors that affect the particle size, e.g., amount of surfactant (or emulsifier) used and the level of agitation, are indirectly important to the molecular weight of the polymer. Temperature also affects the particle size and the kinetics of the various free-radical steps in polymerization. Higher temperatures normally result in lower-molecular-weight polymers.

For emulsion polymerization, increased levels of initiator result in higher rates of polymerization and in somewhat reduced molecular weights. The molecular weights obtained in such a case are often still as high as those obtained when suspension- or bulk-polymerization techniques are used.

Low-molecular-weight compounds, including unreacted styrene, ethylbenzene from the styrene feed, and low polymers (dimers, trimers, etc.), are always present in rather significant amounts in the polystyrene product from the reactor. Low polymers (or oligomers) are present at about 0.3 and 2.5 percent for thermal-initiated runs at 80 and 230°C, respectively (29). In many cases, these low-molecular-weight materials are separated from the polystyrene by subjecting the hot and semimolten polystyrene to high vacuums. The flashing of these compounds is relatively slow and often incomplete because of their low vapor pressure when dissolved in polystyrene.

Fewer oligomers are normally formed for low-temperature polymerizations or for ionic-type polymerizations, compared with high-temperature free-radical polymerizations. These low-molecular-weight materials may be present in sufficient quantities to have a significant effect on the properties (6).

1. These materials act as plasticizers, reducing the strength and the already low heat-distortion temperature.

2. Discoloration may result, since styrene and certain other unsaturated compounds are attacked by oxygen at higher temperatures or by exposure to sunlight. Unreacted initiator left in the polymer sometimes promotes color formation.

3. Slow evaporation of styrene and ethylbenzene from the surface may have detrimental effects. Crazing, primarily a strain-relaxation problem (21), seriously impairs the appearance of the polystyrene surface. Evaporation may sometimes contribute to the problem.

Antioxidants, lubricants, or pigments are frequently added in fairly small amounts to the final polystyrene. These materials must be chosen carefully so that the physical or chemical properties are not seriously affected.

Morphology of the polystyrene molecules can be varied by the stereo-specificity of the repeating units in the polymer chain. Since each repeating unit has an asymmetric carbon atom, the phenyl groups can be attached to the polymeric chain in isotactic, syndiotactic, or atactic configurations (5).

Atactic polystyrenes always are noncrystalline, or amorphous, and at atmospheric conditions are in the glass state. Although amorphous, the molecules do have a semiordered arrangement (28), and there may be a rather high degree of parallel packing of the polymer molecules. Polystyrene has relatively large interchain distances because of the bulky aromatic rings, and the linear molecule has a helical conformation.

Isotactic polystyrenes can be produced using Ziegler-type solid catalysts (25), and the resulting polymers sluggishly crystallize to about 50 percent crystallinity if cooling of the molten polymers is very slow. Crystalline polystyrene has characteristics similar in many respects to those of other crystalline polyolefins. It is opaque because of the differences in refractive indexes between the crystallites and the amorphous regions. The mechanical properties of this polymer are similar to those of conventional atactic polystyrene at temperatures less than 80°C, the glass-point temperature (5), but the crystalline polymers have higher softening temperatures (2). If isotactic polystyrene is quenched from above its first-order transition temperature, about 230°C, the polymer is optically clear and almost completely amorphous. In such a case, the physical properties of the polymer are similar to those of atactic polystyrene.

Unsaturation occurs to some extent in polystyrene because of disproportionation or chain-transfer steps with styrene. Double bonds are of course relatively susceptible to oxygen attack.

Branching of the polystyrene molecule generally is thought to occur to only a very small extent (3, 28). It has been suggested that polymer particles that stick to the reactor walls may eventually branch or even cross-link to form gels (6).

Kinetics of Polymerization

Mathematical models that predict both the kinetics of polymerization and the molecular-weight characteristics of the resulting polystyrene are complicated since the overall reaction is complex. Terms must be included in such a model for all important initiation, propagation, termination, and transfer steps. Several industrial companies have developed proprietary models.

Recently, kinetic models have been published for free-radical polymerizations using 2,2'-azo-bisisobutyronitrile as an initiator. Models were presented for continuous-flow stirred-tank reactors (13, 14, 19) and for batch reactors (16). The five reaction steps assumed to be important were thermal decomposition of the initiator, propagation, termination by coupling (or combination of two free radicals), and chain transfers with the monomer and the solvent. Several somewhat arbitrary assumptions were made in order to obtain an adequate fit between the experimental and predicted results.

It was assumed in the above models that 60 percent of the initiator fragments generated polymer chains. In addition, both solvent- and viscosity-correction factors were applied to the kinetic rate constant for the coupling step. Increased concentrations of solvents and/or higher viscosities are thought to be effective in decreasing the mobility of the reactants and hence in decreasing the effective kinetic rate constants. Although the predicted results agree quite well with the experimental results, the resulting kinetic model appears to be at best only semitheoretical. Caution should be employed in using these models for predicting mechanistic details of the polymerization.

Platt (28) has reported kinetic information for thermally initiated polymerization of styrene. At a constant temperature, the rate of polymerization decreases steadily with increased conversions and hence at decreased concentrations of unreacted styrene, as shown in Fig. 8-1. At conversons of about 90 to 95 percent, the rate becomes very low and difficulty is experienced in getting essentially complete polymerizations. Two techniques are frequently used to obtain significant rates of polymerizaton at high conversions: the temperature is often raised to perhaps 220 to 280°C as the conversions increase, and certain initiators can be

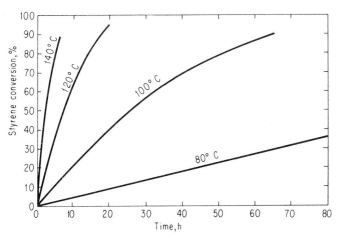

Fig. 8-1 Thermally initiated polymerization of styrene as a function of time.

used to promote polymerization at higher conversions (and higher temperatures) (1).

There is still the need for a better understanding of the reaction mechanism and for more kinetic information. Kinetic models of the type that have been published are most useful for investigating the effect of operating variables on both the kinetics and type of polystyrene produced. Tailoring a polystyrene product for a given molecular-weight distribution and hence specific applications is greatly simplified with such a model.

Transport Phenomena Occurring during Polymerization

Temperature control and flow patterns of the mixtures of polystyrene and styrene are critical in many polymerization reactors. These mixtures have viscosities as high in some cases as 1 million cP (10); they are also nonnewtonian fluids, for which the viscosity decreases with increased shear rates. The viscosity is also affected by the molecular weight of the polystyrene and to at least some extent by pressure. Pressure drops in some flow reactors are relatively high because of high viscosities of the reaction mixture. Thermal conductivities of the mixtures are often very low, frequently varying from about 0.2 to 1.2 Btu/(h)(ft^2)($^\circ$F/ft) and depending to a considerable extent on the level of polymerization. Heat-transfer rates from such mixtures are very low.

Flow and temperature patterns in flow-reactor systems may be quite complicated. Since the flows are generally laminar, and since shear rates in the nonnewtonian fluids vary significantly in a direction normal to the flow, the average residence times of differential volumes (or portions) of the reaction mixture vary significantly as they pass through a tubular or tower reactor. Brasie (10) indicates that 10 percent of the material may have residence times that are 4 to 10 times the mean residence time. Large temperature gradients in a direction normal to flow also occur, but agitation helps minimize them. Brasie (11) has shown that radial temperature variations of about 15 to 80°C can be expected in 1.5-in tubes as polystyrene concentration increases up to 30 percent. Even larger temperature variations occur at higher concentrations.

From electron-microscope observations, Grancio and Williams (15) have reported evidence for the complicated transfer steps occurring inside the monomer-polymer particles during emulsion polymerization. The centers of the particles are polymer-rich while the shells are monomer-rich.

As indicated earlier, diffusion or transfer of polymeric free radicals in the reaction mixture may be a controlling step in the polymerization sequence, especially when the polymerization is almost complete.

II. COMMERCIAL PROCESSES FOR POLYSTYRENE

Several processes are used for the commercial production of homopolymers (general-purpose or crystal polystyrenes) and of copolymers of styrene (6, 26, 37, 61). Major commercial processes for homopolymers employ techniques of bulk (or mass), solvent (actually a modified bulk), and suspension polymerization. These processes and emulsion-polymerization processes are also employed for the production of important copolymers of styrene. Other polymerization techniques have been considered, including cationic and anionic polymerizations, but the latter are of little or no commercial importance. Commercial techniques for the production of homopolymers will now be considered in detail. Since the same processes and often the same equipment are used for certain copolymers, some of the examples refer to copolymer production.

Bulk and Solvent Polymerization Processes

Styrene has been polymerized commercially for over 30 years by bulk- (or mass) polymerization processes operated both batchwise and by continuous-flow techniques. A modified bulk process that involves the addition of a small amount of solvent, such as ethylbenzene, is sometimes referred to as a solvent or solution process. These two types of process are of importance for the production of homopolymers of styrenes and some copolymers including impact (rubber-modified) polystyrenes. In general, for comparable levels of polymerization, reaction mixtures for the production of homopolymers have lower viscosities than mixtures for impact polystyrenes because of the dissolved rubber in the latter products. Heat transfer and temperature control are generally easier for homopolymer production, permitting differences in the reactor design and operation.

Flow Processes for Bulk or Solvent Polymerizations Flow processes like those used by The Dow Chemical Co., Badische Anilin- und Soda-Fabrik (BASF), and Union Carbide Corp. produce large quantities of polystyrenes. The processes of the first two companies were reported in 1970 to produce 1.8 billion lb of polymer per year whereas the Union Carbide process yielded 0.25 billion lb (61). These processes are similar in many respects to the well-publicized one (Fig. 8-2) used at Ludwigshafen, Germany, in the early 1940s (6, 26, 47).

For production of homopolymers of styrene (crystal polystyrenes), somewhat less agitation is required than for the production of impact polystyrenes, and so somewhat simpler reaction systems are frequently permissible. Polymerization vessels suitable for impact polystyrenes are also suitable for the homopolymers, however. In some plants, economics dictates that the equipment be designed for production of both types of polystyrenes, permitting more flexibility in the plant operation.

Flow processes using bulk- or solvent-polymerization techniques employ two or more reactors in a series. In the initial reactor, often referred to as the prepolymerizer, up to about 11 to 50 percent of the styrene is polymerized. Polymerization to almost 100 percent then occurs in the later reactors in the series. Stirred autoclaves and tower reactors are employed in these processes. In some processes discussed later, suspension-polymerization reactors operated batchwise are employed for the final stages.

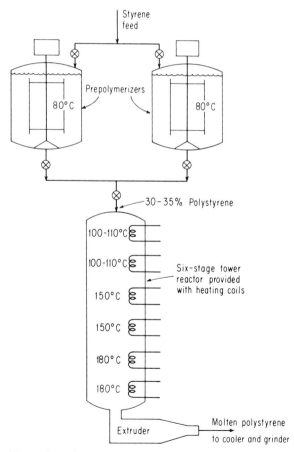

Fig. 8-2 Polystyrene tower process used in Germany in the early 1940s.

Stirred Autoclaves Used as Polymerization Vessels: Stirred autoclaves are used for the polymerization of styrene in both continuous-flow and batch processes. They can be designed either as prepolymerizers or as polymerization vessels to be used when high concentrations of polystyrene are present.

When autoclaves are used as prepolymerizers, rather conventional mixing equipment is normally adequate, since the reaction mixtures have fairly low viscosities. Heat transfer to the internal coils or tubes is in general good. Such autoclaves are frequently used as prepolymerizers in plants producing both crystal and impact polystyrenes. For the polymerization of high-viscosity mixtures, such as those with high polymer concentrations, the autoclaves must be provided with low-speed heavy-duty agitators that require high power inputs. Considerable backmixing of the reaction mixture generally occurs in these autoclaves, resulting in relatively constant temperatures and compositions throughout the vessel. With high-viscosity mixtures, however, mixing is less complete, and some temperature and composition gradients may occur.

The prepolymerizers used at Ludwigshafen in the early 1940s were stirred autoclaves constructed of aluminum and provided with internal heat-transfer coils (6, 26, 47). Flat paddles rotated at 50 to 60 r/min provided agitation in the 530-gal autoclaves, and the packing gland for the agitator shaft was water-cooled. A stream of nitrogen was introduced to flush styrene from the region where the agitator shaft entered the bottom of the packing gland. Consequently adsorption of styrene and the eventual accumulation of polymer there were prevented. This nitrogen also served as an inert blanket over the reaction mixture. Exit gases from the prepolymerizer vessel were cooled in a small water-cooled pipe-coil condenser. The condensed styrene was later recycled.

In the early German plants, approximately 30 percent of the styrene was polymerized in the prepolymerizers operated at about 80°C with an average residence time of 60 to 70 h. Because of the low rates of polymerization, two prepolymerizers were used in series in each polymerization line (see Fig. 8-2). Refractive-index measurements of the product stream were taken at frequent intervals to determine the concentration of the polystyrene.

Considerably faster rates of polymerization are obtained if higher temperatures are used. The average molecular weight of polystyrene is significantly affected by the operating temperatures, however. At 80°C the average is estimated at 370,000 by ultracentrifuge methods (47) or 880,000 by osmometric methods (37). At 120°C, a polymer with a lower molecular weight is produced.

The product mixtures from these early prepolymerizers were processed in two ways. Generally the mixture flowed continuously to a tower reactor (discussed later) for additional polymerization.   At other times, the polymeric syrup containing 30 percent polystyrene was fed to a vacuum two-roll drum dryer (6, 26, 47, 61).   The unreacted styrene was evaporated, recovered, and then recycled.   The chrome-plated dryer rolls, rotated at 1.5 to 2.0 r/min, were approximately 20 in in diameter

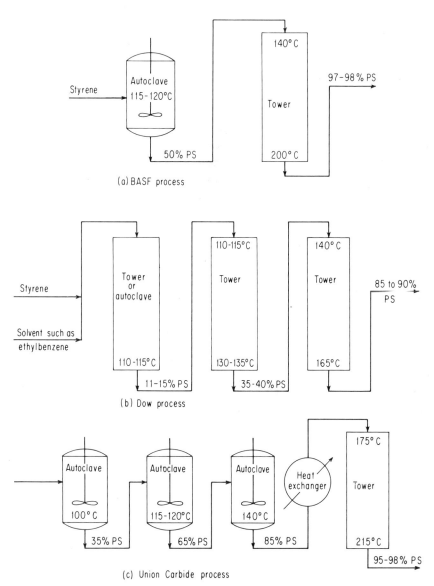

Fig. 8-3 Summary of three major processes for bulk polymerization of polystyrene: (a) BASF, (b) Dow, and (c) Union Carbide process.

and 48 in long. The system was maintained at an absolute pressure of 10 to 15 mm Hg using a three-stage steam ejector. The film thickness of the polymer on the rolls was about 0.06 in. In the German plant, two prepolymerizers and two drum dryers were used for production of 30 to 35 lb/h of polystyrene with especially high molecular weight. Such a small unit is no longer economical.

BASF still uses a reactor system similar to the one shown in Fig. 8-2, namely an autoclave employed as the prepolymerizer and a tower reactor for the final stages of polymerization. Their current prepolymerizer is operated at a higher temperature (about 115° to 120°C), however, as shown in Fig. 8-3a. With a residence time of about 4 to 5 h, the polystyrene concentration of the reaction mixture is about 50 weight percent.

The Dow Chemical flow process (see Fig. 8-3b) has a prepolymerizer that is normally a stirred autoclave. In one unit, a jacketed Pfaudler glass-lined autoclave equipped with an S-shaped blade stirrer was provided as the prepolymerizer (63). In an example reported for production of an impact polystyrene, the vessel was operated at 112°C with an agitator speed of 67 r/min. The feed rate was 2,425 lb/h of a mixture containing mainly styrene, the remainder consisting of dissolved rubber, ethylbenzene, and various additives. About 7,500 lb of the reaction mixture, equivalent to approximately 1,000 gal, was present in the reactor. The exit mixture contained 11 percent of polymeric materials based on an average residence time calculated to be slightly greater than 3 hr. For homopolymer production, somewhat higher polymer concentrations (and hence higher temperatures) can be allowed with adequate temperature control.

In an autoclave with rather conventional mixing and with heat transfer through only the jacket, a 30 percent level of polystyrene in a polystyrene-styrene mixture at 80°C is approximately the maximum allowable. Temperature control tends to become inadequate at higher polymer concentrations because of the resulting high viscosities, and then runaway reactions may occur. To permit autoclaves to be operated at higher concentrations of polymers, several modifications are employed for improving heat removal:

1. Smaller reactors with higher surface-to-volume ratios are used.
2. Internal heat-transfer surfaces, such as coils or tubes, are used.
3. Part of the styrene is vaporized and possibly refluxed.
4. More agitation is provided to increase the heat-transfer rates and to minimize temperature and composition gradients.
5. Small amounts of a solvent such as ethylbenzene are added to reduce the viscosity of the reaction mixture, causing higher heat-transfer

coefficients.   The concentration of the styrene is also reduced somewhat by dilution with the solvent.

The viscosity of the mixtures can also be reduced by increased operating temperatures.   The increased rates of heat transfer then possible are balanced in part (and sometimes entirely) by the higher rates of heat release caused by the faster rates of polymerization.

In the Union Carbide Corp. process (Fig. 8-3c) three or more autoclaves are used in series (31).   The styrene monomer at 25°C is charged to the first autoclave, where at 100°C approximately 35 percent of the styrene is polymerized.   Further polymerization up to a 65 percent level is obtained in the second autoclave, operated at 115 to 120°C.   A polystyrene level of 85 percent is obtained in the third vessel, maintained at 140°C.

The Union Carbide autoclaves are jacketed vessels provided with a vertical agitator shaft.   Several impellers on each shaft cause the reaction mixture to recirculate with an upward flow near the shaft.   In addition, the bottom impeller has blades that "scrape the mass from the sides and force it inwardly toward the shaft and out the outlet."   Because of the high viscosities of the polymer mixtures formed, the agitator structure is "rugged and heavy."   The rate of rotation of each impeller shaft is relatively slow.

A portion of the styrene from the reaction mixture in the Union Carbide reactors is allowed to vaporize and then is condensed and recirculated.   The temperature level in each autoclave is regulated by the level of vacuum applied, which is in the range of 18 to 22 in Hg.

Based on the rates of styrene polymerization (see Fig. 8-1), the three Union Carbide reactors are of different size.   The rates of polymerization are estimated to be approximately 2, 4, and 7.5 percent polymerization per hour in the first, second, and third reactors, respectively.   The amounts of polystyrene production in these three reactors are then calculated to be about 1,1, 2.2, and 4.2 lb/h per cubic foot of reaction mixture.   The first reactor apparently has a volumetric capacity several times greater than the third reactor; the second reactor is about twice as large as the first reactor.   The size of the reactors is dictated to a considerable extent by the need to maintain adequate temperature control during the critical stages of polymerization.   Higher surface-to-volume ratios in the smaller reactors expedite heat transfer when the viscosities of the reaction mixtures are high and when the rates of polymerization are large.   The agitators in the three reactors are probably quite different in design.

The partially polymerized mixture from the Union Carbide autoclaves is pumped by a screw conveyor to a pressure of almost 2 atm before entering the final stage of polymerization. This mixture, containing about 85 percent polystyrene, first flows through a preheater that is jacketed and provided with a mandrel at its axis. The mixture flows in a narrow annular passageway and is heated to approximately 175°C before entering a tower reactor (discussed later). Some polymerization of course occurs in the preheater.

Tower Reactors: Tower (or tube) reactors are widely used for the production of homopolymers and copolymers of styrene. They can be designed for all levels of polymerization, including reaction mixtures with high viscosities. Tower reactors are often custom-built for specific types of polystyrene or for a specific plant. Aluminum, stainless steel, or stainless-steel-clad metals are common materials of construction.

Large tower reactors provide complicated heat-transfer surfaces; outer jackets, multiple tubes or coils in either vertical or horizontal positions, internally cooled agitators, and extended heat-transfer surfaces, such as special fins, are commonly employed. Reactors used for production of impact polystyrenes especially are generally equipped with agitators to increase heat-transfer coefficients and also to minimize channeling of the nonnewtonian fluids in the reactor. Power requirements of these agitators are high—sometimes are high as 1 hp of energy per pound of impact polystyrene produced per hour. Oil, water, and air have been used as heat-transfer agents in the jacket, tubes, or coils.

In the process used at Ludwigshafen in the 1940s (Fig. 8-2) the tower reactors had no mechanical agitation, and the polystyrene concentrations in them increased from about 30 to 95 percent. These units were relatively small, with diameters of 0.8 m and heights of 6 m (6, 47). Each tower was constructed of stainless-clad steel and was divided into six sections of approximately equal size. Temperatures were controlled by a jacket, internal coils, and an external electric-resistance heater. Temperatures in the six sections of the reactor were approximately 100 to 110, 100 to 110, 150, 150, 180, and 180°C, respectively. Since mechanical agitation was not provided in this vessel, rather large variations of temperature were present in each section. The exit product was electrically heated, and almost pure (about 95 percent) polystyrene was obtained at about 200°C.

Each small German reactor was operated with average residence times of about 60 to 65 h and produced polystyrene at about 44 kg/h, equivalent to about 800,000 lb/year. The rate expressed as 11 kg/h per cubic meter of tower volume [or 0.7 lb(h)(ft³)] is low relative to modern units. In the early German plant, no real attempt was made to devolatilize the 3 to 5 percent of volatiles, mainly styrene, which is the current practice.

A large and modified tower reactor employed in the current process

(61) used at Ludwigshafen (Fig. 8-3a) is patented (36). The feed mixture to this reactor contains 45 to 60 percent polystyrene, often about 50 percent, and the remainder styrene. This mixture at 100 to 130°C enters the top zone of the reactor, which is maintained at 140°C. Later zones are at higher temperatures, and the bottom zone is at approximately 200°C. The exit product from the tower contains 97 to 98 percent polystyrene when the residence time in the tower is 3 to 4 h. In this tower, some styrene is allowed to vaporize and is removed from the top of the tower to help maintain temperature control. This styrene is condensed and then recycled to the prepolymerizer.

Current BASF towers are capable of higher yields of polystyrene on a given volume basis than the earlier German units. This improvement is caused primarily by improved methods of temperature control that allow somewhat higher temperatures and hence higher kinetics of polymerization. Approximately 7 to 8 lb/h of polystyrene is produced per cubic foot of reactor volume.

A tower reactor 15 ft in diameter and 40 ft high has been reported for the production of approximately 40 million lb/year of polystyrene (69). Such a tower appears to be essentially a scale-up of the BASF tower just mentioned, at least relative to its production capacity per unit of reactor volume.

Tower reactors like those used in Ludwigshafen must be provided with means for controlling the temperature. Several methods are available, including numerous tubes and coils used for heat-transfer purposes. Details on the internal construction of the BASF reactors have not been published, but a patent indicates a suitable arrangement for placing the tubes (34). A vertical tower reactor is filled with numerous banks of horizontal tubes arranged so that the tubes in each bank are parallel to the others. The tubes in alternating banks are horizontally perpendicular to each other.

Details of the commercial tower reactors currently used in the Union Carbide process (Fig. 8-3c) have not been reported. A small tower reactor described in a patent (31) contained two sections and was not mechanically agitated. A torpedo provided between these sections produced a narrow annular region for flow. For a larger tower reactor, more sections might be desired to promote plug flow of reactants in the system. An average residence time of 8 h completes the polymerization of styrene in the tower reactor from about 85 up to 95 to 97 percent polystyrene using an exit temperature of approximately 215°C.

Tower reactors of The Dow Chemical Co. are often agitated, and they are used for production of both crystal and impact polystyrenes (42, 54, 59, 63, 67). Figure 8-3b is the simplified flowsheet for a solvent-, or modified bulk-, polymerization process. Three reactors are frequently connected in series. The solvent reduces the viscosity of the reaction mixture so that temperature control is easier.

Ruffing et al. (63) have reported many details of the three reactors used in series for the production of about 18 million lb/year of high-impact polystyrene. These reactors can also be used for the homopolymerization of styrene. The first reactor, an autoclave, which contains approximately 1,000 gal (7,500 lb) of reactants, was discussed earlier in this chapter. The second unit is a special tower reactor with a cross-sectional area of 20 ft² and height of 18 ft. This vessel was constructed from two circular (48-in-diameter) columns 16 ft high. One-quarter of the circumference or shell was cut from each column, and the two columns were then welded together at the open sections. The cross-sectional area of the final tower resembles a figure eight. The reactor is provided with two vertical impeller shafts; one shaft is located at the axis of each of the two cylindrical sections of the figure-eight column.

Heat-transfer surfaces are provided for this reactor by two means: the entire reactor is jacketed, and the two impellers have internal cooling systems. Each impeller consists of a shaft and attached crossarms. The shafts are constructed of 6-in piping. The crossarms are about 20 in long and constructed of 3-in piping; and three cross arms are welded 120° apart in a given horizontal plane or bank on the shaft. The banks of cross arms are positioned so that the axes of these arms are 4 in apart. Provisions are made with tubing inside the hollow shaft and crossarms so that coolant (probably water or oil) is pumped inside the impeller to and from the extremities of each arm.

The two impeller shafts are rotated in a counter direction, providing an overlapping action and effective agitation. The crossarms on the two shafts do not collide although they make a close approach as they rotate. Rotating the arms at approximately 10 r/min is used for production of impact-type polystyrene.

This second reactor has a capacity of about 16,000 lb of reaction mixture and hence provides a residence time of about 7 h. The temperatures employed in the reactor for an impact polystyrene range from about 112°C in the top section to approximately 133°C in the bottom. The polymeric concentration increases from about 11 to 37 percent in this vessel with a resulting significant increase in the viscosity of the reaction mixture.

The third reactor of the three-reactor sequence (Fig. 8-3b) is a jacketed vertical tower 48 in in diameter and 16 ft high. This reactor is provided with a single impeller shaft to which are attached 20-in bars about 2 in in diameter. Banks of three bars are positioned about

12 in apart. Since heat transfer is considerably less critical in this third reactor, the impeller system is not cooled and the impeller rotates at a much slower rate of about 2 to 3 r/min. Temperatures increase from about 140°C at the top, to 155°C at the midsection, to 165°C or higher at the bottom. This reactor has a capacity of about 10,000 lb and hence provides a residence time of 1 h or slightly greater to cause the polymer concentration to increase from about 37 to 85 percent. Since about 10 percent of nonreactive solvent is present, the polymerization is essentially completed in this reactor.

The three-reactor unit described by Ruffing et al. (63) has a fairly high rate of polystyrene production. The prepolymerizer (or reactor 1), second reactor (a tower reactor), and the final tower reactor have polystyrene rates that have been calculated as approximately 2.1, 2.1, and 6.3 lb/h per cubic foot of reactor volume, respectively. The overall rate is about 3.4. The process using three reactors in series is said to be completely dependent on the solution of heat-transfer problems (42).

The level of backmixing in a reactor is important for polymer quality and properties. Some backmixing normally occurs, but most tower reactors, including those of the Dow process, are built to minimize it. Brasie (41) has patented a jacketed reactor; it is essentially a tower reactor in which little backmixing occurs and which is highly effective for high-viscosity mixtures. A complicated series of agitators is employed, and the reaction mixture flows in a very serpentine manner past numerous cooled tubes and the large finlike surfaces connected to the tubes. Because of the large heat-transfer surfaces, the amount of heat transfer in this reactor is high. This reactor would presumably be very effective for mixtures containing high concentrations of polystyrene.

The capacity of polystyrene units and lines has increased significantly in the last few years, resulting in a decrease in the operating cost per given weight of product. Table 8-1 indicates the raw materials and the utilities required per metric ton of polystyrene for the process of Mitsui Toatsu Chemicals, Inc. In their unit, four reactors are employed in series. Utility costs are estimated to be about 0.25 cent per pound of polystyrene, and such costs are probably fairly typical for modern processes.

Reactor lines capable of producing 130 million lb/year are currently employed (61), and even larger ones are technically possible. There may be little economic advantage in large units, however, since switching from one type of polystyrene to another is relatively slow, causing production of some off-grade material during the interim. These units are generally employed only for the production of the most important polymers, such as general-purpose (crystal) or impact polystyrenes. Hence, few switches of product are ever made. A given plant often has more than one unit or line, resulting in total plant capacities of 200 million lb/year and greater (12).

Postpolymerization Steps for Bulk Processes   In modern processes, the product mixture leaving the last polymerization vessel is devolatilized, perhaps mechanically worked, and then pelletized. Devolatilization is necessary to remove most of the unreacted styrene, solvent, and low polymers, including dimers and trimers, which have a detrimental effect on the properties of polystyrene.

A technique used by The Dow Chemical Co. involves a combination of a heat exchanger and a finishing tower (35, 63). The molten product stream from the final tower reactor is heated in a heat exchanger to 250 to 280°C. During heating, some additional polymerization of styrene occurs, and with impact polystyrenes some cross-linking results.

The heated polystyrene in the Dow process is then fed to the finishing column, where the melt collects in a pool or a cup above a horizontal plate located near the top of the column. The melt then passes downward through a plurality of ¼-in holes in this plate, and the resulting strands of polymer fall at least 3 ft before they coalesce in a pool in the bottom of the tower. Aluminum-clad steel is used on the inside walls to minimize dark surface deposits. An absolute pressure of about 10 mm Hg is provided, and the diffusion and vaporization of the volatile compounds are affected by the diameter and the length of the falling

TABLE 8-1   Materials and Utilities Required per Metric Ton of Polystyrene for Polystyrene Process of Mitsui Toatsu Chemicals, Inc.*

Materials:

Styrene, kg	997
lb	2,200
Solvent, kg	4
lb	8.8
Additives, kg	23.4
lb	51.5
Blueing dye, g	0.2

Utilities:

Electric power, kWh	340
Steam, kg	400
lb	880
Process water, m³	2.5
gal	660
Fuel (C heavy oil), liters	31
gal	8.3
Nitrogen, Nm³	1.5

* *Hydrocarbon Process.*, November 1971, p. 202.

strands and by the temperature. Temperature affects the diffusion rates in the melt and the flow characteristics of the strands. Design and operation of this unit are aided by complicated mathematical models of the flow phenomena, concentration gradients of the volatile compounds, and the temperature in the falling strands. The residence time of the polystyrene in this unit should be less than 10 to 20 min to minimize thermal degradation and cross-linking reactions.

A column of this design used in a production line of about 18 million lb/year of high-impact polystyrene has a diameter of 5 ft and a height of 10 ft (63). About 220 lb/h of vapor is volatilized, then condensed, purified by distillation, and recycled. The devolatilized melt is extruded and pelletized by conventional techniques.

A roll mill operated with a vacuum of about 29 in Hg has also been used for the devolatilization operation (31). In this case, the temperature rises to about 225°C because of the mechanical working; some of the residual styrene polymerizes; and most remaining styrene, ethylbenzene, and highly volatile materials are vaporized to yield polystyrene of 98.5 to 99 percent purity or even higher. Subsequently the molten polystyrene is sometimes stretched, often up to about 300 percent, as the polymer is cooled. The stretched polystyrene is then pelletized by a rotating cutter. Such polystyrene is said to mold more easily than the unstretched material.

The devolatilizers sold by the Marco Development Co. are used for processing several polymers, including polystyrenes, high-density polyethylenes, and polypropylenes (65). For polystyrene units (see Fig. 8-4), the volatile content can be reduced from as high as 53 percent to 1.5 to 2.0 percent when the unit is operated at 80 mm Hg absolute pressure. With a second unit operated at 9 mm Hg, the concentration can be reduced to 0.1 to 0.2 percent. Units are provided with a rotating vertical screw that has a diminishing diameter at its lower end. The melt as it flows downward is worked between the screw and the housing, exposing new surface to the vacuum. Polymer retention time is usually less than 1 min, thus minimizing thermal degradation. Units with vented turnscrews are used if a lower volatile content is required, but such units are considerably more expensive. It is claimed that a conventional unit with a 4,000 lb/h capacity has been used for over 4 years without serious mechanical problems.

A four-screw self-cleaning devolatilizing extruder that is used for processing 2,200 to 2,600 lb/h of polystyrene is shown in Fig. 8-5. Extruders of this design have capacities up to 10,000 to 12,000 lb/h. In these units, the polystyrene melt passes horizontally through a series of compartments or stages. Pressures in these stages vary from essentially atmospheric to high vacuums. Unreacted styrene and any solvent present are removed from the top of each stage. The devolatilized polystyrene can then be extruded and pelletized at the exit of the extruder.

Thin-film devices have also found commercial application for devolatilizing various high-viscosity liquids (68). Rotary blades that just clear the walls of the vessel are operated at high speeds, thus causing high shear gradients in the thin films on the wall. This shearing action significantly reduces the apparent viscosities of the nonnewtonian fluids in the film. Pilot-plant data have been successfully used for the design of commercial units, and thin-film units appear to be of interest to polystyrene producers.

Bulk Polymerization Processes Operated Batchwise Two relatively different batch processes for bulk polymerization of styrene are operated commercially, but their importance is considerably less than that of the flow processes just discussed. The basic techniques of the batch processes are the same as developed in the 1930s and 1940s; i.e., styrene and possibly an initiator are mixed and heated until polymerization is completed.

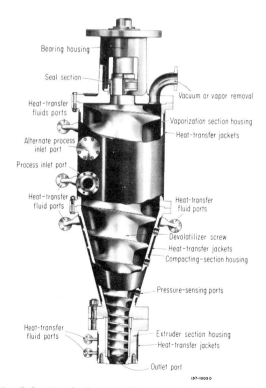

Fig. 8-4 Devolatilizer. (*Marco Development Co.*)

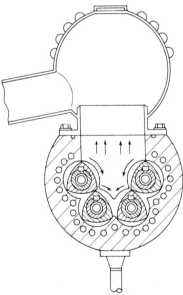

Fig. 8-5 Photograph and cross-sectional drawing of four-screw self-cleaning de-volatilizing extruder for polystyrene.   (*Werner and Pfleiderer Corp.*)

The batch processes employed by at least four American companies, including one of the largest polystyrene producers, use a chamber-type reactor (see Fig. 8-6) sold by T. Shriver and Co. (53).   No mechanical agitation is employed in these reactors; they contain numerous cavities in which polymerization occurs.   The equipment is relatively simple and includes storage tanks for the styrene and possibly for comonomers; pumps for the reactant olefins; proportioning control devices for controlling the amounts of styrene, comonomer, and initiator to be fed to the mixing tank; an agitated tank for mixing the reactants; often a prepolymerization reactor; the chamber-type reactor; and finally equipment for crushing and pulverizing large blocks of polystyrene. A supply of steam and cooling water is also needed.

The prepolymerizers, if used, are operated batchwise. Polymerization in an autoclave reactor is at approximately 80°C until about 30 to 35 percent of the styrene has polymerized. The syruplike mixture is then transferred to a chamber-type reactor.

The chamber-type reactors are similar to plate-and-frame filter presses. Each reactor consists of several combinations of plates and frames. Two cavities or molds are provided in each frame, allowing the production of two blocks of polystyrene. The platens and frames are made of cast aluminum, and the surfaces contacting the polystyrene are machined. Each platen is hollow and is provided with an inlet and an outlet line. Steam and water lines are connected to the inlet line using Teflon-lined hoses attached to suitable manifolds. Hot water can be produced by mixing cold water and steam.

The assembly of platens and frames is clamped together with a hydraulic ram before filling the reactor. The platens and frames are assembled without gaskets since styrene quickly polymerizes to produce its own gasketing material, which remains in place after the reactor is emptied.

The operation of the reactor is basically quite simple, at least in theory. Styrene or a partially polymerized (or prepolymerized) styrene with the consistency of honey is used to fill the cavities, or frames,

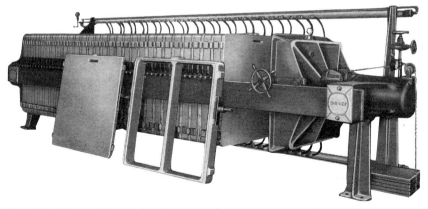

Fig. 8-6 Plate-and-frame batch reactor for polystyrene production. Reactor has 30 chambers, each producing two 24 × 52-in slabs 5 in thick. (*T. Shriver and Co., Inc.*)

of the reactor. Steam or hot water is then used to heat the platens and the styrene until polymerization starts. When a 28 percent prepolymerized mixture is allowed to heat steadily from 90 to 200°C over a period of 5 h and then is held at 200°C for 1 h, essentially complete polymerization is obtained (1). Initiators (discussed later) are sometimes added to promote high polymerization levels. Finally the reactor is cooled with water, the reactor is opened, and the polystyrene blocks are removed. Sticking of the blocks in the frames and to the platens has been a problem, and devices to remove the blocks have been developed.

The Shriver reactors are generally designed with 28 to 56 frames. The polystyrene blocks obtained are approximately 24 by 52 by 5 in and weigh about 200 lb. A 56-frame reactor produces 112 blocks per batch, or about 22,000 lb (11 metric tons) per batch. Figure 8-6 shows a reactor with 30 frames.

Temperature control is a critical factor in the Shriver reactor, and temperature gradients that occur in the reaction mixture during polymerization can be minimized if relatively low rates of polymerization are maintained. Operating variables for controlling the reaction, and hence the type of polymer obtained, include the temperature of the platens (temperature level and temperature as a function of the batch cycle), the amount and type of initiator used, and the level of prepolymerization of the feedstock. Approximately one batch of product per day can be obtained if pure styrene is used as a feedstock whereas up to two batches are possible with a partially polymerized styrene. Hence, a 56-frame reactor can produce about 7.5 to 15 million lb/year of polystyrene.

If thinner frames, and hence thinner and smaller blocks of polystyrene, were used in the reactor, somewhat faster rates of polymerization would be allowable. Then increased number of batches per day would be possible.

An important advantage of this bulk polymerization process is the modest capital cost for the installed equipment. The reactor is the most expensive piece of equipment, and Shriver indicated in the early 1970s a price of about $117,000 for a 53-frame reactor (53).

Crystal polystyrene was previously produced in large amounts in unagitated polymerization vessels containing bundles of cooling coils and possibly other heat-transfer surfaces in addition to the jacketed walls. These reactors are currently used to a more limited extent. A typical reactor of this type contains a large number of tubes, such as the one described by Platzer.* All portions of the reaction mixture are within several inches (probably generally 2 in or less) of a heat-transfer sur-

* See Fig. 1 of Ref. 61.

face. Somewhat better temperature control can often be realized in these reactors than in plate-and-frame reactors. The degree of temperature control depends to a considerable extent on the distances between the tubes and the other heat-transfer surfaces in the reactor.

Liquid styrene or a prepolymerized mixture is added to the reactor, and the contents are then heated to start polymerization. The molecular weight of the polystyrene is controlled primarily by the temperature-time program, which is carefully followed. When polymerization is essentially completed, the reactor temperature is raised to reduce the viscosity of the reaction mixture to permit rapid draining. The polymer is then drained and devolatilized.

Improved control of the temperature in these batch reactors can be realized if mechanical agitators are also used. The time at reaction conditions and the temperature of the reactants can then be very carefully controlled, often better than in flow-reactor systems. As a result, polystyrenes with better-controlled molecular-weight distributions can often be produced.

The time to drain the polystyrene product from the batch reactor, to devolatilize, and to pelletize the polymer should preferably be fairly short. Otherwise the level of polymerization and, to some extent, thermal degradation may change, especially with the last portion of the polystyrene drained from the reactor.

Removal of the volatile compounds in the liquid-polystyrene product stream can be accomplished rather easily by methods similar to those employed for flow processes; but when solid polystyrene blocks are obtained, as in the chamber-type reactor, devolatilization may be harder. If the solid blocks are pulverized, ground, and then pelletized, the volatile compounds can be removed while the polystyrene is in the molten state during the pelletizing operation.

A Monsanto patent (1) indicates a method for obtaining higher levels of polymerization during batch polymerizations such as those in a chamber-type reactor. When small amounts of alkanoic acids and an initiator that is either dialkyl peroxide, alkyl perbenzoate, or alkyl peracetate are added to the styrene feed, the residual styrene content is reduced to about 0.35 to 0.5 weight percent. A mixture of organic hydroperoxides and a peroxy compound used in conjunction with the alkanoic acid, however, reduces the residual styrene content to less than 0.15 percent.

Labor costs for batch processes, and especially the one with the chamber-type reactor, tend to be high compared with flow processes. Reactor assembly and opening and polymer handling result in high labor costs.

Design of Bulk Polymerization Reactors Bulk (or solvent) polymerization reactors are designed to obtain as high a rate of polymerization as possible, commensurate with adequate control of the temperature. Inadequate temperature control may result in runaway reactions and hot spots, causing styrene to vaporize, building up pressure. At least two serious explosions have occurred in commercial-flow reactors. Furthermore, temperature has a significant effect on the molecular weight of the polystyrene and also on the viscosity of a given reaction mixture.

Increased levels of agitation and of solvent and more heat-transfer surface all tend to give better temperature control, but they also increase the cost of operation. Hence it is uneconomical to provide more than a safe minimum of each. Both the capital and operating costs for the agitator system in flow processes are often significant portions of the overall costs. The electric motors and the gear reducers for reducing the motor speed down to the desired low levels for the agitator shaft are frequently relatively large. Combining several reactors in series for bulk-polymerization flow processes offers the important advantage that significant variations in the operating conditions can be provided rather easily in the various reactors containing materials of different composition.

For the design of flow reactors, including tubular reactors, complicated mathematical models solved with digital computers are often used to predict temperatures, pressures, flow velocities, polymer concentrations, and polymer characteristics, including molecular weight (11, 56, 57). Fairly theoretical information is needed about the physical properties and the kinetics of polymerization in order to solve these models. Part of this information is in the literature, but considerably more is in the proprietary files of various companies that produce polystyrene. These models have proved helpful for evaluating new reactor concepts or for modifying operating conditions. Some models also predict the characteristics of the product, including the molecular weight.

Suspension-Polymerization Processes

Suspension-polymerization processes are preferred by several companies for the production of homopolymers of styrene, impact polystyrene, expandable polystyrene, styrene-acrylonitrile copolymers, and some ABS copolymers. Plants having annual capacities of 100 million lb/year or greater have been built (38). In the suspension-type polymerizers, the organic phase is dispersed as droplets throughout the continuous water phase.

Polymerizer Design and Operation In a typical polystyrene plant, at least several polymerization reactors are operated batchwise, and they are similar in design to those used in other suspension processes, including polyvinyl chloride production. The reactor temperature is controlled using a jacket and sometimes with internal-cooled baffles (38).

The reactor vessels range in size from about 2,000 to at least 15,000 gal; even larger vessels are currently being designed. Both glass-lined and more recently stainless-steel reactors are employed commercially (see Fig. 8-7 for a picture of one such reactor). The latter type, having inner walls of polished 304 stainless steel, provides significantly higher heat-transfer coefficients than glass-lined reactors. The larger reactors are generally of the stainless-steel type and often contain cooled baffles.

Mechanical agitators, frequently turbine type, and baffles provide the desired agitation, and the organic reaction mixture, initially primarily styrene, is dispersed as droplets of about 0.1 to 1.0 mm throughout the water phase. Polymerization occurs in these droplets within several hours' time, and the water maintains excellent temperature control in the droplets.

The feedstocks and process conditions employed depend on the type of polystyrene required. A water-to-styrene ratio varying from about 1:1 to 3:1 is currently common (28). Grim (50) in a 1955 patent, however, has reported examples in which even lower ratios, such as 1:3, were used in relatively small reactors. Smaller ratios tend to give higher production capacities for a given reactor, but temperature control by heat transfer with the jacket becomes more difficult because of the greater amounts of polymer formed. At lower ratios, polymer sticking or deposition on the walls of the reactors increases. Such depositions are to some extent affected by the wall temperatures of the reactors.

Fig. 8-7. Electropolished interior of 15,000-gal stainless-steel-clad autoclave used for suspension polymerization of styrene. (*Brighton Corp.*)

Initiators soluble in styrene are commonly used in suspension processes, since the reaction temperatures are generally too low for adequate thermal initiation of polymerization. Now more than one initiator is frequently used, and a programmed increase of temperature is provided for the batch run. Selecting the initiators so that free radicals are produced over a relatively wide range of temperatures is recommended for effective operation of a reactor (45).

The exact recipes used for suspension processes are generally considered proprietary. Suspending agents and protective colloids are always used to stabilize the dispersed organic droplets in the water. Several inorganic salts, including calcium, barium, and magnesium phosphates, are often employed in conjunction with a small amount of an organic surfactant such as dodecyl benzene sulfonate. Grim (50) has presented considerable information for polymerizations with benzoyl peroxide as the initiator and with calcium phosphates. He reported several runs in which the first 6 to 7 h of the run were at 90°C. Some calcium carbonate was then added as a buffer, and the remaining 3 to 5 h of the run were at 110 to 115°C. Other recipes often use about 1 to 5 percent zinc oxide (based on the styrene monomer) or magnesium silicate (37, 28). These inorganic salts are relatively insoluble in the water, and they accumulate at the interface between the two liquid phases. Organic polymers that are relatively water-soluble are also used in many recipes. Polymers that also accumulate at the interfaces include polyvinyl alcohol, methyl cellulose, and pectin.

Suspension processes for polymerization of styrene are similar to those described earlier for production of polyvinyl chloride, and flow sheets of the polystyrene processes have been published (28, 33, 43). Temperature control during styrene polymerization is considerable easier to achieve because of lower heats of polymerization than for polyvinyl chloride processes. As a result, large polymerization vessels are frequently used for polystyrene processes, causing the production cost per unit of polymer to be decreased.

An inert gas such as purified nitrogen is frequently used as a blanketing agent in the polymerizers in order to maintain a positive pressure at all times during the cycle and prevent air leaks. At temperatures less than about 93°C, the sum of the vapor pressures of water and styrene is less than atmospheric. Furthermore, the partial pressure of the styrene drops as polymerization progresses.

Cosden Petroleum Co. have described their method for adding the reactants to the polymerizer (33). Warm, deaerated, and highly purified water is first added. Agitation is then started, and the liquid styrene and any comonomers are next added. Sometimes the stabilizers, surfactants, and initiator are premixed and added with the styrene, and sometimes they are added separately. The reaction suspension is then heated to start polymerization. The use of warm water as a feedstock minimizes the time required for heating the mixture to reaction temperatures. The time to complete polymerization varies particularly with the amount and type of initiator(s) used and the temperature.

The critical stage of polymerization normally occurs at polymer concentrations between 30 and 70 percent. In this range, premature agglomeration of the soft, semisolid, and highly sticky droplets of polymer is always a matter of concern (37). Normally a relatively constant power input is required by the agitator motor, but sudden increases, as measured by a wattmeter, indicate that agglomeration has started. Additional suspension agent should be added immediately since further agglomeration may cause the motor to stall. In severe cases, a large amount of inhibitor, e.g., sulfur dissolved in styrene, is added to stop polymerization. Failure to follow these steps may result in excessive agglomeration of the organic reactants, followed by a runaway reaction. Pressure-relief valves and exhaust lines are provided for such emergency situations.

Insufficient agitation of the suspension causes poorly formed beads or droplets (37), and too much agitation may result in deformed beads or entrapped gas (from the gas phase) in the beads. A momentary failure of the agitator often results in irreversible agglomeration.

With two or more initiators and with increased temperatures during the run, more uniform concentrations of free radicals and more uniform rates of polymerization are frequently possible. Reactor design and operation are based in part on the maximum rate of heat removal. Thus more uniform rates of polymerization permit higher production capacities of a reactor. As polymerization progresses, and as the styrene concentration decreases, a sharp drop in the rate of polymerization eventually is reached. Residual-styrene concentrations at the end of a run are frequently as low as 0.1 percent. Even with carefully controlled polymerizations, the ratios of the maximum to the average rates of polymerization vary from perhaps 2 to 3.5 or even 4. The range of this ratio has decreased as improved operating techniques have been developed.

Water is normally used as a coolant during the exothermic portions of the batch run. The relatively cool water is pumped through the jacket at a rapid rate to promote high rates of heat transfer. Part of the exit water from the jacket is recycled with a pump, and the remainder is discarded. Either normal cooling water, refrigerated water, or occasionally a refrigerated brine solution is used as makeup to the jacket. The makeup liquid must of course be at a significantly lower temperature than the liquid in the jacket if reasonably low water usage is to be realized. During the start-up or the final stages of the batch run, either hot water or steam is used in the jacket. Hot water is prepared by adding steam directly to the water stream being recirculated.

Calculations are shown in Table 8-2 for a polymerization in a 5,000-gal glass-lined reactor. The operating conditions and variables shown are quite typical, and the water provided in the jacket has an average temperature of 107°F (42°C). Normal cooling water is often adequate for the above run. When the ratio of the maximum to the average rates of polymerization is reduced to less than 2.5 (as shown in Table 8-2), the average temperature of the coolant is higher than 107°F. With larger reactors that have lower heat-transfer surface-to-volume ratios, lower temperatures are required for the coolant. With water used in the jacket, glass-lined reactors are generally limited to 10,000 gal and less, but larger stainless-steel reactors can be used.

Upon completion of the polymerization, the polymer-water slurry in the polymerizer is generally cooled to 185°F (85°C) or even less. Agglomeration of the polymer particles is then minimal as they are transferred to the hold tank.

Cosden Petroleum Corp. indicated that in 1958 they were using four 2,600-gal reactors for the production of about 20 million lb/year of polystyrene (33). This production rate is equivalent to 0.23 lb/h per gallon of reactor volume. Such a value is probably quite typical for commer-

TABLE 8-2 Coolant-Water Temperature for Suspension-Polymerization Run Using Styrene and Glass-lined Reactor

Nominal reactor size, gal	5,000
Available heat-transfer surface, ft^2	350
Working volume (90% full), gal	4,500
Volume ratio of water to styrene	1.1:1
Volume of styrene in charge, gal	2,150
Styrene (7.5 lb/gal), lb	16,000
Conversion fraction during run	1.0
Total reaction time, h	6
Total heat of polymerization during run (300 Btu/lb), million Btu	4.8
Average rate of heat release, Btu/h	800,000
Reaction temperature	203°F (95°C)
Ratio of maximum to average rate of polymerization	2.5
Maximum rate of heat release, million Btu/h	2
Maximum temperature difference for reactor jacket, if $U = 60$ Btu/(h)(ft^2)(°F) $$\Delta T = \frac{2,000,000}{(350)(60)} = 96°F$$	96°F (53°C)
Coolant temperature required in jacket, $203 - 96 = 107°F$ (43°C)	107°F (43°C)
Temperature of water feed stream	$<107°F$ (43°C)

cial reactors operating with a total time for a batch run of 14 h. About 5 h of that time is required on the average in each run for loading (or charging) reactants, evacuation of air, heating, cooling, discharging, scheduling, and cleaning (38). When the actual time for polymerization is reduced to about 5 h, the total time for a run approximates 10 h. In such a case, the production rate increases to about 0.32 lb/(h)(gal). Although the time required for polymerization is the single largest one in the overall batch cycle, process improvements to allow shorter times for heating or cleaning, for example, would be highly important in reducing the overall time cycle and hence in increasing the overall production capacity of a reactor.

From a technical standpoint, continuous-flow processes involving suspension polymerization techniques seem feasible. Stark (66), for example, describes a system using several stirred autoclaves in series. Cocurrent flow of an inert gas and of the liquid–liquid suspension in the connecting lines between the autoclaves is provided to minimize plugging and separation of the liquid phases in the transfer lines. The economic advantages of such a flow process are apparently minimal, compared with batch systems, especially when several relatively different polystyrenes must be manufactured in a single reactor train.

Polystyrene Recovery Operations for Suspension Processes The polystyrene-water slurry from a polymerizer is normally pumped to a hold tank provided with an agitator in order to maintain dispersion of the polymer particles in the water. Cosden Petroleum Co. (33) used one 2,600-gal hold tank for every two suspension polymerizers (also 2,600 gal each). Their hold tanks were glass-lined and jacketed for temperature control. The function of hold tanks is at least threefold:

1. The polymer-water slurry is cooled below the heat-distortion temperature of the polymer, generally to 122 to 140°F (50 to 60°C).
2. Chemicals such as HCl are added to promote solubilization of the suspension agents in the water.
3. The tank serves as a storage tank until the slurry can be centrifuged.

Both batch and continuous-flow centrifuges have been used. Three-step centrifugation was used by Cosden Petroleum Co. (33). The slurry was first centrifuged in a continuous solid-bowl centrifuge to separate the water and solid phases. The solids were then washed with water in the second step. In the last step, the wash water was separated from the solids and discarded. The solid polymer particles normally retain 1 to 5 percent water.

The wet polymer particles should be dried to less than 0.5 percent water, thus avoiding problems when the polymer is later extruded or pelletized. Cosden Petroleum Co. used a hot-air rotary dryer that could handle up to 4,500 lb/h of wet polymer, but it was normally operated at lower rates. Their single dryer was employed for the products obtained from their four polymerization vessels, and so careful scheduling of the polymerization runs was required to prevent mixing the different products from these runs. The temperature of the air used for drying was regulated so that the temperature of the solid polymer particles was 185°F (85°C) or less. Higher temperatures cause polymer degradation and sticking or agglomeration of particles. In other plants, counter-current steam-tube dryers and rotary-vacuum dryers have been employed.

After drying, the polystyrene particles are pneumatically transferred to hold or storage tanks for blending and testing purposes. From here, they are generally transferred to an extruder, where they are pelletized. The pellets are next stored and prepared for shipment.

The design and operation of the hold tanks, centrifuges, dryers, pelletizers, and storage and solid-transfer facilities for the polystyrene products are similar to those for polyvinyl chlorides produced in a suspension process. The Cosden polystyrene equipment is in a building maintained under a slight positive pressure of air in order to minimize dust. The air to the building is water-scrubbed and filtered.

Suspension processes can be adapted for the production of expandable polystyrene beads used to make foamed-polystyrene materials. An organic solvent (often pentane) is then added to the reaction mixture, where it accumulates in the organic phase. Following polymerization, the resulting beads containing adsorbed solvent are filtered or centrifuged, water-washed, and dried to produce beads containing 5 to 7 percent pentane. Eight additional suspension reactors were recently used by BASF in their plant at Jamesbury, New Jersey, to increase the plant capacity by 55 million lb of expandable product per year.

For production of the expandable polystyrene, a high-molecular-weight crystal-grade polystyrene is required to produce the proper structure when the beads are expanded. The solvent used always diffuses to some extent out of the organic phase whether in the manufacturing process (and particularly drying) or during storage. Since pentane forms explosive mixtures, precautionary measures must be taken whenever it is used. Other solvents have been used, including fluorocarbons, or Freons, which are nontoxic and nonflammable. Expandable polystyrenes can also be produced in solvent-type bulk processes. In such a case, the solvent, often pentane, is added to the styrene feedstock, and the solvent remains dissolved in the final product. Commercial development of these latter processes is probably relatively small compared with the suspension type.

Rather wide variations in the physical properties can be realized for the general-purpose polystyrenes produced by suspension- or by bulk-polymerization processes by varying the operating conditions employed. Suspension, continuous-flow bulk, and some batch bulk processes have been perfected so that essentially identical polymers can be produced by each method, and curves of the molecular weight distribution can be essentially duplicated. Products obtained with the plate-and-frame type of reactors often have rather wide range of molecular weights (69); data furnished by Jones (55) give one such comparison:

Manufacturing method	Number-average molecular weight M_n	Weight-average molecular weight M_w	M_w/M_n
Suspension................	103,000	299,000	2.9
Continuous bulk...........	103,000	274,000	2.7
Batch bulk (unagitated)....	71,000	340,000	4.8

Emulsion Polymerization Processes

Most ABS copolymers, other copolymers, and several specialty homopolymers of styrene are produced by emulsion processes. The relative importance of emulsion processes has decreased significantly for homopolymers of styrene in the last 30 years, and most homopolymers (crystal or general-purpose polystyrene) are now produced more cheaply by bulk or suspension processes. Several homopolymers of especially high molecular weight and hence high softening temperatures are still made by emulsion processes. In addition, latices obtained by emulsion polymerization find more or less direct use in some surface coatings.

The manufacturing techniques used for ABS copolymers, other copolymers, and impact polystyrene will be discussed later. The emulsion processes for conventional polystyrene and closely related polymers will now be described next.

Process Details for Emulsion Polymerization   The polymerization vessels used for suspension processes can generally also be used for emulsion processes, but the preferred type of impeller and baffles may be different for these two types of polymerization (48). Almost all reactors are glass-lined since scaling or polymer-coating problems are frequently severe during emulsion polymerization. Polymerizers as large as 10,000 gal are suitable and use water as the coolant.

The flowsheet and some operational details of the unit employed several years ago by BASF have been published (26, 37). The feedstocks for a specific polymer were as follows: 3,100 kg (6,800 lb) of deaerated water; 130 kg (286 lb) of 14% soap solution; 3 kg (6.6 lb) of sodium pyrophosphate; 2.8 kg (6.2 lb) of potassium persulfate (used as an initiator); and 1,500 kg (3,300 lb) of styrene. This mixture was introduced with agitation into the polymerization vessel, which had an apparent volume of approximately 1,500 gal. The pH of the mixture was adjusted to 10 to 11 by adding sodium hydroxide. Air was flushed from the reactor using nitrogen, and the reactor was heated with hot water or steam (in the jacket of the reactor) to almost 70°C. After a short induction period, the polymerization started as the temperature rose to 75°C, at which time cooling water was provided to the jacket. The temperature continued to rise, however, to 105°C within about 0.5 to 1.0 h causing a reactor pressure of approximately 2 atm. The temperature then slowly decreased to 90°C. Nitrogen was then used to flush out any styrene vapors left, the system was partially cooled, and the emulsion was discharged to the hold tank. An emulsion containing 30 percent solids was obtained in this run.

The BASF reaction vessels were sometimes provided with reflux condensers as an aid to temperature control (45). The preferred initiator, potassium persulfate, was added as a water solution that was freshly prepared to minimize decomposition during standing. In many current cases, the initiator solution is added during most of the run or during the initial stages, perhaps the first 1 to 2 h of the run. On other occasions, it is all added at the start of the run. Frequently the emulsifiers (or surfactants) are all added at the start of the run. Adding styrene, emulsifiers, and initiators during the run, although complicating the process, is often effective in improving the quality of the polymer and in reducing the amounts of surfactants required.

An azeotropic mixture of water and styrene refluxes at about 93.5°C (37). When the reaction is completed, the reflux temperature rises to 100°C. This technique is frequently used to signal the completion of the polymerization. Unreacted styrene is recovered at the end of a run by steam stripping, and the styrene is condensed and recovered using the overhead condenser of the reactor.

Many questions concerning the mechanism and role of the micelles still exist for all emulsion polymerizations. The level of solubility of the monomer in the water phase is important since the monomer transfers from dispersed droplets of monomer through the water phase to the micelles. Styrene is only slightly soluble in water, whereas vinyl chloride and (especially) vinyl acetate are much more soluble. The importance of the steps occurring during emulsion polymerization may then be quite different for these three monomers. Disagreements have been noted in the literature over the location of initiation. Claims have recently been made that initiation occurs in the water phase (62); others say it occurs in the micelles (51, 64); and still others (26) claim dual-phase polymerization.

The products obtained using continuous-flow emulsion processes that employ either one or three (in series) stirred reactors have been com pared to the products from batch runs (46, 59). Broader ranges of particle sizes are obtained in the flow system. Information concerning the rates of polymerization and the molecular-weight characteristics of the polymer have also been reported. Such information will be helpful if commercial flow processes are to be developed.

Recovery of Polystyrene for Emulsion Process The BASF method used in the 1950s for polymer recovery involved discharging the emulsion to an agitated hold tank, where the emulsion was cooled to perhaps 30°C. The emulsion then flowed to the first of four agitated vessels connected in series. In the first vessel, coagulation occurred as a 0.5% solution of formic acid was added. In the second vessel, the slurry was heated to 60 to 90°C, causing partial agglomeration of the polystyrene solids. In the third vessel, ammonium hydroxide was added, and the mixture was heated. Finally in the last vessel, the slurry was cooled to approximately 30°C.

BASF transferred the slurry with a centrifugal pump to a vacuum filter. The resulting cake was washed with water, partially dried on a drum dryer, and then dried in a tray dryer. With a larger operation and with more modern equipment, centrifuges and a rotating tubular dryer would probably be used now.

Fluidized-bed dryers have also been used for recovery and drying of the solids from emulsions. Uniroyal, Inc. currently uses large flash dryers for drying their wet ABS solids.

Cost of recovery of the dry polystyrene is obviously relatively high for emulsion processes. Furthermore, removal of all the surfactants is not readily possible except by expensive operations. As a result, the color and to some extent the stability of the resulting polymers tend to be relatively poor.

Choice of Processes

Choice of the best process for production of general-purpose polystyrenes involves consideration of many factors but especially the cost of operation and the quality of the product. Emulsion processes in general cannot compete with bulk or suspension processes on either of the above two bases. High-molecular-weight polystyrenes however, can, be produced by emulsion processes rather easily.

Suspension processes and bulk (or modified bulk) flow processes can be operated to produce polymers of similar quality. Economic considerations often favor one process for specific types of products. Suspension processes, compared with bulk-flow processes, have relatively simple and cheap reactors, excellent versatility for the production of numerous types of products, and relatively standard operation. Disadvantages of suspension processes include more difficulty in obtaining high-purity products since water, stabilizers, and unused initiators tend to be contaminants. The costs of stabilizers and initiators are also significant. Labor costs are higher for the suspension processes.

For a 50 million lb/year plant, total production costs, excepting only the cost of the styrene, are estimated for general-purpose polystyrenes to be 1.75 to 2.12 cents/lb for suspension processes, compared with 2.14 cents/lb for processes using plate-and-frame reactors (40). The rather wide range of costs for suspension processes depends mainly on the size of reactors used, number of products to be manufactured, and degree of flexibility to be incorporated in the plant. For small plant capacities, up to perhaps 50 million lb/year, operating costs per given weight of polystyrene produced are often lower for suspension processes than continuous-flow bulk processes. The reverse is generally true with larger plants, however.

The suggestion has been made that molten polystyrene from a reactor or devolatilizer be immediately molded or extruded to produce the desired end-use items (39). In essence, a single line or process would be used to polymerize styrene and to do the desired molding or extrusion. Development and use of such a combined operation are potentially interesting since it eliminates several intermediate steps that are rather costly, including pelletizing, packaging, storage, shipping, etc. Success of such an operation will depend on the ability to coordinate the polymerization and molding operations. The reactor (probably using bulk-polymerization techniques) will often need to be small but highly flexible in operation. The operators for such a unit will need to understand the factors that affect the quality and physical properties of polystyrene.

The suggestion has been made of using a combination of cationic and suspension polymerizations (44). A low-molecular-weight polystyrene is produced by cationic polymerization; it is then dissolved in styrene, and the mixture is polymerized in a suspension process to produce a polystyrene that has improved molding characteristics. This or similar techniques of blending polymers appear to be practical for production of at least certain speciality polystyrenes.

Although currently not of commercial importance, at least two unique types of polystyrene may find applications. First, isotactic or stereospecific polystyrenes have unique properties, and the polymerization processes involving coordination catalysts are of at least latent commercial interest (28). Second, for the preparation of polystyrenes with very narrow molecular-weight distributions, certain anionic polymerizations can be used (32).

Some research is currently in progress to develop new processes. A unique process developed by BASF employs a falling-drop principle in the polymerization vessel (61). The liquid styrene containing dissolved sodium naphthenate and tetrahydrofuran is sprayed downward into a large towerlike reactor. As the droplets settle, polymerization occurs and part of the styrene vaporizes. Good temperature control is maintained because of the partial vaporization of the styrene, and this process may find commercial application.

Polymerization technology and processes are available for sale or licensing from at least nine companies (60).

III. COMMERCIAL PROCESSES FOR STYRENE COPOLYMERS

Copolymers for which styrene is the major comonomer are produced in large quantities since their physical properties are better for numerous applications than those of the homopolymers of styrene (6, 9, 22, 28, 103). Both random and graft copolymers are of importance, and several olefins are used as comonomers with the styrene. Considering first the random copolymers, acrylonitrile is employed to produce copolymers with improved heat-distortion temperatures, improved strengths, and improved resistances to hydrocarbons. These copolymers are also transparent, have better craze resistance, and are relatively tough. Copolymers (or terpolymers) with α-methylstyrene have higher deformation temperatures because of stiffening of the polymeric chain. Glasslike copolymers with good light stability are prepared using methyl methacrylate as a comonomer. Maleic anhydride is used to produce copolymers having acid groups, and the resulting copolymers are employed in epoxy and coating resins. Latex paints sometimes contain copolymers prepared from styrene and butadiene.

Impact polystyrenes and ABS plastics are styrene-type copolymers of major importance, some graft copolymerization often occurs, and each material contains two solid phases. An elastomeric or rubber phase is dispersed throughout a solid matrix consisting primarily of polystyrene or a styrene copolymer (see Fig. 8-8 for micrographs). Factors affecting the physical properties of a two-phase plastic are considerably more complex than those for a single-phase one (101). Variables of importance for the continuous phase include the average molecular weight, molecular-weight distribution, and the composition if it is a copolymer. Variables of importance for the elastomeric phase are the number, size, distribution of sizes, and shapes of the dispersed particles. In addition, the composition (since it is often a copolymer), degree of

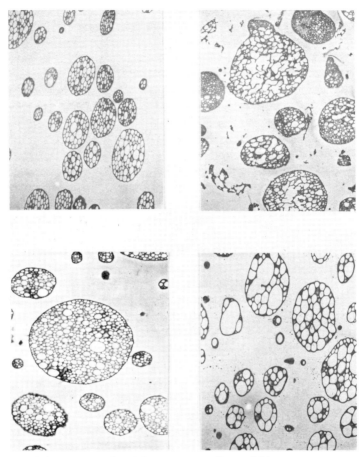

Fig. 8-8 Micrographs of four commercially available impact styrenes. (*The Dow Chemical Co.*)

grafting on the elastomeric molecules, and amount of cross-linking all have a significant effect on the properties.

For random-type copolymers, the two or more comonomers react, forming a chain molecule in which the repeating groups are distributed randomly along the chain. Several features of the molecular structure affect the physical properties. These include the average molecular weight, molecular-weight distribution, composition (or ratio) of repeating units in chain, range of compositions (since the compositions may change as polymerization progresses), and possible branching and cross-linking of polymer chains. Manufacturing techniques employed to produce these many copolymers are unusual in several cases, especially for the two-phase polymers.

Production of Impact Polystyrenes

Impact polystyrenes, or, as they are sometimes called, rubber-modified polystyrenes, are produced by several techniques, but the main ones are outlined in Fig. 8-9. Mass (or bulk) polymerization is always employed for the prepolymerization or the initial stages of polymerization of the processes shown in Fig. 8-9. The final stages of polymerization may be by either bulk (or modified-bulk) or suspension methods. In these processes, some styrene is graft-polymerized on the rubber, and less rubber is needed for the desired impact properties compared with other processes (not shown in Fig. 8-9), in which the rubber and polystyrene are mechanically blended. The impact polystyrenes always consist of two phases (see Fig. 8-8), a polystyrene phase (preferably the continuous one) and a rubber phase, which is dispersed. The properties

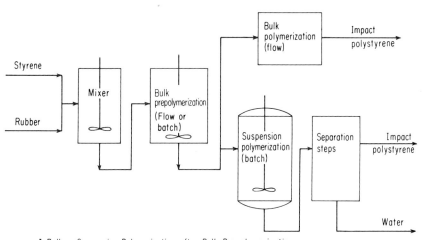

I. Bulk or Suspension Polymerization after Bulk Prepolymerization

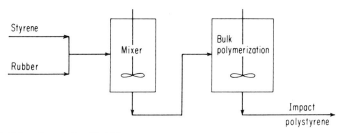

II. Bulk polymerization (Batch)

Fig. 8-9   Process outlines for impact polystyrenes.

of the final polymer vary significantly depending on the geometrical configuration of the two phases (89).

Impact polystyrenes are sometimes characterized as medium-impact, high-impact, or even extra-high-impact polystyrenes, depending on the impact strength of the polymer. In general, a higher strength is obtained when more rubber is incorporated in the polystyrene. High-impact polystyrenes generally contain about 5 to 10 percent rubber, whereas extra-high-impact ones contain up to 25 percent (22). The method employed for incorporating the rubber in the product is also important.

Impact Polystyrenes Containing Graft Copolymers   A solution of rubber dissolved in styrene is the feedstock used for production of impact polystyrenes containing graft copolymers, and it is prepared in a mixing tank (see process outline I or II of Fig. 8-9). Rubber crumbs are added to the styrene, and a solvent such as ethylbenzene is also sometimes added in relatively small amounts. The mixture is then agitated at room temperature for several hours until most if not all of the rubber is dissolved; and finally the solution, which looks like mineral oil and has the same viscosity, is filtered to remove any undissolved rubber (92). The rate and degree of dissolution of course depend on the amount and type of rubber used and the condition and size of the rubber crumbs.

Polymerization of Solutions of Rubber Dissolved in Styrene: The phenomena occurring during polymerization of styrene-rubber solutions are complicated since two phases form after about 6 to 10 percent of the styrene has polymerized (92). A polystyrene phase (consisting of polystyrene and dissolved styrene) separates as small droplets and is dispersed throughout the rubber phase (initially a liquid phase of rubber dissolved in styrene), as shown in the upper left micrograph of Fig. 8-10. As polymerization continues, styrene reacts by one of the following methods:

1. It polymerizes to form polystyrene, causing the polystyrene phase to grow in volume. This method is normally the predominant one.

2. It copolymerizes to form grafts or branches on the rubber molecules. The resulting graft copolymers presumably accumulate mainly at the interface between the two phases, and they act to some extent as emulsion or dispersion stabilizers (28).

As polymerization occurs, some of the styrene transfers from the rubber phase to the polystyrene phase, which is growing in size. Eventually all or at least most of the styrene polymerizes to form polystyrene or graft copolymers.

FORMATION OF RUBBER PARTICLES BY
PHASE INVERSION OF A POLYMERIC OIL-IN-OIL EMULSION

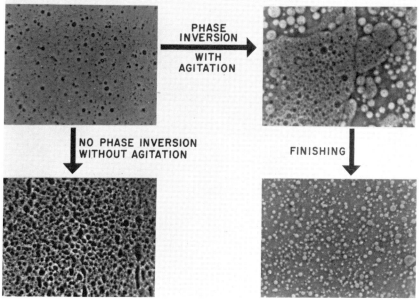

Fig. 8-10 Micrographs indicating two-phase behavior as a styrene-rubber mixture is polymerized, with and without agitation. (*The Dow Chemical Co.*)

The polystyrene phase increases in weight (and volume) rather rapidly, especially during the initial stages of polymerization. The rubber molecules also increase somewhat in weight because of graft polymerization. When a feedstock containing 5 percent rubber and the remainder styrene is used, the ratio of the rubber phase to the polystyrene phase is about 4 for a styrene conversion of 5 percent (92). This ratio decreases with further polymerization and is essentially 0.5 for styrene conversions of 15 to 30 percent.

A key phenomenon that has a most important effect on the properties of final impact polystyrenes can be made to occur after about 7 to 12 percent of the styrene has polymerized. Phase inversion occurs with sufficient agitation, as shown in the upper right micrograph of Fig. 8-10, so that the rubber-phase particles are dispersed throughout the polystyrene phase (28, 92). The minimum level of agitation (or shear rate) required for such an inversion increases as the rubber concentration in the reaction mixture increases (81). The type of agitator used is important since shear rates tend to be quite nonuniform throughout the reaction mixture. After inversion and with continued polymerization, the micrograph of the final impact polystyrene is like that in the lower right corner of Fig. 8-10.

The level of polymerization for phase inversion varies somewhat with the amount and type of rubber used in the feedstock and with the operating conditions. When the phases invert, the apparent viscosity of the emulsion suddenly decreases since the viscosity of the polystyrene phase is considerably less than that of the rubber phase.

With little or no agitation, the rate of phase inversion is negligibly slow because of the high viscosity of the rubber phase. In such a case, the rubber phase then remains primarily as the continuous phase with a spongy network (see Fig. 8-10). The desired physical properties of the final polymer are realized for most applications only if phase inversion has occurred. Some multiple inversions also form, especially if only moderate agitation is provided; in this case, small particles of polystyrene remain occluded in the dispersed rubber phase.

Immediately after phase inversion, the dispersed droplets of rubber plus dissolved styrene have a relatively large size. With continued polymerization, some dissolved styrene polymerizes to form polystyrene that remains occluded in the rubber phase, but most of it transfers to the polystyrene phase. The dispersed rubber droplets then decrease in size with continued polymerization, and they are gradually transformed into a solid rubber phase. Some cross-linking occurs in the rubber, especially when polymerization is almost completed.

Chain-transfer agents, sometimes called polymerization regulators, are generally added to the reaction mixture in order to obtain a polystyrene that has a lower molecular weight. The dimer of α-methylstyrene is often used for this purpose. This dimer, when added in amounts less than 1 weight percent, significantly lowers the viscosity of the reaction mixture especially after phase inversion occurs (28). Phase inversion occurs in addition at slightly lower styrene conversions when the dimer is added. For a styrene solution containing 7 percent polybutadiene, phase inversion was experienced at a conversion of 10 percent (and a viscosity of about 50 P) when no dimer was used. When, however, 0.6 weight percent dimer was used, phase inversion occurred at about 9 percent conversion; the viscosity decreased from about 30 to 15 P because of inversion.

Chain-transfer agents are distributed between the two phases formed (28). The relative amounts in each phase vary as polymerization progresses and as the relative volumes and composition of the phases change. Changes in the concentrations of the transfer agents in each phase affect the levels of chain transfer in that phase. Ethylbenzene and other solvents sometimes added to the reaction mixture also act to at least some extent as chain-transfer agents.

Grafting of the polystyrene branches on the rubber polymers probably occurs by at least two methods (28, 80). First, styrene copolymerizes to some extent with the unsaturated groups of the rubber molecule. Such copolymerization leads to some cross-linking and three-dimensional polymers in the rubber phase. Cross-linking is more common at higher temperatures, frequently employed during the later stages of bulk polymerization. Second, grafting can occur after a chain-transfer step with the rubber in which the free radical is transferred to a carbon atom alpha to an unsaturated group; the carbon-hydrogen bonds of the methylene groups in the alpha positions are relatively weak, allowing easy chain transfer. The length of polystyrene-type graft depends on the concentrations of the styrene and chain-transfer agents in the rubber phase.

The dissolved rubber in styrene has more effect than just dilution relative to the rate of the styrene polymerization (28). The rubber often decreases the overall rate, presumably because of stabilizers or impurities in the rubber or because of degradative chain-transfer steps with the rubber. Hence the type of additives used in the feedstock recipes may have a significant effect.

Bulk-Polymerization Processes for Impact Polystyrenes: Both continuous-flow and batch processes involving bulk-polymerization techniques have been used for commercial production of impact polystyrenes, as outlined in Fig. 8-9. These processes are similar to those used for general-purpose (or crystal) polystyrenes.

The continuous-flow process employed by The Dow Chemical Co. (42, 54, 63, 70) uses a combination of three reactors, arranged in series (Fig. 8-3b). Small amounts of solvent are added, and the process is sometimes referred to as a solvent process. It is used by several American companies and at least one British one.

The feedstock reported for one Dow unit was 2,200 lb/h of a solution containing 5.5 percent polybutadiene and the remainder styrene. In addition, 220 lb/h of ethylbenzene (as a solvent to lower viscosity) plus small amounts of white mineral oil (as a lubricant and plasticizer), 2,6-di(*tert*-butyl)-4-methylphenol (as an antioxidant), and the dimer of α-methylstyrene (as a polymerization regulator) were continuously introduced to the reactors. With their three-tower arrangement, polymerization was completed within about 14 to 15 h residence time as the temperature of the reaction mixture increased from about 112 to 165°C in the reactors (63). At these temperatures, polymerization is initiated by thermal reactions of the styrene, and a free-radical initiator is not needed.

The product mixture from the last tower reactor passed through a tubular heat exchanger. Here the mixture was heated to 220 to 285°C for about 10 to 30 min, resulting in some cross-linking in the rubber phase but not enough to render the polymer thermosetting (or nonthermoplastic). The rubber is in a sense vulcanized by this technique. In an example reported, the temperature was raised to 235°C. Following heating, the polymeric material was devolatilized to remove ethylbenzene, unreacted styrene, and other volatile compounds. The rubber-modified polystyrene produced by this process has the rubber dispersed in particles ranging in size from about 2 to 25 μm (63).

A Czechoslovakian process also employs three polymerization vessels arranged in series (73) and is somewhat similar to the Dow process. Feedstocks for this process contain 6.5 to 8.5 percent rubber dissolved in styrene. Small amounts of required additives are also introduced continuously into the feedstream. Utilities required per metric ton of product are 2,000 kWh of electricity, 1,750 lb of steam at 6.5 atm pressure, 13,250 gal of cooling water, 10,000 gal of pure water, nitrogen, and compressed air. Costs of these utilities probably total about 1 cent per pound of product.

Batch reactors were used several years ago for the production of most impact polystyrenes, and they still are used to a limited extent. A cylindrical reactor that proved to be highly successful is provided with several internal pancake coils so that good temperature control is maintained during the entire batch run. These coils are positioned with their centers at the axis of the reactor, and they are arranged parallel to each other. Heavy-duty agitators are provided in the spaces between coils, and a batch run is normally completed in 12 to 18 h. In general, batch processes can no longer compete economically with the modern flow processes, especially since the amounts of product are now so large.

Suspension-Polymerization Processes for Impact Polystyrenes: Large amounts of impact polystyrenes are produced using a bulk-polymerization technique for the prepolymerization stages of the reaction and a suspension-polymerization technique for the remainder of the reaction (Fig. 8-9).

Patents (98, 99) assigned to Monsanto Company give examples in which styrene-rubber mixtures containing 6 weight percent rubber are used as feedstocks. Small amounts of an initiator such as di(*tert*-butyl) peroxide, an antioxidant, a C_{12} mercaptan used as molecular-weight modifier, and a refined hydrocarbon oil used as a lubricant and plasticizer are also added to the feedstock. In this process, a solvent is not normally added as in the Dow process. The prepolymerization is accomplished in stirred autoclaves probably in most cases batchwise. Sufficient conversion (probably in 10 to 18 percent range) of styrene occurs and adequate agitation is provided so that phase inversion occurs; i.e., the polystyrene phase becomes the continuous one. A semicontinuous process has also been patented (71) in which only part of the prepolymerized mixture is removed from the autoclave for each suspension run. Because of the initiator used, prepolymerization is completed in several hours.

Following prepolymerization, the prepolymerized solution is added to a suspension-polymerization reactor, where a water-continuous suspension is formed with the prepolymerized solution dispersed throughout it. A recipe reported contains 100 parts (by weight) of prepolymerized solution, 200 parts of water, 0.25 part of surfactant, 0.15 part of calcium chloride granules that dissolve in the water, and 0.25 part of a copolymer of acrylic acids and other acrylate comonomers (98). The copolymer lowers the pH to the preferred range of 3 to 4.

As little as 100 parts of water per 100 parts of prepolymerized solution has been used in other recipes (28). The water is often heated before being added to the reactor, and the method of adding the reactants is important. The prepolymerized mixture, initiator, and water are added to the reactor, and agitation is started with temperatures up to at least 110°C. Then the dispersing agent and the copolymer of acrylic acids and acrylate comonomers are added next. Polymerization then continues. In an example presented, the reaction mixture is first maintained at 130°C for 3 h and then at 140°C for 5 h. When the polymerization mixture is centrifuged, the solid polymer particles contain only 7 percent water.

The suspension recipes used commercially are generally highly proprietary since recipe changes often result in small but significant changes in the polymer properties or the cost of polymerization. Polyvinyl alcohol has also been used as a stabilizer (100), and various types and levels of initiators are used to affect the rates of polymerization. The temperature of course also affects the rates of initiator decomposition and hence the overall rates of polymerization. At the temperatures normally employed, thermal initiation is also a factor. Temperatures from about 90 to 140°C have been reported with times required for the batch polymerization varying from 2 to 8 h.

When impact polystyrenes containing relatively high amounts of rubber, perhaps 10 percent and greater, are to be produced, suspension-polymerization methods are probably preferred. Reaction mixtures for such products do not handle easily in the bulk-flow process since they tend to have higher viscosities unless larger amounts of solvent are used.

Impact Polystyrenes Using Mechanical Blending Two mechanical-blending techniques are used to a limited extent for the production of impact polystyrenes. In the first, polystyrene and rubber emulsions, each prepared by emulsion polymerization, are mixed for at least several minutes (96). The mixed polymers are then precipitated, sometimes using hot ethanol, and the solid polymers are recovered by filtration or centrifuging. The solids are then washed, dried, and pelletized or molded. In the second method, polystyrene and a rubber are mechanically mixed using a Banbury mill, blending rolls, or extrusion equipment.

When the rubber and polystyrene are mechanically blended, higher rubber contents up to perhaps 20 to 30 percent are required in order to obtain the desired impact properties. Because of the larger amounts of rubber employed, surface characteristics of the polystyrene are only fair and the polystyrene has poorer aging characteristics. The dispersed rubber particles in this case have rather irregular and noncircular cross sections (22, 89). The particle sizes vary with the amount of thermomechanical working provided during mixing. Care must be taken during working to prevent overheating, which would degrade the polymers.

The cost of preparing mechanical blends of polystyrene and rubber are generally rather high. Furthermore the physical properties of these blends are only fair as a rule. As a result, it is not surprising that relatively small amounts of impact polystyrenes are currently prepared by mechanical blending.

Types of Elastomers for Impact Polystyrenes   Several types of rubbers have been used for production of impact polystyrene including styrene-butadiene random and block (93) copolymers, polybutadiene, nitrile-type rubbers (primarily copolymers of butadiene and acrylonitrile), and recently EPDM (ethylene-propylene-diene copolymers) rubbers (76). Properties of impact polystyrenes are affected by the characteristics and chemical structure of the rubber (63, 80), including its color, residual amounts of unsaturated groups, Mooney viscosity (related to the average molecular weight), molecular-weight distribution, and relative amounts of cis, trans, and vinyl unsaturated groups in the rubber. As discussed earlier, the rubber should preferably be readily soluble in the styrene. EPDM rubbers are often quite insoluble, but a high degree of unsaturation promotes solubility. Special techniques that have not been published are probably required when EPDM rubbers are used.

The moisture, ash, and organic acid contents of the rubber affect the color, clarity, and stability of the final polymer (80). Because of the improved quality required for the rubbers, considerable attention must be given to their manufacture. As a result, the costs of the rubbers used for rubber-modified polystyrene are higher than rubbers used for tires and most other purposes.

Costs for producing rubber-modified polystyrenes are higher than those for general-purpose polystyrene because the rubber itself is fairly expensive and the feedstock solution of rubber in styrene must be prepared. Operating costs for the impact polystyrenes are generally 1.5

to 3.0 cents/lb greater than the general-purpose grades. The additional cost depends to some extent on the amount and type of rubber used. Especially for plants with capacities of over 50 million lb/year, total operating costs for continuous-flow process are often 0.5 cent/lb cheaper than those of batch processes, including suspension processes.

Production of SAN Copolymers and ABS Plastics

SAN (or PSAN)* copolymers and ABS plastics are related since SAN copolymers are the matrix (or continuous phase) in ABS plastics. A rubber phase is the dispersed phase in ABS plastics. Certain portions of the manufacturing procedures are often similar for the two types of polymers.

SAN copolymers are the polymerization products of styrene and acrylonitrile in which the styrene content varies from about 63 to 90 percent. The range for obtaining the best physical properties is generally from 65 to 80 percent (103). Resistance of the SAN copolymers to solvents including motor oils, kerosene, gasoline, and greases is good in this range. Significant improvements are realized for the flexural, impact, and tensile strengths over those for general-purpose (crystal) polystyrenes. Maximum values are obtained for the flexural, impact, and tensile strengths when the styrene contents of the copolymers are about 70, 70, and 66 percent, respectively. The copolymers are transparent but have a slightly yellow cast that becomes darker as the acrylonitrile content increases.

SAN copolymers and ABS plastics account for slightly more than 7.5 percent of the styrene currently used in the United States (12). Production of these two plastics is growing rapidly and is considerably over 0.6 billion lb/year. Because of their excellent physical properties they find wide uses for piping and plumbing, automotive parts, furniture, appliances, eating utensils, telephone housings, and sporting gear, including football and motorcycle helmets.

Bulk, emulsion, and suspension processes are used commercially for the copolymerization of styrene and acrylonitrile (9).

Bulk Processes for SAN Copolymers   Bulk processes (or modified bulk processes sometimes called solvent processes since a small amount of solvent is added) are used for production of SAN copolymers. These processes are related to some extent to processes previously described for general-purpose polystyrenes. Bulk copolymerization of most mixtures of styrene and acrylonitrile has two complicating features, however.

1. The relative reactivities of the two comonomers vary except for mixtures that contain about 72 weight percent styrene (103). This com-

---

* Styrene-acrylonitrile (or polystyrene-acrylonitrile).

position depends to some extent on the temperature and other reaction conditions. In general for mixtures containing more styrene, the acrylonitrile is more reactive. The reverse is true for mixtures with less styrene. Hence as the level of polymerization increases for mixtures other than those containing 72 percent styrene, the compositions of the unreacted comonomer mixture and of the copolymers change.

2. SAN copolymers tend to have rather high softening temperatures and hence high viscosities compared with homopolymers of styrene. Sufficiently low viscosities are obtained by the use of fairly high temperatures, about 130 to 180°C; the addition to the reaction mixture of fairly large amounts of inert solvents such as ethylbenzene, chlorobenzene, benzene, toluene, or methyl ethyl ketone (88, 104); and/or only partial polymerization of the reaction mixture. In the latter case, the unreacted comonomers act as solvents for the copolymers. With copolymer contents in the reaction mixture up to 40 percent, the viscosities at reaction temperatures are generally quite low.

Ease of molding and the physical properties of the SAN copolymers are affected to a considerable extent by the polymerization temperatures, usually 130 to 180°C (103). At lower temperatures, the molecular weight of the copolymers is very high, often too high for easy molding. Furthermore, the rate of polymerization is low. At higher temperatures, the product is weak and brittle since it has a low molecular weight.

The effectiveness of obtaining chain-transfer steps with common chain-transfer agents varies significantly as the composition of the styrene-acrylonitrile mixture changes (28). When the mixture becomes richer in acrylonitrile, the molecular weight of the copolymer often decreases rapidly. As indicated earlier, the composition of the mixture often changes during batch polymerizations, resulting in a broad spectrum of molecular weights.

A continuous-flow process for the production of numerous copolymers including SAN copolymers has been reported (84, 85, 103), and a simplified flowsheet in which a tubular loop is used as the main reactor is shown in Fig. 8-11. The reaction mixture, containing perhaps 40 percent SAN, is recirculated by a pump many times around the reactor.

Two streams are continually withdrawn from the reactor. One is the product stream, which is sent to the devolatilizer (discussed later). The second stream is combined with the feed and recycle mixtures of comonomers, then passed through a heat exchanger, and eventually returned to the reactor. Key features in the process are to maintain minimum differences of the temperature and composition of the mixture in the reaction system that includes the reactor proper, heat exchanger, and connecting lines. Such restrictions are necessary in order to obtain copolymers with relatively uniform compositions and molecular weights since copolymerization occurs throughout the entire system. To obtain minimum temperature and composition differences, a recycle loop is provided around the heat exchanger. A major operating problem is to obtain rapid mixing of the viscous copolymer-rich solution with the low-viscosity mixtures of comonomers. For normal pumping and mixing, viscosities of 5,000 cP or less are desired for the reaction mixture. Process controls for this process have been described (84, 104).

Two other reactor modifications have been reported (85). In one modification, a heat exchanger was employed with provisions to recirculate the reaction mixture on the average of about 800 passes per hour using a recycle pump. In yet another modification, a stirred and jacketed autoclave was found satisfactory.

For several examples reported for this continuous-flow process (85), several hours were required to reach steady-state operation (including a constant-composition copolymer). The time to reach equilibrium depends on the average residence time of the reactants in the reaction portion of the system. Such residence times depend on the rates of polymerization, and with the styrene-acrylonitrile mixtures normally used, about 1 h is common (103). Increased amounts of acrylonitrile in the reaction mixture increase the rates of polymerization. Design of a suitable reactor requires careful considerations of the flow and recycle patterns and of adequate heat transfer.

The product mixture from the reactor is pumped to a devolatilizer to separate the substantial amounts of unreacted styrene and acrylonitrile from the SAN copolymer. The unreacted comonomers are condensed and recycled. Provisions are normally required to purify these comonomers and to remove by continuous distillation inert diluents that may

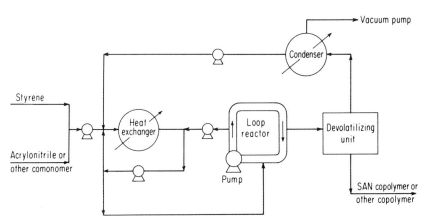

Fig. 8-11   Process for production of SAN or other styrene copolymers.

be introduced in the feedstock. Devolatilizers suitable for the SAN copolymers are similar to those previously described for general-purpose polystyrenes. Recovery and pelletizing the devolatilized copolymer involve conventional processing steps.

SAN copolymers produced by this continuous-flow process can be made with wide variations in composition and yet with excellent homogeneity (or constant compositions) for all polymeric molecules of a specific product. Examples were presented for copolymers with styrene contents varying from about 70 to 92 weight percent (85). Since these copolymers are relatively hygroscopic and are capable of adsorbing up to several percent moisture, special precautions are necessary to prevent moisture pickup during storage. Sometimes the pellets are stored in Saran-lined drums. If the pellets do become wet, they should be dried before fabrication since otherwise streaks will appear on the fabricated parts.

Emulsion and Suspension Processes for SAN Copolymers Emulsion and suspension processes, including both the equipment and operating procedures, for copolymerization of styrene and acrylonitrile are similar to the batch processes described earlier for production of general-purpose polystyrenes.

For suspension copolymerizations, a recipe of Mino (91) specifies the following weights of reactants: comonomers, 100; water, 200; 1-azo-2 bis-1-phenylethane (as initiator), 0.1; n-decylmercaptan, 0.1; and calcium hydroxyapatites, 0.2. For SAN copolymers with styrene compositions of commercial interest, about 6 to 7 h is required to complete the copolymerization. Grim (50) in example XVIII of his patent has also outlined the recipe used for production of an SAN copolymer containing 10 percent acrylonitrile.

A complicating feature of these two processes (each of which has water-continuous phases) is that acrylonitrile is soluble, often up to several percent, in the water. The styrene, however, is only very slightly soluble. The actual amount of acrylonitrile dissolved increases of course as the relative amount of water in the two-phase system is increased. It decreases, however, with smaller concentrations of acrylonitrile in the organic phase and hence at higher polymerization levels. As a result, the styrene-to-acrylonitrile ratio in the organic phase is always higher, and especially so at the start of a polymerization run, than the overall ratio of these two comonomers in the polymerization vessel. Since the ratio in the organic phase controls the reaction and hence the composition of the copolymers produced, it is obvious that the compositions and molecular weight of copolymers tend to vary as the batch polymerization progresses, and the copolymer molecules are generally more heterogeneous than comparable bulk-polymerization runs. Mino (91) has shown that the acrylonitrile content of copolymers produced by suspension technique may vary by at least several percent, depending on the conversion level. If a specific copolymer is too heterogeneous in character, the copolymer molecules may be sufficiently incompatible with each other to result in partial loss of light transmission, resulting in haziness (28).

Recovery of SAN copolymers that have been prepared by emulsion or suspension processes is similar to that for general-purpose polystyrenes produced by the same method.

Production of ABS Plastics   Several methods for production of ABS plastics (77, 94) are outlined in Fig. 8-12. In older and now less important methods, an SAN copolymer is mechanically blended with a rubber, often a nitrile rubber; generally water emulsions of these two copolymers are prepared and then blended. Newer ABS production procedures involve the polymerization of a mixture of rubber (often polybutadiene), styrene, and acrylonitrile using emulsion, bulk, and suspension-polymerization techniques. Part of the styrene and butadiene polymerizes to form grafted branches on the polybutadiene chain, and part reacts to form SAN copolymers. Additional SAN copolymer is sometimes mechanically blended to form the final ABS plastic.

Rubber Latices for ABS Plastics: Rubber latices used for production of the ABS plastics are normally prepared by relatively standard batch-operated emulsion-polymerization processes. Details of such processes are presented by Whitby (102), who has presented recipes for the preparation of nitrile rubbers that often contain up to 35 percent acrylonitrile and the remainder butadiene. The effect of operating conditions on the reaction and on the properties of the rubber was also discussed. Latices containing perhaps 30 to 35 percent rubber are common, and they are produced at about 5 to 50°C.

When butadiene polymerizes by a free-radical mechanism, as during emulsion polymerization, the butadiene can add to the growing polymer chain by trans-1,4, cis-1,4, or 1,2 additions. In the latter case a side vinyl branch is present on every other carbon atom of the chain. The relative fraction of these three addition methods has a considerable effect on the properties of the polybutadiene. Binder (71) has reported such information for a large number of polybutadienes polymerized by standard emulsion-polymerization techniques at temperatures from 5 to 70°C. The initiator and surfactant used were found to affect the addition of butadiene to the chain, but temperature generally had a greater effect. The cis-1,4 additions generally increased from about 5 to 20 percent as the temperature was raised, the trans-1,4 additions decreased from approximately 80 to 60 percent, and 1,2 additions increased slightly from 15 to 20 percent.

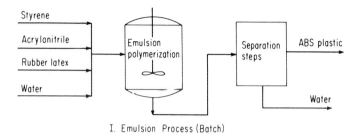

I. Emulsion Process (Batch)

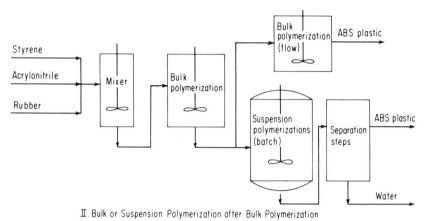

II. Bulk or Suspension Polymerization after Bulk Polymerization

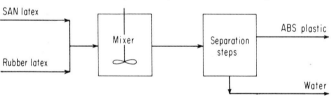

III. Mechanical Mixing of Latices

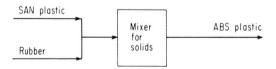

IV. Mechanical Mixing of Solids

Fig. 8-12   Process outlines for ABS plastics.

A recipe used by Union Carbide Corporation (82) for production of polybutadiene in their ABS process was as follows: 100 parts butadiene, 200 parts water, 3 parts sodium oleate, 10 parts carbon tetrachloride (used as a molecular-weight modifier), and 0.5 part $K_2S_2O_8$.

Considerable modifications have occurred within the last several years in emulsion-polymerization techniques for butadiene and other dienes. New recipes have been developed that allow rather large changes in the kinetics of polymerization and the chemical characteristics (and hence physical properties) of the rubber molecules. Some cross-linking and ring formation always occur during polymerization, particularly at higher levels of polymerization. For this reason, polymerization is frequently stopped at conversion levels from about 70 to 90 percent.

The size of the dispersed particles of rubber in the latex is of importance, especially when two or more latices are to be blended in the preparation of an ABS plastic (or of an impact polystyrene). The size of particles in the latices will affect the size of rubber particles in the final product. Hence in preparing a rubber latex, it is important to consider those factors, including the recipe and the operating variables, which affect not only the polymerization steps but also the characteristics of the latex formed. The amount of surfactant, for example, is highly important in this respect.

Although continuous-flow polymerization processes are employed for the manufacture of rubbers for tires, a batch system is undoubtedly always used for the rubbers incorporated into ABS plastics. The production capacity required and the relatively large number of latices needed do not justify flow processes.

Emulsion Processes for ABS Plastics: Emulsion polymerization processes are widely used for the production of newer and often improved varieties of ABS plastics in which styrene-acrylonitrile copolymers are grafted to the rubber (see Fig. 8-12). Styrene and acrylonitrile are added to the rubber, often a polybutadiene, latex. The reactants used by Borg-Warner Corp. (72, 87) for one of their plastics are as follows on a weight basis: rubber latex on a polybutadiene equivalent, 30; acrylonitrile, 25; styrene, 45; cumene hydroperoxide (an initiator), 0.75; sodium salt of hydrogenated disproportionated rosin, 2.0; sodium pyrophosphate, 0.5; sodium hydroxide, 0.15; dextrose, 1.0; ferrous sulfate, 0.01; and total water, including that in original latex, 160. The mixture polymerized within about 6 h when heated to about 65 to 85°C.

The ABS plastic from the above latex is recovered by coagulation with dilute brine, such as sodium chloride, and with sulfuric acid. The slurry is then heated to 95°C to promote granulation (or agglomeration) of the particles. The solids are then separated by filtration (or possibly centrifuging), are water-washed, and finally dried. The equipment and operation procedures are in general similar to those for production of homopolymers of styrene by emulsion-polymerization processes.

Relatively large variations of the recipes used for the final copolymerization step are possible. Such changes affect the rates of polymerization, the operating variables required, and/or the methods of ABS plastic recovery to at least some extent.

Bulk Processes for ABS Plastics: Bulk processes similar to those described earlier for the production of impact polystyrenes are also suitable for the production of ABS plastics (9) (see Fig. 8-12). The modified bulk process using small amounts of solvents such as ethylbenzene (70), described earlier for the production of general-purpose (crystal) polystyrenes and impact polystyrenes, is such a process. In this case, the rubber, usually polybutadiene, is dissolved in a mixture of styrene, acrylonitrile, and often some ethylbenzene. The combined mixture is prepolymerized to 20 to 40 percent with agitation to obtain phase inversion. The polymerization is completed in agitated tower reactors.

An important advantage of this bulk process, compared with the emulsion process, is that a wider range of rubbers, including *cis*-polybutadiene and other rubbers produced by solution processes using Ziegler or other solid catalysts, can be employed.

Suspension Processes for ABS Plastics: Grafted ABS plastics can also be produced by suspension processes (9) (Fig. 8-12) similar to those used for impact polystyrenes. A solution of rubber in styrene and acrylonitrile is first prepolymerized in a bulk system with agitation to obtain phase inversion, and then polymerization is continued in a suspension with water as the continuous phase. The types of ABS plastic that can be produced are similar to those produced by the bulk processes.

Mitsui Toatsu Chemicals have recently developed a suspension process for producing a plastic that is related to an ABS plastic but is not one since it apparently contains no butadiene (76). For the rubber, they use EPDM terpolymers (of ethylene, propylene, and a diene that is presumably a nonconjugated one). EPDM is highly resistant to oxygen and weather attack, compared with rubbers produced primarily from butadiene and isoprene. The claim has been made that the key point was to learn how to graft a branch onto the linear rubber molecule, but the method used has not been published. Part of the acrylonitrile and styrene forms a graft copolymeric branch on the terpolymer, and the remainder forms free SAN copolymers. The new plastic is said to have better weather resistance than conventional ABS plastics.

Mechanical Blending of Rubber and SAN Plastics: The most common method for blending rubber and SAN copolymer is to employ latices of each, as shown in Fig. 8-12. Uniroyal at their Baton Rouge plant use large head tanks constructed of 304 stainless steel. The level of agitation, the temperature, and the time for mixing are all important in blending these two latices. Such conditions affect the dispersion of the rubber phase in the finished plastic. The method of coagulation is also important for the same reason. Uniroyal uses a large flash dryer constructed of 304 stainless steel for drying the wet product after it has been washed.

The first ABS plastics produced in the late 1940s were manufactured by blending SAN plastics and nitrile rubbers on rubber mills in a ratio of approximately 65:35 at temperatures of 150 to 200°C (79). It is doubtful if such milling operations (Fig. 8-12) are of much current importance for production of the basic ABS plastic. However, melt mixing of graft ABS plastics with additional SAN copolymers is still important as a method for varying the physical properties of the final ABS plastic.

Process Considerations for ABS Plastics: The ratios, or relative amounts, of comonomers to be used for ABS plastic production vary over wide ranges, depending on the specific plastics being produced. The relative amounts of acrylonitrile (A), butadiene (B), and styrene (S) employed often are in the ranges of 20 to 30 percent, 6 to 35 percent, and 45 to 70 percent, respectively. Other comonomers sometimes used include methacrylonitrile, isoprene, and α-methylstyrene (9, 22).

Manufacturing variables, as already indicated, affect the geometry of the dispersed rubber phase in the ABS plastics. Key parameters describing the chemical and physical characteristics of the plastics are discussed by Weaver (101). As a rule, the plastics have an excellent blend of toughness, tensile strength, and creep resistance. In general, more rubber or butadiene content results in a higher impact resistance but lower hardness, tensile strength, and heat-distortion temperature. More acrylonitrile improves chemical resistance, tensile strength, and heat-distortion point.

Compatibility or chemical similarity of the resin and rubber is a key factor in obtaining desirable impact strengths for ABS plastics (101). Graft copolymerization of styrene-acrylonitrile branches on the rubber molecules tends to make the rubber phase much more compatible with the SAN phase. For ABS plastics formed by mechanical-blending techniques, in which no grafting occurs, a nitrile rubber is normally preferred since it is much more compatible with the SAN resin than a polybutadiene rubber.

The properties of ABS plastics are sometimes modified and even improved, particularly in the case of impact resistance, by blending ABS plastics with other polymers including polyvinyl chlorides, polycarbonates, or polyurethanes to produce the so-called ABS alloys.

Production of Styrene-Butadiene Copolymers

Copolymers of styrene and butadiene containing 50 percent or more styrene are of considerable commercial importance. The estimated production capacities of 13 American companies was 440 million lb/year by 1969 (12), and it is probably larger at present. These copolymers have properties that are often intermediate between a plastic and an elastomer.

Copolymers containing about 60 percent styrene find large uses in latex paints. The latices are prepared by batch emulsion-polymerization techniques similar to those used for making styrene-butadiene synthetic rubbers, which are of major importance. Sometimes other comonomers such as alkyl acrylates or acrylic acid are also added to the reaction mixture. These latter comonomers are highly polar and improve the adhesive characteristics of the resulting copolymers.

Block copolymers of the type S—B—S are produced from styrene and butadiene by Shell Chemical Company (86, 97); B refers to a block of polybutadiene and S to one of polystyrene.

In a Shell process, a solvent such as benzene and part of the styrene to be used are added to a stirred autoclave. The catalyst is added, and the styrene polymerizes. With a sec-butyllithium catalyst, a temperature of 40°C is suitable. Then butadiene (or isoprene for certain other copolymers) is added. Polymerization is continued until the diene is essentially all reacted. The remaining styrene is next added, and polymerization is continued until the styrene is almost depleted. By such a polymerization technique, the butadiene polymerizes to form blocks that are largely cis-polybutadiene. In another manufacturing procedure, a dianion is used to initiate the polymerization of butadiene. The resulting B block has an active site on each end. When styrene is added, it polymerizes to form the S—B—S block copolymer.

Block copolymers of styrene and butadiene vary in properties from soft rubbery to hard, semirigid materials. The physical properties vary with the relative amounts of styrene and diene used and with the sizes of the block. In general, more styrene makes a more rigid and harder material. Hydrogenation of various block copolymers is reported to yield tough, clear, heat-resistant polymers (95).

A major use of styrene is for the production of copolymers used as synthetic rubbers, the major one being a copolymer of butadiene and styrene in essentially a 3:1 weight ratio. Discussion of the processes employed for production of these elastomers is beyond the scope of this book.

Production of Speciality Styrene Copolymers

Styrene is used for the preparation of numerous speciality copolymers. In general, the polymerization processes are operated batchwise for these copolymers because of the relatively small production capacities needed. The processes are as a rule very similar to processes already described.

Chlorostyrene is a comonomer that has recently become available on a developmental basis (78). Copolymers of styrene and chlorostyrene have a reduced flammability. Various other chlorinated and brominated comonomers also find use as flame retardants (90). Mechanical compounding with inorganic salts, glass fibers, or organic compounds containing phosphorus, chlorine, or bromine atoms can also be used for this purpose.

Vinyltoluenes, which are closely related to styrene, are employed to prepare copolymers for paint, varnish, and polyester applications. Copolymers using divinylbenzene find important applications in preparing ion-exchange resins; a suspension-polymerization reaction is part of the overall operation (83). Divinylbenzene is also sometimes used in small amounts as a comonomer to increase the heat resistance and strength of the final plastic.

Several copolymers which generally contain more than 50 percent styrene and which find rather large-scale uses are discussed next.

Copolymers Containing α-Methylstyrene   Copolymers of styrene and α-methylstyrene are usually produced by emulsion-polymerization techniques (22). Such a process very effectively controls the temperature, which is important since α-methylstyrene increases the heat-deformation temperature of the plastic.

Hanson and Zimmerman (85), however, have reported success in producing copolymers with α-methylstyrene contents from about 10 to 39 percent in their recycle bulk-copolymerization process, described earlier for styrene-acrylonitrile copolymers.

Copolymers with Methyl Methacrylate   Copolymers of styrene and methyl methacrylate are used for crystal-clear plastics with good light stability. The copolymers often contain up to 60 percent by weight of methyl methacrylate. Bulk, solvent, suspension, or emulsion techniques can all be used for this copolymerization, but a conventional batch suspension process is probably most common.

Copolymers with Maleic Anhydride   Maleic anhydride and styrene are normally dissolved in a solvent such as acetone, benzene, or xylene, and then batch-polymerized at about 50°C in the presence of benzoyl peroxide (75). The comonomers tend to react on essentially an equimolar and alternate basis to form the linear molecules.

The resulting copolymers are rather brittle but are useful in latex paints, pigment dispersants, floor polishes, and epoxy resins. The anhydride groups can be reacted with glycols or diamines to produce cross-linked resins.

Literature Cited

 1. Barrett, G. R., and A. F. Harris (to Monsanto Co.): Process for Polymerizing Styrene, U.S. Pat. 3,222,341 (Dec. 7, 1965).
 2. Basdekis, C. H.: Styrene Polymers, W. M. Smith (ed.), in "Manufacture of Plastics," vol. I, pp. 402–436, Reinhold, New York, 1964.
 3. Bevington, J. C., G. M. Guzman, and H. W. Melville: Self-Branching in the Polymerization of Styrene, Proc. R. Soc., A221: 453 (1954).
 4. Bevington, J. C., H. W. Melville, and R. P. Taylor: The Termination Reaction in Radical Polymerization: Polymerization of Methyl Methacrylate and Styrene at 25°, J. Polym. Sci., 12: 449 (1945).
 5. Billmeyer, F. W.: "Textbook of Polymer Science," 2d ed., Interscience, New York, 1971.
 6. Boundy, R. H., and R. F. Boyer: "Styrene," Reinhold, New York, 1952.
 7. Boyer, R. F.: Styrene Polymers (Physical Properties), in N. M. Bikales (ed.), "Encyclopedia of Polymer Science and Technology," vol. 13, pp. 251–277, Interscience, New York, 1970.
 8. Boyer, R. F., and H. Keskkula: Styrene Polymers (Characterization), in N. M. Bikales (ed.), "Encyclopedia of Polymer Science and Technology," vol. 13, pp. 206–232, Interscience, New York, 1970.
 9. Boyer, R. F., H. Keskkula, and A. Platt: Styrene Polymers, in N. M. Bikales (ed.), "Encyclopedia of Polymer Science and Technology," vol. 13, pp. 128–135, Interscience, New York, 1970.
10. Brasie, W. C.: Fluid Flow in Polymer Handling Systems, Chem. Eng. Prog., 58(1): 45 (1962).
11. Brasie, W. C.: Elements of Polymer Engineering, 1968 Professors' Workshop, Midland, Mich., May 1968.
12. Styrene Forecast, Chem. Eng. News, Sept. 22, 1969, pp. 22–25.
13. Duerksen, J. H., and A. E. Hamielec: Polymer Reactors and Molecular Weight Distribution: Free-Radical Polymerization in Steady-State Stirred-Tank Reactor Trains, J. Polym. Sci., C25: 155 (1968).
14. Duerksen, J. H., A. E. Hamielec, and J. W. Hodgins: Polymer Reactors and Molecular Weight Distribution: Free Radical Polymerization in a Continuous Stirred-Tank Reactor, AIChE J., 13: 1081 (1967).
15. Grancio, M. R., and D. J. Williams: Morphology of Monomer-Polymer Particle in Styrene Emulsion Polymerization, J. Polym. Sci., (A-1)8: 2617 (1970).
16. Hamielec, A. E., J. W. Hodgins, and K. Tebbens: Polymer Reactors and Molecular Weight: Free Radical Polymerization in a Batch Reactor, AIChE J., 13: 1087 (1967).
17. Henrici-Olive, G., and S. Olive: Kettenübertragung bei der radikalischen Polymerisation, Fortschr. Hochpolym. Forsch., 2: 496–577 (1961).
18. Henrici-Olive, G., S. Olive, and G. V. Schulz: Selbstverzweigung und Ubertragungsreaktion am Polymeren bei Polystyrol, Z. Phys. Chem., N.F., 20: 176 (1959).
19. Hui, A. W., and A. E. Hamielec: Polymer Reactors and Molecular Weight Distribution: Free-Radical Polymerization in a Transient Stirred-Tank Reactor Train, J. Polym. Sci., C25: 167 (1969).

20. Johnson, R. B.: Polystyrene, *Plast. World*, August 1969, pp. 64–67.
21. Kambour, R. P.: The Role of Crazing in the Mechanism of Fracture of Glassy Polymers, *Appl. Polym. Symp.*, 7: 215 (1968).
22. Keskkula, H.: Styrene Polymers (Plastics), in N. M. Bikales (ed.), "Encyclopedia of Polymer Science and Technology," vol. 13, pp. 395–425, Interscience, New York, 1970.
23. Keskkula, H., A. Platt, and R. F. Boyer: Styrene Plastics, in A. Standen (ed.), "Kirk-Othmer Encyclopedia of Chemical Technology," 2d ed., vol. 19, pp. 85–134, Interscience, New York, 1969.
24. Lang, J. L. (to Dow Chemical Co.): Process of Polymerizing Monovinyl Aromatic Compounds with Rubber, U.S. Pat. 2,646,418 (July 21, 1953).
25. Natta, G., F. Danusso, and D. Sianesi (to Montecatini): Polymerization of Styrene, U.S. Pat. 3,161,624 (Dec. 15, 1964).
26. Ohlinger, H.: "Polystyrol," Springer-Verlag, Berlin, 1955.
27. Pennwalt Corporation, Lucidol Div.: *Bulletins*, Buffalo, N.Y., 1970.
28. Platt, A. E.: Styrene Polymers (Polymerization), in N. M. Bikales (ed.), "Encyclopedia of Polymer Science and Technology," vol. 13, pp. 156–206, Interscience, New York, 1970.
29. Stein, D. J., and H. Mosthaf: Oligomer Formation in the Thermal Polymerization of Styrene, *Angew. Makromol. Chem.*, 2: 39–50 (1968).
30. Wehr, H. W., and Nagle, F. B. (to Dow Chemical Co.): Alpha Alkyl Styrene Dimer Polymerization Modifiers, U.S. Pat. 2,732,371 (Jan. 24, 1956).
31. Allen, I., W. R. Marshall, and G. E. Wightman (to Union Carbide Corp.): Continuous Bulk Polymerization of Styrene, U.S. Pat. 2,496,653 (Feb. 7, 1950).
32. Altares, T., and E. L. Clark: Pilot Plant Preparation of Polystyrene of Very Narrow Molecular Weight Distribution, *Ind. Eng. Chem. Prod. Res. Dev.*, 9: 168 (1970).
33. Anderson, E. V., R. Brown, and C. E. Belton: Styrene: Crude Oil to Polymer, *Ind. Eng. Chem.*, 52: 550 (1960).
34. Amos, J. L., J. C. Frank, and K. E. Stober (to Dow Chemical Co.): Process of Conducting Exothermic Bulk Polymerization, U.S. Pat. 2,714,101 (July 26, 1955).
35. Amos, J. L., and A. F. Roche (to Dow Chemical Co.): Removal of Volatile Ingredients from Thermoplastic Polymers, U.S. Pat. 2,849,430 (Aug. 26, 1958).
36. Badische Anilin- und Soda-Fabrik (by S. Stadelmann, H. Ohlinger, and H. Mostof): Continuous Bulk Polymerization of Styrene, German Pat. 1,112,631 (Aug. 10, 1961).
37. Basdekis, C. H.: Styrene Polymers, in W. M. Smith (ed.), "Manufacture of Plastics," vol. I, Reinhold, New York; 1964.
38. Basel, L., and J. Papp: Polymerization Procedures, Industrial, in N. M. Bikales (ed.), "Encyclopedia of Polymer Science and Technology," vol. 11, pp. 280–304, Interscience, New York, 1969.
39. Bishop, R. B.: From Styrene to End-Use Item in One Easy Operation, *163d Am. Chem. Soc. Meet., Boston, Mass., 1972.*
40. Bishop, R. B.: Find Polystyrene Plant Costs, *Hydrocarbon Process.*, November 1972, pp. 137–140.
41. Brasie, W. C. (to Dow Chemical Co.): Heat Exchanger Agitator, U.S. Pat. 3,280,899 (Oct. 25, 1966).
42. Production of Polystyrene by Continuous Solvent Process, *Br. Plast.*, January 1957, pp. 26–27.
43. Polystyrene via "Natural" Ethyl Benzene, *Chem. Eng.*, Dec. 1, 1958, pp. 98–101.

44. Cleland, W. J., H. G. Henshall, A. N. Roper, J. M. Waring, and E. Seijo (to Shell Oil Co.): Polymerization of Vinyl Aromatic Compounds, U.S. Pat. 3,052,664 (Sept. 4, 1962).

45. Coulter, K. E., H. Kehde, and B. L. Hiscock: Styrene and Related Monomers, in E. C. Leonard (ed.), "Vinyl and Diene Monomers," pt 2, pp. 479–576, Wiley-Interscience, New York, 1971.

46. DeGraff, A. W., and G. W. Poehlein: Emulsion Polymerization of Styrene in a Single Continuous Stirred-Tank Reactor, J. Polym. Sci. (A-2)9: 1955 (1971).

47. Dunlop, R. D., and F. E. Reese: Continuous Polymerization in Germany, Ind. Eng. Chem., 40: 654 (1948).

48. Farber, E.: Suspension Polymerization, in N. M. Bikales (ed.), "Encyclopedia of Polymer Science and Technology," vol. 13, pp. 552–571, Interscience, New York, 1970.

49. Gerrens, H., and K. Kuchner: Continuous Emulsion Polymerization of Styrene and Methyl Acrylate, Brit. Polym. J., 2(1–2): 18–24 (1970).

50. Grim, J. M. (to Koppers Co.): Suspension Polymerization Using Synthetic Calcium Phosphate, U.S. Pat. 2,715,118 (Aug. 9, 1955).

51. Harkins, W. D.: A General Theory of the Mechanism of Emulsion Polymerization, J. Amer. Chem. Soc., 69: 1428 (1947).

52. Hauser, E. A., and E. Perry: Emulsion Polymerization of Styrene, J. Phys. Colloid Chem., 52: 1175 (1948).

53. Hutton, J. L., and R. K. Jessen (T. Shriver and Co., Inc., Harrison, N.J.): personal communications, 1970, 1972.

54. Continuous Thermal Polymerization of Styrene, Ind. Chem., January 1957, pp. 11–14.

55. Jones, E. M. (Monsanto Co., St. Louis, Mo.): personal communication, 1970.

56. Liou, D. W.: Process Simulation and Design of Continuous Flow Reactor, Ind. Eng. Chem. Div., 161st Meet. Am. Chem. Soc., Los Angeles, 1971.

57. Lynn, S., and J. E. Huff: Polymerization in a Tubular Reactor, AIChE J., 17: 475 (1971).

58. Mack, W. A. (Werner and Pfleiderer Corp., Walwich. N.J.), personal communication, 1972.

59. McDonald, D. L., K. E. Coulter, and J. L. McCurdy (to Dow Chemical Co.): Process for Mass Polymerization in Vertical Unmixed Strata, U.S. Pat. 2,727,884 (Dec. 20, 1955).

60. Miller, R. L.: Process Technology for License or Sale, Chem. Eng., Apr. 20, 1970, pp. 114–144.

61. Platzer, N.: Design of Continuous and Batch Polymerization Processes, Ind. Eng. Chem., 62(1): 6 (1970).

62. Roe, C. P.: Surface Chemistry Aspects of Emulsion Polymerization, Ind. Eng. Chem., 60(9): 20 (1968).

63. Ruffing, N. R., B. A. Kozakiewcz, and B. B. Cave (to Dow Chemical Co.): Process for Making Graft Copolymers of Vinyl Aromatic Compounds and Stereospecific Rubbers, U.S. Pat. 3,243,481 (Mar. 29, 1966).

64. Smith, W. V., and R. H. Ewart: Kinetics of Emulsion Polymerization, J. Chem. Phys., 16: 592 (1948).

65. Sparrow, R. F. (Marco Development Co., New Castle, Del.): personal communication, 1971.

66. Stark, A. H. (to Dow Chemical Co.): Continuous Polymerization Process, U.S. Pat. 3,007,903 (Nov. 7, 1961).

67. Stober, K. E. (to Dow Chemical Co.): Method for Polymerizing Styrene, U.S. Pat. 2,530,409 (Nov. 21, 1950).
68. Widmer, F.: Behavior of Viscous Polymers during Solvent Stripping or Reaction in an Agitated Thin Film, *143d Amer. Chem. Soc. Meet., Boston, Mass.,* 1972.
69. Wohl, M. H.: Bulk Polymerization, *Chem. Eng.,* Aug. 1, 1966, pp. 60–64.
70. Amos, J. L., J. L. McCurdy, and O. R. McIntire (to Dow Chemical Co.): Method of Making Linear Interpolymers of Monvinyl Aromatic Compounds and Natural or Synthetic Rubber, U.S. Pat. 2,694, 692 (Nov. 16, 1954).
71. Binder, J. L.: Microstructures of Polybutadienes and Butadiene-Styrene Copolymers, *Ind. Eng. Chem.,* **46:** 1727 (1954).
72. Borg-Warner Corp.: Blends of Polymeric Products, British Pat. 841,889 (July 20, 1960).
73. High Strength Polystyrene, *Brit. Chem. Eng. Process Supp.,* November 1967, pp. 78–79.
74. Brown, R. P., R. H. Dyer, W. M. Mayes, and P. D. Meek (to Cosden Oil and Chemical Co.): Masterbatch Prepolymerization Feed for Suspension Polymerization, U.S. Pat. 3,188,364 (June 8, 1965).
75. Brownell, G. L.: Acids, Maleic and Fumaric, in N. M. Bikales (ed.), "Encyclopedia of Polymer Science and Technology," vol. 1, pp. 67–95, Interscience, New York, 1964.
76. EPDM Resins Show High Weather Resistance, *Chem. Eng. News,* Dec. 7, 1970, p. 55.
77. Chopey, N. P.: Three Polymerization Lines Yield ABS Plastics Array, *Chem. Eng.,* Sept. 17, 1962, pp. 154–156.
78. Coulter, K. E., H. Kehde, and B. L. Hiscock: Styrene and Related Monomers, *Professors' Workshop, Midland, Mich., May 1968.*
79. Daly, L. E. (to United States Rubber Co.): Composition of Butadiene-Acrylonitrile Copolymer and Styrene-Acrylonitrile Copolymer, U.S. Pat. 2,439,202 (Apr. 6, 1948).
80. DeLand, D. L., J. R. Purdon, and D. P. Schoneman: Elastomers for High-Impact Polystyrene, *Chem. Eng. Prog.,* **63**(7): 118 (1967).
81. Freeguard, G. F., and M. Karmarkar: Production of Rubber-modified Polystyrene, *J. Appl. Polym. Sci.,* **15:** 1619–1655 (1971).
82. Fremon, G. H., and W. N. Stoops (to Union Carbide Corp.): Shock-resistant Rigid Polymer, U.S. Pat. 3,168,593 (Feb. 2, 1965).
83. Guccione, E.: A Look at the Synthesis of Ion-Exchange Resins, *Chem. Eng.,* Apr. 15, 1963, pp. 138–140.
84. Hanson, A. W., and J. S. Best: Polymerization Method, U.S. Pat. 2,989,517 (June 20, 1961).
85. Hanson, A. W., and R. L. Zimmerman: Continuous Recycle Copolymerization, *Ind. Eng. Chem.,* **49:** 1803 (1957).
86. Higginbottom, B., and W. R. Hendricks: Styrene-Butadiene, in "1970–1971 *Modern Plastics* Encyclopedia," pp. 217–218, McGraw-Hill, Inc., New York, 1970.
87. Irvin, H. I. (to Borg-Warner Corp.): Blends of Graft Copolymers of Polybutadiene, Styrene, and Acrylonitrile with Interpolymers of α-Methylstyrene and Acrylonitrile, U.S. Pat. 3,010,936 (Nov. 28, 1961).
88. Jones, C., B. Harris, and F. L. Ingley (to Dow Chemical Co.): Styrene-Acrylonitrile Copolymers U.S. Pat. 2,739,142 (Mar. 20, 1956).
89. Keskkula, H., and P. A. Traylor: Microstructure of Some Rubber-reinforced Polystyrenes, *J. Appl. Polym. Sci.,* **11:** 2361 (1967).

90. Lindemann, R. F.: Flame Retardants for Polystyrene, *Ind. Eng. Chem.*, **61**(5): 71, 1969.
91. Mino, G.: Copolymerization of Styrene and Acrylonitrile in Aqueous Dispersion, *J. Polym. Sci.*, **22**: 369 (1956).
92. Molau, G. E., and H. Keskkula: Mechanism of Rubber Particle Formation in Rubber-modified Vinyl Polymers, *J. Polym. Sci.*, (A-1)4: 1595 (1966).
93. Molau, G. E., and W. M. Wittbrodt: Colloidal Properties of Styrene-Butadiene Block Copolymers, *Macromolecules*, **1**: 260 (1968).
94. Nelb, R. G.: ABS Plastics, in W. M. Smith (ed.), "Manufacture of Plastics," vol. 1., pp. 437–455, Reinhold, New York, 1964.
95. Pendleton, J. F., D. F. Hoeg, and E. P. Goldberg: Novel Heat Resistant Plastics from Hydrogenation of Styrene Polymers, *163d Am. Chem. Soc. Meet., Boston, Mass.*, 1972.
96. Seymour, R. B. (to Monsanto Co.): Plasticized Polystyrene, U.S. Pat. 2,574,439 (Nov. 5, 1951).
97. Shell International Research (to G. Holden and R. Milkovich): Rubberlike Block Copolymers, Belg. Pat. 627,652 (July 29, 1963).
98. Stein, A., and R. L. Walter (to Monsanto Co.): Suspension Process for the Polymerization of Vinylidene Aromatic Hydrocarbons, U.S. Pat. 2,862,906 (Dec. 2, 1958).
99. Stein, A., and R. L. Walter (to Monsanto Co.): Suspension Process for Polymerization of Vinylidene Aromatic Hydrocarbons Having Rubbery Conjugated 1,3-Diene Polymers Dissolved Therein, U.S. Pat. 2,862,907 (Dec. 12, 1958).
100. Styrene Products Ltd. (by H. G. Henshall and E. G. Barber): Polyvinyl-Aromatic Compounds, British Pat. 854,238 (Nov. 16, 1960).
101. Weaver, E. P.: Acrylonitrile-Butadiene-Styrene, "1968 *Modern Plastics* Encyclopedia," pp. 111–114, McGraw-Hill, Inc., New York, 1967.
102. Whitby, G. S.: "Synthetic Rubber," Wiley, New York, 1954.
103. Ziemba, G. P.: Acrylonitrile-Styrene Copolymers, in N. Bikales (ed.), "Encyclopedia of Polymer Science and Technology," vol. 1, pp. 425–435, Interscience, New York, 1964.
104. Zimmerman, R. L., J. S. Best, P. N. Hall, and A. W. Hanson: Continuous Recycle Copolymerization Design for Effective Heat Exchange and for Handling Inert Diluents, *Adv. Chem. Ser.*, **34**: 225 (1962).

Index